Chapman & Hall/CRC Mathematical Biology and Medicine Series

KNOWLEDGE DISCOVERY IN PROTEOMICS

CHAPMAN & HALL/CRC
Mathematical Biology and Medicine Series

Aims and scope:
This series aims to capture new developments and summarize what is known over the whole spectrum of mathematical and computational biology and medicine. It seeks to encourage the integration of mathematical, statistical and computational methods into biology by publishing a broad range of textbooks, reference works and handbooks. The titles included in the series are meant to appeal to students, researchers and professionals in the mathematical, statistical and computational sciences, fundamental biology and bioengineering, as well as interdisciplinary researchers involved in the field. The inclusion of concrete examples and applications, and programming techniques and examples, is highly encouraged.

Series Editors

Alison M. Etheridge
Department of Statistics
University of Oxford

Louis J. Gross
Department of Ecology and Evolutionary Biology
University of Tennessee

Suzanne Lenhart
Department of Mathematics
University of Tennessee

Philip K. Maini
Mathematical Institute
University of Oxford

Shoba Ranganathan
Research Institute of Biotechnology
Macquarie University

Hershel M. Safer
Informatics Department
Zetiq Technologies, Ltd.

Eberhard O. Voit
The Wallace H. Couter Department of Biomedical Engineering
Georgia Tech and Emory University

Proposals for the series should be submitted to one of the series editors above or directly to:
CRC Press UK
23 Blades Court
Deodar Road
London SW15 2NU
UK

Chapman & Hall/CRC Mathematical Biology and Medicine Series

KNOWLEDGE DISCOVERY IN PROTEOMICS

Igor Jurisica
Dennis Wigle

CRC Press
Taylor & Francis Group
Boca Raton London New York

CRC Press is an imprint of the
Taylor & Francis Group, an **informa** business
A CHAPMAN & HALL BOOK

CRC Press
Taylor & Francis Group
6000 Broken Sound Parkway NW, Suite 300
Boca Raton, FL 33487-2742

First issued in paperback 2019

© 2006 by Taylor & Francis Group, LLC
CRC Press is an imprint of Taylor & Francis Group, an Informa business

No claim to original U.S. Government works

ISBN-13: 978-1-58488-439-2 (hbk)
ISBN-13: 978-0-367-39217-8 (pbk)

Library of Congress Cataloging-in-Publication Data

Catalog record is available from the Library of Congress

**Visit the Taylor & Francis Web site at
http://www.taylorandfrancis.com**

**and the CRC Press Web site at
http://www.crcpress.com**

Contents

List of Tables

List of Figures

Symbol Description

ℓ The average shortest path length of scale-free networks.

γ The power-law exponent in scale-free degree distributions $P(k) \approx k^{-\gamma}$.

$a.m.u.$ Atomic mass units.

C The clustering coefficient of a network.

C_v The clustering coefficient of node v in a networks.

$C(k)$ The average clustering coefficient of degree k nodes.

$C(p)$ The clustering coefficient of small-world networks with rewiring probability p.

C_{vw} Mutual clustering coefficient, a variable describing the topology of a protein-protein interaction network close to proteins v and w. This variable expresses the probability that the number of mutual neighbors observed for pair (v, w) would be matched or exceeded by chance.

$d(u, v)$ The shortest path length between nodes u and v.

$D(G, H)$ Relative graphlet frequency distance between two networks G and H.

Da Dalton

ΔCn The normalized difference between the best match and the second highest scoring match.

E A set of edges, or links of a graph.

$|E|$ Number of edges in a graph.

$E(G)$ A set of edges, or links of graph G.

G A graph, or a network.

$G(V, E)$ A graph, or a network.

$G_{n,p}$ Erdös and Rényi random graph on n nodes with each possible edge being present with probability p.

I_{vw} A variable indicating the absence or presence $(0/1)$ of interaction between proteins v and w, as detected by small-scale screens of interaction, as detected by small-scale screens.

$<k>$ The average degree of a graph.

K_n A complete graph on n nodes.

kV Kilo volt

$L(p)$ Characteristic path length, i.e., the shortest path length between two nodes averaged over all pairs of nodes, for small-world networks with rewiring probability p.

m Number of edges in a graph.

m/z Mass to charge ratio.

n Number of nodes in a graph.

nm Nanometer.

$N(v)$ The set of neighbors of node v (called the *neighborhood* of v).

$N[v]$ The *closed neighborhood* of v, which is defined as $N[v] = N(v) \cup \{v\}$.

$N_i(G)$ The number of graphlets of type i in network G.

ppm Parts per million.

P_k A path with k nodes.

$P(k)$ The probability that a randomly selected node of a network has degree k.

$T(G)$ The total number of graphlets in network G.

V A set of nodes, or vertices of a graph.

$|V|$ Number of nodes in a graph.

$V(G)$ A set of nodes, or vertices of graph G.

$Xcorr$ Cross-correlation score, which is based on the spectral fit between recorded and generated spectrum.

Y_{vw} A variable indicating the absence or presence (0/1) of interaction between proteins v and w, as detected by yeast 2-hybrid.

Preface

Who knows useful things, not many things, is wise.
Aeschylus (ca. 525-456 BC)

The nascent fields of bioinformatics and computational biology are currently an odd amalgam of everything from biologists with a computational bent, through physicists and mathematicians, to computer scientists and engineers sifting through the myriad of data and grappling with biological questions. Much of the excitement comes from a collective sense that there is something truly new evolving. Hardware and software limitations are declaring themselves as major challenges to managing and interpreting the avalanche of data from high-throughput biological platforms. This "drinking from the fire hydrant" sensation continues to spark interest and draw technical skill from other domains. As we move forward to true systems biology experimentation, it is increasingly obvious that experts in robotics, engineering, mathematics, physics, and computer science have become key players alongside traditional molecular biology.

Life sciences applications are typically characterized by multimodal representations, lack of complete and consistent domain theories, rapid evolution of domain knowledge, high dimensionality, and large amounts of missing information. Data in these domains require robust approaches to deal with missing and noisy information. Modern proteomics is no exception. As our understanding of protein structure and function becomes ever more complicated, we have reached a point in time where the actual management of data is a major hurdle to knowledge discovery. Many of the browse-through applications of yesterday are clearly not useful for computational manipulation. If the data was not created having data mining and decision support in mind, how well can it serve that purpose?

We felt this book was a timely discussion of some of the key issues in the field. In subsequent chapters we discuss a number of examples from our own experience that represent some of the challenges of knowledge discovery in high-throughput proteomics. This discussion is by no means comprehensive, and does not attempt to highlight all relevant domains. However, we hope to provide the reader with an overview of what we envision as an important and emerging field in its own right by discussing the challenges and potential solutions to the problems presented. We have selected five specific domains to discuss: (1) Mass spectrometry based protein analysis; (2) Protein–protein interaction network analysis; (3) Systematic high-throughput protein crystallization; (4) A systematic and integrated analysis of multiple data repositories using a diverse set of algorithms and tools; and (5) Systems biology. In each of these areas, we describe the challenges created by the type of data produced, and potential solutions to the problem of data mining within the domain. We hope this stimulates even more discussion, and newer and better ways to deal with the problems at hand.

Igor Jurisica and Dennis A. Wigle

Acknowledgments

The authors would like to thank the Hauptman-Woodward high-throughput screening lab in Buffalo, NY for providing crystallography image data, and especially Angela Lauricella for her meticulous hand-scoring of the images. We highly benefited from a long term collaboration with George T. deTitta and Joe R. Luft, which led to the development of this system. We are also grateful to Thomas Acton, Rong Xiao, Gaetano Montelione, and other members of the NESG (North-East Structural Genomics Consortium) protein production team for providing the many protein samples that have made the crystallography work possible. Earlier work on case-based reasoning and protein crystallography also highly benefited from a long term collaboration with Janice Glasgow at Queen's University.

Lukas Jurisica deserves mention for his help with Lego robotics and photography. Although too numerous to list individually, we are indebted to many research groups generating and providing high-throughput data both locally and via the web from a distance. We also extend thanks to those who continue to stimulate, encourage, provoke, and inspire us with new ideas and new ways of looking at a problem.

Last but not least, we would like to thank for the research support from the National Science and Engineering Research Council, the Institute for Robotics and Intelligent Systems, National Institutes of Health, and the Fashion Show fund from Princess Margaret Hospital Foundation. We also gratefully acknowledge the hardware and software support from IBM Life Sciences through a Shared University Research Grant and IBM Faculty Partnership Award.

Contributing Authors

Igor Jurisica is a Scientist at the Ontario Cancer Institute, University
Health Network since 2000, Assistant Professor in the Departments
of Computer Science and Medical Biophysics, University of Toronto,
Adjunct Assistant Professor at School of Computing Science, Queen's
University, and a Visiting Scientist at the IBM Centre for Advanced
Studies. He earned his Dipl. Ing. degree in Computer Science and
Engineering from the Slovak Technical University in 1991, M.Sc. and
Ph.D. in Computer Science from the University of Toronto in 1993 and
1998 respectively.

Dr. Jurisica's research focuses on computational biology, and representation, analysis and visualization of high dimensional data generated by
high-throughput biology experiments.

Of particular interest is the use of comparative analysis for the mining of
integrated datasets such as protein–protein interaction, gene expression
profiling, and high-throughput screens for protein crystallization.

Ontario Cancer Institute, Princess Margaret Hospital/UHN
Division of Signaling Biology
610 University Avenue, Rm 8-413
Toronto, Ontario M5G 2M9
juris@ai.utoronto.ca, http://www.cs.utoronto.ca/~juris

Dennis A. Wigle is the Program Director of Cancer Genomics for the Thoracic Oncology Site Group at Princess Margaret Hospital in Toronto. He
earned his Ph.D. from Queen's University in 1994 and his M.D. from
the University of Toronto in 1997. Dr. Wigle is currently completing his
clinical training in Thoracic Surgery at the University of Toronto. His
research interests focus on the molecular biology of lung development
and disease.

Division of Thoracic Surgery
University Health Network
200 Elizabeth Street
Toronto, Ontario M5G 2C4
dennis.wigle@utoronto.ca

Kevin R. Brown is a Ph.D. candidate in the Department of Medical Biophysics, University of Toronto, under the supervision of Dr. Igor Jurisica. His research focus has been on expanding the human interactome using model organism data, and integration of various data types to help mine and interpret the interactome and microarray datasets. Mr. Brown has a background in molecular biology, biochemistry and computer science, obtained a M.Sc. from the University of Toronto, and a B.Sc. from the University of Waterloo.

> Ontario Cancer Institute, Princess Margaret Hospital/UHN
> Division of Signaling Biology
> and University of Toronto, Department of Medical Biophysics
> 610 University Avenue
> Toronto, Ontario M5G 2M9
> kbrown@uhnres.utoronto.ca

Christian Anders Cumbaa is a Research Assistant in Jurisica lab. The main focus of his research is machine learning and data mining, and high-throughput image analysis. He earned his M.Math. degree from University of Waterloo in 2001.

> Ontario Cancer Institute, Princess Margaret Hospital/UHN
> Division of Signaling Biology
> 610 University Avenue
> Toronto, Ontario M5G 2M9
> cumbaa@uhnres.utoronto.ca

Andrew Emili is an Assistant Professor in the Program in Proteomics & Bioinformatics and the Banting & Best Department of Medical Research. He is cross-appointed to the Graduate Department of Medical Genetics & Microbiology. Dr. Emili has a B.Sc. in Microbiology and Immunology from McGill University, and an M.Sc. and Ph.D. in Molecular Genetics from the University of Toronto.

Before joining the University of Toronto in 2000, Dr. Emili carried out research training with Dr. Leland Hartwell, noted genetics researcher and 2001 Nobel laureate, at the Fred Hutchinson Research Center in Seattle, and with Dr. John Yates, a leader and pioneer in protein mass spectrometry. Current research interests of the Emili laboratory are focused on proteomics investigation of the complex biochemical circuitry that underlies cell growth, proliferation and development with a particular focus on DNA replication and repair and cell division pathways.

> University of Toronto
> Banting & Best Department of Medical Research
> 112 College St. Rm. 416
> Toronto, ONT M5G 1L6
> andrew.emili@utoronto.ca, http://www.utoronto.ca/emililab/

Thomas Kislinger is a post-doctoral fellow in the laboratory of Dr. Andrew Emili at the University of Toronto since March 2002. Dr. Kislinger works on the development and application of LC-MS based proteomics technologies for the analysis of mammalian proteoms.

Thomas prepared his Ph.D. thesis on the biology and chemistry of advanced glycation endproducts at the Friedrich Alexander University in Erlangen, Germany and the Columbia University in New York, USA. He received his Ph.D. in July 2001.

> University of Toronto
> Banting & Best Department of Medical Research
> 112 College St.
> Toronto, ONT M5G 1L6
> thomas.kislinger@utoronto.ca

Max Kotlyar is a Ph.D. candidate at the Department of Medical Biophysics, University of Toronto, under supervision of Dr. Igor Jurisica. His main research focus is using data mining for predicting protein–protein interactions.

Previously, Mr. Kotlyar was employed at Molecular Mining Corporation, analyzing biological data by machine learning methods. He has a M.Sc. in Computer Science from Queen's University.

> Ontario Cancer Institute, Princess Margaret Hospital/UHN
> Division of Signaling Biology
> and University of Toronto, Department of Medical Biophysics
> 610 University Avenue
> Toronto, Ontario M5G 2M9
> mkotlyar@uhnres.utoronto.ca

Natasa Przulj is a postdoctoral fellow at Samuel Lunenfeld Research Institute in Toronto, working under the supervision of Prof. Jeff Wrana and Prof. Igor Jurisica. Her work involves large network analysis with a particular emphasis on protein–protein interaction networks. Dr. Przulj has received a Ph.D. and an M.Sc. from the Department of Computer Science at the University of Toronto and a B.Sc. in Mathematics and Computer Science from Simon Fraser University. She also studied at the University of Belgrade and worked in industry for two years.

> Samuel Lunenfeld Research Institute
> 600 University Avenue
> Toronto, Ontario M5G 1X5
> natasha@ai.utoronto.ca

Chapter 1

Introduction

As our understanding of biology becomes ever more complicated, we have reached a point in time where the actual management of data is a major stumbling block to the interpretation of results from modern proteomic platforms. We felt this book was a timely attempt to discuss emerging issues in this field, propose suggestions to diminish the bottleneck, and project on future challenges and solutions in proteomics and high-throughput (HTP) biology in general over the coming years.

We unfortunately live and work in an era where just coping with new terms has become an ever-increasing challenge [46, 130, 137, 258, 342, 357, 403]. Let us begin by first defining the terms *knowledge discovery* and *proteomics* to set the context for further discussion within the book.

1.1 Knowledge Discovery

Knowledge discovery (KD) as a term of reference may seem self-evident, but has become an increasingly useful descriptor as biological and other forms of data have become more and more complex. The term itself was coined during the first Knowledge Discovery in Data Bases workshop held in 1989. The process of finding useful patterns in data has been described in many ways, with terms such as data mining, knowledge extraction, information discovery, information harvesting, data archaeology, and data pattern processing widely used [553]. Although it has similar goals to data mining, knowledge discovery emphasizes the end product of the process, which is knowledge. This leads to the following definition: "Knowledge discovery is a (nontrivial) process of identifying valid, novel, potentially useful, and ultimately understandable patterns in data" [165]. Knowledge discovery usually employs statistical data analysis methods but also methods in pattern recognition and artificial intelligence (AI). Data base management systems (DBMSs) ensure that the process is scalable (see Figure 1.1). None of the components alone is sufficient for knowledge discovery in proteomics.

In other words, knowledge discovery stands for the sum of real knowledge that can be gained from the analysis and interpretation of data. It is an iterative process, involving intertwined analysis and interpretation, often using

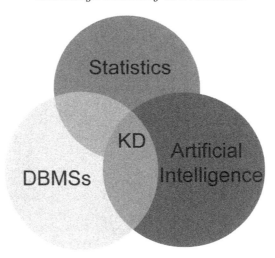

FIGURE 1.1: Knowledge discovery (KD) in complex domains requires combined methods from artificial intelligence and statistics. Scalability of discovery systems is ensured and supported by efficient data base management.

multiple methods, to uncover the "truth" in the data. Patterns described should be novel, and in a form that human users will be able to understand and apply.

Discovered knowledge has three important facets:

1. Form;

2. Representation; and

3. Degree of certainty.

The *pattern* is a statement S in a language L that describes relationships among a subset of facts FS with a certainty C. Such patterns include clusters, sequences or association rules. To be potentially useful, S must be simpler than the enumeration of all the facts in FS. In addition, since the discovered pattern is usually not true across all the data, the certainty factor is required to determine the faith put into a discovery. The *certainty* involves the integrity of the data, the size of the sample on which the discovery was performed, and the degree of support from available domain knowledge. Only interesting patterns are considered to be knowledge. Although a subjective measure, an interesting pattern is nontrivial to compute, novel, and useful.

The main use of discovered knowledge is for hypothesis formulation and verification, building models for forecasting, planning and predicting, decision rule discovery, information cleaning, identifying outliers, information organization and structure determination. However, as we move from manufacturing and marketing domains into sciences, we need to apply mining algorithms not only to data, but to knowledge as well. As a result, we not only aim at discovering patterns, but aim to support knowledge management and evolution. In addition, discovered knowledge can be used for optimization of information systems.

1.1.1   Knowledge–Discovery Process

There are two main groups of knowledge–discovery algorithms:

1. Quantitative discovery, which locates relationships among numeric values;

2. Qualitative discovery, which finds logical relationship among values.

In many domains, including biological and medical, this approach must be extended by incorporating symbolic attribute values and visual attributes from still images. The discovery process involves two main steps: *identification* (see Section 1.1.1.1) and *description of patterns* (see Section 1.1.1.2). In general, several additional steps are involved: from selecting, preprocessing and transforming the data set, through using diverse data-mining algorithms, to visualizing the patterns (see Figure 1.2).

1.1.1.1   Identification of Patterns

Identification of patterns is done by clustering records into subclasses that reflect the patterns (sometimes called unsupervised learning). There are two basic methods:

1. Traditional *numeric methods* use cluster analysis and mathematical taxonomy [18, 265]. These algorithms produce classes that maximize similarity within classes and minimize similarity between classes. Although many different similarity metrics have been proposed, the most widely used is Euclidean distance between numeric attributes. The similarity of two objects can also be represented as a number, the value of a similarity function applied to symbolic descriptions of objects. However, such similarity measures are context free and these methods fail to capture the properties of a cluster as a whole and are not derivable from properties of individual entities. As a result, these algorithms work well on numeric data only. In addition, these approaches cannot use background knowledge, such as likely shapes of clusters [18, 230].

   One of the useful applications of pattern identification is recognizing "outliers," i.e., data outside of a normal region of interest or input. In general, outliers could be unusual but correct or unusual and incorrect. They can be detected using histograms and then removed by threshold filters, or identified by calculating the mean and standard deviation and

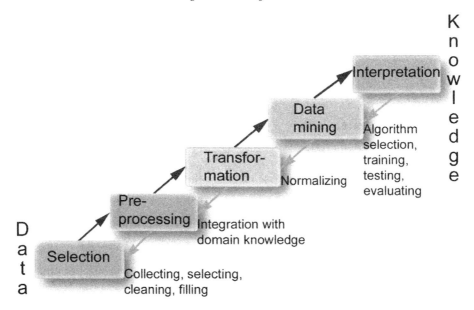

FIGURE 1.2: Knowledge discovery is a process involving several steps, which are iteratively repeated and intertwined. Blue arrows represent sequence of steps from data to knowledge, and green arrows illustrate feedback loops.

removed by specifying a "window," for example two standard deviations from the mean [385].

2. *Conceptual clustering methods* determine clusters not only by attribute similarity but also by conceptual cohesiveness, as defined by background information [504]. Inductive learning is a process of generalizing specific facts or observations. It is a basic strategy by which one can acquire and organize knowledge. For example, generalization can be used to organize existing concepts into taxonomies, referred to as isA or generalization hierarchies. The most general concept is the root of the hierarchy. Inheritance inference rule states that attributes and properties of a class are also attributes and properties of its specializations.

This process can be exploited further, by using two main forms of inductive learning: instance-to-class induction and clustering. During instance-to-class induction the learning system is presented with independent instances, representing class and the task is to induce a general description of the class. This process can result in introducing either a new entity-class or a super-class relationship. If two subclasses share the same properties then it is possible to introduce a new super-class as a generalization of the subclasses. If subclasses of a generalization are associated with an entity-class then the entity-class can be moved to the super-class. For example, after knowing that student, faculty, and technician isA person, and since faculty and technician have salary while student has a scholarship, a new entity-class — employee isA person — may be introduced. Thus, faculty, and technician isA employee, while student and employee isA person.

Clustering problems arise when several objects or situations are presented to a learner and the task is to find classes that group objects. For both numeric and conceptual clustering, there are two main approaches: (1) *flat clustering*, where there is only one level of clusters; and (2) *hierarchical clustering*, where nested clusters are created [230].

The hierarchical clustering process can progress either *top-down* or *bottom-up*:

1. Top-down clustering starts with all items being a member of the cluster, then removing "outliers" into new clusters;

2. Bottom-up approaches start with no items being in the cluster, then moving similar items to clusters.

The process can be interactive, by combining domain knowledge with user's knowledge and visual skills to solve more complex problems. Interactive clustering algorithms prove to be particularly useful in evolving domains [552].

1.1.1.2 Description of Patterns

Description of patterns is achieved by summarizing relevant qualities of the identified classes (also referred to as supervised learning). The task is to determine commonalities or differences among class members by deriving descriptions of the classes. There are two main streams: empirical algorithms that detect statistical regularity of patterns, and knowledge-based algorithms (explanation-based learning) that incorporate domain knowledge to explain class membership. For example, aggregation views patterns as aggregates of their components. In turn, components of a pattern might themselves be aggregates of other simpler components.

Identified and described patterns must also include measures of their quality. Since different tasks may require disparate priorities, it is useful to have multiple measure of quality. There are several standard approaches to quantify quality, including certainty, interestingness, generality and efficiency:

- **Certainty** measures the trust we have in a pattern, i.e., how much the existing data supports it;

- **Interestingness** specifies how novel, useful, nontrivial and understandable is the discovered pattern;

- **Generality** expresses the strength of a pattern in terms of the size of the subset of objects, described by the pattern;

- **Efficiency** measures the time required to compute the pattern and space needed to store it.

1.1.2   Applications of Knowledge Discovery

There are numerous applications of knowledge discovery, but we will focus only on a few that are related to information systems and knowledge management. Symbolic knowledge discovery can be used for three main tasks:

1. System optimization;

2. Knowledge evolution;

3. Evidence creation [283].

Image-feature extraction algorithms can automatically and objectively annotate and describe images [127, 495]. *System optimization* is performed by knowledge representation effective for a given context and task, and organizing knowledge into context-based clusters. The importance of individual descriptors (knowledge properties) may change with context or task, and thus several context-dependent clusters of knowledge may be created. During navigation and reasoning, relevant knowledge sources are thus located in physical proximity, which results in a more efficient access.

Knowledge evolution is supported by adding descriptors to assist knowledge discrimination during prediction and classification, removing redundant knowledge and descriptors, creating hierarchies of descriptors and their values and finding associations [7]. This process is useful in creating hierarchies of and relationships among knowledge sources. Creating hierarchies of descriptors and their values enhances knowledge organization and thus improves system performance and domain comprehensibility.

Evidence-based reasoning is achieved by analyzing created clusters, hierarchies and associations of knowledge sources to identify underlying principles in the domain. Evidence could be a rule discovered by generalization from individual experiences. Although some rules may be obtained from a human expert, deriving them by analyzing an existing knowledge repository provides explanation of the rule and justification of its validity. Furthermore, over time, the support for the rule may change, and it can be automatically removed or modified to reflect the change in the knowledge repository. Thus, this implements a systematic process of knowledge compilation to support evidence-based reasoning. We will return to knowledge discovery in subsequent chapters.

1.2 Proteomics

Recent years have witnessed an explosion in the power of biological experimentation. Although difficult to trace to one specific advance, this appears to have arisen from a convergence of full genome sequencing, developments in robotics, and an overall paradigm shift in how biologists think about and approach experimentation. In the current lexicon, high-throughput (HTP) biological domains comprise relatively recent methods developed in the fields of genomics and proteomics, typically involving robotic platforms at some level, and producing thousands or tens of thousands of data points for each experiment. It is these specific areas where knowledge discovery is a young and emerging field. Many of these experimental techniques have only been developed in the past few years, and produce data of unprecedented complexity in unfamiliar formats.

How would one define proteomics? Although difficult to generate a specific definition, the simplest would characterize it as the study of protein composition in a protein complex, organelle, cell, or entire organism [468]. For the purpose of this text, we characterize high-throughput (HTP) proteomics as any approach to the study of proteins involving robotic platforms and requiring computational support for data collection and analysis.

Proteomic data provide key information regarding the structure and function of biological systems. Much of this information is not coded for in DNA sequence, and hence its acquisition has increased in importance with the publication of full genome sequences for many organisms. Such information includes aspects of the protein composition of a cell, expression levels, alternate splicing products, and posttranslational modifications such as protein phosphorylation. Assays of complex protein mixtures are rapidly improving in the number of proteins that can be analyzed simultaneously, the aspects of these proteins that can be determined, and the cost of doing such experimentation. As proteomic technology continues to improve, it will play an increasing role in both basic and translational research [468].

The development of mass spectrometry is arguably the single greatest accelerant to the creation of modern proteomics as an emerging discipline. The ability to separate molecules based on different size and charge was first described in 1912 by J.J. Thompson, who won the Nobel Prize in 1906 for investigations of the conduction of electricity by gases. Despite years of intense MS development, the goal of analyzing large macromolecules and proteins remained elusive for over 70 years. It was subsequently acknowledged that this could be accomplished using a chemical preparation of charged molecules in the gas phase followed by physical separation of ions in a vacuum. For biomolecules the challenge was to find a viable procedure for the chemical preparation of the sample. The well-defined breakthrough of electrospray ionization (ESI) came in 1988 when John Fenn presented an identification of polypeptides and proteins of molecular weight 40 kDa [168]. Fenn showed that a molecular-weight accuracy of 0.01% could be obtained by applying a signal-averaging method to the multiple ions formed in the ESI process. The findings were based on developments that had started in 1984 in Fenn's laboratory at Yale, when electrospray and mass spectrometry were successfully combined for the first time.

In the 1990's, mass spectrometry definitively crossed the border to biochemistry. The general ways that it provides structural determination, identification and trace level analysis have many applications. It has become an attractive alternative to Edman sequencing, and has an unsurpassed ability to identify posttranscriptional modifications and non-covalent interactions. These developments and the informatics issues they create are discussed in the chapter on mass spectrometry.

In recent history, the development of computational tools to effectively deal with the data produced from HTP platforms has been a critical part of allowing biologists to fully leverage the information contained within. The most prominent example is the development and wide distribution of the BLAST algorithm [15, 16], a pattern recognition tool that has evolved as the most effective way to rapidly search data bases of sequence information for homology. The design and testing of this software by mathematicians and computer scientists has enabled the morass of genome and EST sequence information being produced to be rapidly analyzed and interpreted by biologists. However, as these large biological datasets and portals are being further integrated and analyzed, new ways of storing and accessing this information are emerging, as the old "one gene/protein at a time approach" is clearly not suitable for the new HTP data analysis.

The current HTP revolution is rapidly changing the way we formulate and test biological hypotheses. Advances in protein profiling by mass spectrometry have suggested the potential to simultaneously view all genes expressed, all subsequent protein products, and all the interacting partners of each individual protein within a biological system. Such views are already having an impact on our understanding of human disease, particularly in the realm of cancer biology [136]. However, these data are typically noisy, incomplete, and extraordinarily complex. For these reasons, the development of computational tools necessary to effectively deal with these data has significantly lagged behind the advances in analytical techniques for their generation. Many of the data analysis approaches make unreasonable assumptions about the application from which they are derived, such as the normal distribution of data or independence of individual variables.

Other existing approaches do not scale well to the problems in HTP proteomics, where frequently one requires the capability to deal with tens or hundreds of thousands of attributes for each instance. Unfortunately, we deal with severely over-specified systems, as we usually have hundreds or even thousand time fewer instances described by these attributes. It follows that with this disproportion, one has to be careful what methods are used for the analysis, and how are the results interpreted, because even statistical significance may not guarantee biological relevance. Eliminating some attributes before analysis is also a dangerous process, as we may eliminate our best evidence.

So why is the process of knowledge discovery a critical part of the HTP revolution in biology and the understanding of life itself? The scientific drive for reductionism has made it clear that the understanding of even a seemingly simple signaling network, with all its interconnections under different cellular conditions, cannot be comprehended by a single researcher staring at pieces of paper or simple images on a monitor. If we are to successfully integrate multiple or all system networks that comprise cellular and organism function, and successfully apply this knowledge to the understanding of disease states, we must develop computational methods that are accessible to domain experts in biology.

This is clearly not an artificial or trivial problem. In most examples, data from HTP methods are noisy, the search space is large, and existing small-scale methods are either too slow or break down in attempting to scale to the problem. Computational analysis is and will be a necessity to deal with such data types that are evolving faster than the ability to effectively integrate and comprehend.

In subsequent chapters we discuss a number of examples from our own experience that represent some of the challenges of knowledge discovery in HTP proteomics. This discussion is by no means comprehensive, and does not attempt to highlight all relevant domains. However, we hope to provide the reader with an overview of what we envision as an important and emerging field in its own right by discussing the challenges and potential solutions to the problems presented. We have selected five specific domains to discuss:

1. Mass spectrometry based protein analysis;

2. Protein–protein interaction network analysis;

3. Systematic HTP protein crystallization;

4. A systematic and integrated analysis of multiple data repositories using a diverse set of algorithms and tools;

5. Systems biology.

In each of these areas, we describe the challenges created by the type of data produced, and potential solutions to the problem of data mining within the domain. We hope this stimulates even more discussion, and newer and better ways to deal with the problems at hand.

Chapter 2

Knowledge Management

2.1 Computational Analysis, Visualization, and Interpretation of HTP Proteomic Data

Application of artificial intelligence techniques to the life sciences requires addressing some fundamental issues: effective information representation, its efficient access, analysis and use during decision-making. It is becoming clear that the high dimensionality poses a serious challenge to existing knowledge–discovery and reasoning tools. Most existing tools were developed for relational types of data, which typically have millions of instances but low dimensionality. Life science domains involve tens of thousands of dimensions, which cause most existing tools to either fail or provide outcomes with limited usefulness. Since these domains are characterized by the diversity of representation formalisms used, the complexity and amount of information present, uncertain or missing values, and the evolution of knowledge, it is necessary for an intelligent information system to be flexible and scalable.

Advances in technology for the HTP generation of proteomic data have raced ahead of the development of computational tools to effectively analyze, visualize, and interpret this information. We believe this bottleneck will worsen as the rush to accumulate such data accelerates. Direct application of many existing computational algorithms will fail for two main reasons:

1. Many approaches make unreasonable assumptions about the application, such as normal distribution of data or independence of variables;

2. Many approaches will not scale to the problems in HTP proteomics, such as tens of thousands of attributes for each instance.

In addition, analyzing and interpreting data in isolation will also not be sufficient, as we further aim at integrated analysis, described in Chapter 6.

Life sciences applications are characterized by multimodal representations, lack of complete and consistent domain theories, rapid evolution of domain knowledge, high dimensionality, and large amounts of missing information. Computational analysis of data in these domains requires an iterative and interactive approach, as representation and organization of knowledge evolves with an increased understanding of the domain. Considering the complexities, researchers prefer to use multiple data generation and computational techniques to boost performance and to provide a more complete analysis, followed by scalable visualization techniques. Through systematic knowledge management, analysis and intelligent use of information, we can significantly increase our understanding of biology in general and health and disease processes in specific. This requires new generation of computational tools and their integrative use. The emerging fields of systems biology and integrative computational biology provide some answers to these problems.

Naturally, both knowledge discovery and HTP proteomics have to be linked to the management of data, information, and resulting knowledge. There is evidence that applying intelligent computational systems first and then statistics provides more systematic identification of all patterns. Combining diverse intelligent computational techniques to the analysis of large, multidimensional datasets provides an increased confidence in the utility of the discovered patterns and will enlarge the overall number of sets of discovered patterns. Applying unbiased learning methods will enable us to identify novel information. The main goal is to design and use an effective and efficient computational system for the management, analysis, visualization, interpretation and use of high-dimensional biological knowledge. The focus is on complex domains, where iterative and interactive environments are required.

Knowledge management is concerned with the representation, organization, acquisition, creation, use, and evolution of knowledge in its many forms. To build effective technologies for knowledge management, we need to understand how the knowledge is going to be used. The most fundamental characterization is to determine if it is going to be used by other computational tools, or directly by humans. This has been discussed on many occasions [258], but is frequently forgotten. The next chapter discusses relevant issues of knowledge management.

2.2   Introduction to Data Management

The largest proportion of bioinformatics with direct use by biologists is in data storage and access. A data model is a mathematical formalism comprising a notation for describing data and a set of operations used to manipulate that data. Many techniques have been proposed for representing and managing codified knowledge, using approaches from computer science, software engineering, data bases, information systems, and artificial intelligence. The main goal is to *formalize* information representation so effective storage, access and use can be supported.

The most prevalent model of data and information representation is a relational data model with a corresponding relational data base management system (RDBMS). An RDBMS is a system to store and manage data bases, including setting up storage structure, loading data, accepting and performing updates and data requests, and controlling access to data through security provisions.

2.2.1   Data Base Management Systems

A data base is a collection of files whose content and structure are defined in a data dictionary. A relational data base comprises a set of tables, each storing records (or instances). A record is represented as a set of attributes, which define a property of a record. Attributes are identified by their name and store a value. All records in a particular table have the same number and type of attributes.

One of the most fundamental aspects of the data base record is a unique identifier. Although technically it is indisputable that being able to unambiguously identify individual records is necessary, biology provides many examples when one name is used to refer to multiple entities, or even more commonly, when one entity has multiple names. One such example relates to protein names; ppif, cyclophilin 3, cyclophilin III and cyp3 are all valid references to the same protein. Although different, one could see the way these names have been constructed; sometime, protein names are completely different, such as SARA, which is also known as ZFYVE9, NSP, hSARA, SMADHIP and MADHIP. Most proteins in existing data bases have at least one synonym (see Table 2.1). Implementing robust protein name recognition requires that the system has access to all available synonyms. A good start is to integrate accepted synonyms from Swiss-Prot[1] and LocusLink[2] data bases. Although this provides 15 synonyms per protein on average, it still does not provide full coverage. This creates an enormous problem for data retrieval. An example of severity of this problem is the HPRD data base,[3] which originally contained 10,500 entries of human-curated human protein–protein interactions (PPIs), later listed almost 20,000 entries, but then had to discard over 6,000 due to potentially incorrect identifiers and names (personal communication).

However, even proper naming convention will not solve all the problems for two main reasons: we have vast amount of legacy data bases, and we have to deal with inherent biological variations and human biases. We continuously deal with evolving gene and protein annotations, existence of mutations, single nucleotide polymorphisms and posttranslational modification, which suggest to use sequence as an ultimate identifier. This is further reinforced by diversity of gene and protein constructs placed on individual high-throughput

---

[1] http://us.expasy.org/sprot/

[2] http://www.ncbi.nlm.nih.gov/LocusLink

[3] http://www.hprd.org

TABLE 2.1: Protein names and synonyms from Swiss-Prot data base.

Swiss-prot ID	P48349
Protein name	14-3-3-like protein GF14 lambda
Synonym	General regulatory factor 6
Synonym	14-3-3-like protein RCI2
Synonym	14-3-3-like protein AFT1
Gene Name	GRF6
Synonym	AFT1
Synonym	RCI2

(HTP) platforms, which unless properly handled result in frustration from nonoverlapping results from different experiments [363].

Another important property of data bases is the ability to efficiently access large sets of data. Many biological data repositories provide public access, but only for human users and in a form not suitable for computer systems. Depending on the overall goal, this may be the most appropriate form, but systems biology approaches require automated access to such repositories. Examples include GeneCards[4], the RIKEN Expression Array data base[5], HPRD[6], etc. [71, 440, 470]. Clearly, graphical navigation through the repository viewing one item (or a few items) at a time is suitable for small-scale manual interpretation. However, large-scale computational analysis requires access to thousands of items at a time, querying them with complex constraints. Any large-scale integration of diverse data sets will require this type of access.

In addition, having consistent and accurate identifiers of biological entities is required for data integration and interoperability. These topics will be further developed in Chapter 6.

2.2.2 Data Base Design Process

The data base design process has several important stages. Identifying mission objectives defines purpose and provides a development focus. Identifying, expressing, and analyzing requirements leads to specifying functional requirements for the application. Functional requirements, in turn, are used during conceptual design of the system.

Conceptual design is used to describe the data requirements of the application. It also identifies and describes entities and relationships [105], and defines business rules (i.e., describes integrity constraints) and views. Entities and relationships are usually described by entity-relationship diagrams (ER

---

[4]http://bioinfo.weizmann.ac.il/cards/

[5]http://read.gsc.riken.go.jp

[6]http://www.hprd.org

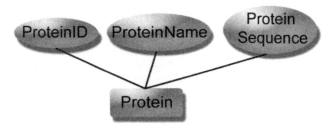

FIGURE 2.1: Entity-relationship diagram of a protein. As a basic entity, proteins have their identifiers (PID), name (Pname) and a sequence (Psequence).

diagrams). An example ER diagram of a an entity "protein" is presented in Figure 2.1.

Proteins are molecules composed of long chains of 20 amino acids, forming linear heteropolymers of fixed length. Proteins constitute more than half of the nonwater weight of human cells. Proteins are characterized by: type, structure, function.

Proteins can be fibrous or globular. Fibrous proteins are soluble in water, such as keratin (hair) or fibroin (silk). Globular proteins are insoluble, e.g., hemoglobin.

Structural classes of proteins include simple proteins, i.e., those containing only amino acids, or conjugated proteins, i.e., those containing other components, e.g., glycoproteins (containing sugars), lipoproteins (containing lipids), metalloproteins (with metal ions), nucleoproteins (bind to nucleic acids, e.g., DNA or RNA), and hemoproteins (contain the group heme, which includes both hemoglobin and myoglobin), etc.

Protein structure can be:

- *Primary* — linear sequence of monomers;

- *Secondary* — defined by the phi and psi angles of the backbone atoms of the amino acid residues, and the hydrogen bonds between main chain atoms;

- *Supersecondary* (motifs) — certain arrangements of 2–3 consecutive secondary structures (alpha-helices or beta-strands);

- *Tertiary* — the linear chains fold into specific three-dimensional conformations;

- *Quaternary* — describes how polypeptide chains associate to form a native protein structure.

Functional classes of proteins include enzymes, transcription factors, hormones, membrane proteins, structural proteins, etc. Transcription factors regulate gene expression, structural proteins create structure of individual cellular compartments. Membrane proteins transport material in and out of cells. Enzymes are catalysts that speed up the breaking apart and putting together of molecules; some enzymes function as regulators, controlling the process. Antibodies are proteins with special shapes that recognize and bind to foreign substances, e.g., bacteria or viruses. Hormones are communicating proteins that act as chemical messengers.

Integrity constraints define either constraints on values for a given application or may define triggers, i.e., rules that govern data base updates (adding, modifying, or deleting records). The most basic constraint in any data base is that each record is uniquely identified by its key, i.e., no identifier can contain an empty value (usually referred to as null). Other constraints may be domain dependent. For example, although a protein sequence can be represented by a string, we need an additional constraint that limits it to a combination of 20 amino acids only. As another example, if we want to add a PPI to a data base, the system must ensure that involved proteins are already available. In addition, it may be required that the interaction is labeled as bidirectional or a specific direction of interaction is defined, otherwise the system would not add it to the data base.

Logical design involves the actual implementation of the data base using a DBMS. Schema refinement involves checking data integrity, table validity, and the relational schema for redundancies and inconsistencies. This process is called normalization and will be covered in the next section.

Physical design and tuning considers typical workloads and refines the data base design accordingly. During this process, we optimize table design, implement indexing and clustering approach [455]. The usual goal is to speedup response time to frequent queries, even at the cost of some redundancy in the data base. In such cases, integrity constraints need to be implemented to ensure that the data base integrity is preserved during updates.

Data modeling determines what we need to represent and how. A *data model* is a collection of concepts that describes data and operators that manipulate that data. We can distinguish three models — conceptual, representational, and physical, as defined bellow:

- *Conceptual data model* defines concepts that describe how the data are perceived. An example includes entity-relationship modeling. During a conceptual design phase, we identify entities and relationships for an application, as well as business rules, i.e., integrity constraints, triggers, and stored procedures. The conceptual schema does not provide any operations for data manipulation.

- *Representational data model* provides concepts that describe how the data are organized. Example of models include relational, hierarchical, network, and object-oriented data models.

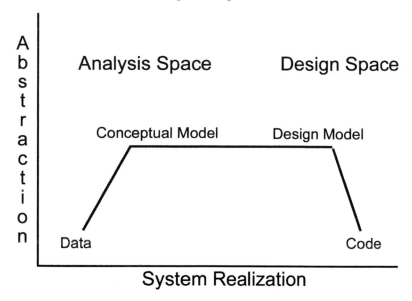

FIGURE 2.2: During design and implementation of a relational data base, we increase and decrease abstraction as we move from data to conceptual design, relational data model, to physical model and implementation.

- *Physical data model* provides concepts that describe how the data are stored, i.e., it describes record formats, a collection of files and indices. In the next section we discuss a relational data model.

The relationship between analysis and design, modeling, and system implementation is presented in Figure 2.2. As we move from actual data to conceptual models, we increase abstraction. Expressing the conceptual design in representational model provides design, which can then be physically implemented.

2.2.3   Relational Data Model

The relational data model [114] is the most accepted data model. It is based on relational algebra and provides a declarative query language to define and manipulate data.

Entity represents a distinguishable real-world object, with described constraints on the permitted values. *Entity set* is a collection of "similar" entities that share attributes. Each entity set has a unique identifier, called *key*.

Attributes describe properties of entities. Individual attributes may have a value or be null. An attribute that identifies each entity in an entity set must have a unique value, and, thus, cannot be null. Attributes can be simple or composite. Examples include the number of amino acids in a sequence,

```
experimental design  (author, type, ...)
array design  (array, element, ...)
samples  (sample source, treatment, labeling, ...)
hybridizations  (protocol, ...)
measurements  (raw data, scanner information, image, ...)
controls  (normalization, control elements, ...)
```

FIGURE 2.3: Designing tables to represent microarray experiments according to MIAME standard.

intrinsic protein properties, or crystallization conditions. Attribute values can be stored or derived, i.e., sequence or sequence length.

A primary key uniquely identifies a record. The most important properties of key columns are persistence, uniqueness and stability:

- *Persistence* — the value must always be present, and it cannot accept null values;

- *Uniqueness* — there is no duplication of values;

- *Stability* — the value should not change over time.

Thus, keys should be defined using attributes that seldom change and are familiar to users. The primary key should contain the fewest number of attributes. One can define a simple primary key, which has only one attribute or a composed primary key that uses a combination of attributes. For example: **Expressed (GID: integer, TID: string, Level: real)**.

A foreign key is a set of fields in one relation that is used to refer to a record in another relation. It is a logical pointer that points to a primary key in the second relation. Referential integrity is achieved when all foreign key constraints are enforced.

As an example, a MIAME (Minimal Information About Microarray Experiment)[7] [80] defines the tables as shown in Figure 2.3.

Since HTP proteomics has started after microarrays, many of the standardized data representations are being adopted to support proteomics experiment representation. For example, Proteome Experimental Data Repository (PEDRo) [187] was designed and implemented to support proteomics data sharing. The HUPO (Human Proteome Organization)[8] PSI (Proteomics Standards Initiative)[9] integrated these efforts and proposed the MIAPE (Minimum Information About a Proteomics Experiment) standard to describe the

---

[7]http://www.mged.org/Workgroups/MIAME/miame.html

[8]http://www.hupo.org

[9]http://psidev.sourceforge.net/

experiment and support interpretation of experimental results [424] and PSI format for PPI data interchange [244].

Relationships among entities can have multiple forms. It can be *one-to-one mapping*, where each entity in entity set E1 is related to, at most, one entity in entity set E2 and vice versa. For example, a clone `hasA` sequence. *One-to-many relationship* allows each entity in E1 be related to, at most, one entity in E2, but an entity in E2 can be related to many entities in E1. For example, a gene `isLocatedOn` a chromosome. *Many-to-many relationship* allows each entity in E1 be related to many entities in E2 and vice versa. For example gene `isExpressedIn` tissue.

In order to develop and use a relational data base, we deploy data definition language (DDL), which defines a conceptual schema, and data manipulation language (DML), used to modify and query data. Different data base systems will have different interfaces that implement these languages, ranging from Structured Query Language (SQL), to visual design and query tools. Further details can be found in related books and user manuals [455].

2.2.4 Data Base Normalization

The normalized design enhances integrity of the data by minimizing redundancy and inconsistency, but at some possible performance cost for certain retrieval applications. The relational data model also defines certain constraints on the tables and attributes, and defines operations on them.

The normal forms defined in relational data base theory represent guidelines for record design. The normalization rules are designed to prevent update anomalies and data inconsistencies. Under first normal form, all occurrences of a record type must contain the same number of fields. Under second and third normal forms, a nonkey field must provide a fact about the key, and use the whole key. In addition, the record must satisfy first normal form. Second normal form is violated when a nonkey field is a fact about a subset of a key. It is only relevant when the key is composite, i.e., consists of several fields. Third normal form is violated when a nonkey field is a fact about another nonkey field. A record is in second and third normal forms if every field is either part of the key or provides a (single-valued) fact about exactly the whole key and nothing else. Under fourth normal form, a record type should not contain two or more independent multivalued facts about an entity. In addition, the record must satisfy third normal form. Fifth normal form deals with cases where information can be reconstructed from smaller pieces of information that can be maintained with less redundancy. Further suggested reading includes [455].

2.3   Knowledge Management

Knowledge management (KM) focuses on the development of concepts, methods, and tools supporting the management of human knowledge. KM tools support the representation, organization, acquisition, creation, use, and evolution of knowledge in its many forms. The goal is to explicitly and formally deal with the semantics of representations and manipulations, to support multiple representations and views, enable and achieve competence, scalability, and extensibility. To realize this, diverse tools and techniques can be used: information repositories, decision-support systems, data analysis and knowledge–discovery algorithms.

The identification of the right concepts for modeling the world for which one would like to do computations is known as ontology. Ontology is a branch of philosophy concerned with the study of what exists.

Over the years, many approaches for representing and managing codified knowledge have been proposed [400]. Although there is a trend to design more flexible, expressible, scalable and robust representations, the choice is highly dependent on the problem domain. In biology, the main challenge is to handle available knowledge effectively, given its ever-increasing complexity and volume. An important aspect in choosing the representation scheme is also consideration of how it will be accessed and used.

As one of the simplest approaches, *controlled vocabularies* provide a standardized dictionary of terms for representing and managing information. Dictionaries can be organized using relations to form taxonomies. *Taxonomies* provide organized dictionaries. *Ontologies* specify domain semantics in terms of conceptual relationships and logical theories. Some of the well known examples include CYC [334], KIF [193], UMLS [328], CIF [544], and GeneOntology [26, 234].

Handling the organization of evolving domains, such as biology and medicine, is challenging because vocabularies are not predefined, taxonomies change, and no single ontology can handle all required semantics. For example, one classification of all organisms on Earth explores similarities of anatomical structures, while molecular biology can define similarities among organisms by defining homologies at the gene or protein levels.

Biology introduces additional problems by relying only on humans to create relationships among individual knowledge sources, since it is not scalable and may be subjective. In order to support systematic knowledge management we need to complement traditional knowledge management techniques with approaches that automate some steps in the process. Tools for automated text processing, information extraction, and information quality control help us to find missing, unexpected, incorrect or incomplete information. In addition, introducing knowledge–discovery systems helps to automate organizing, utilizing, and evolving large knowledge repositories. We aim to use these tools for knowledge to have $4C$ attributes: clean, correct, comprehensive, and current. Through supporting domain knowledge evolution, KM can assist "cleaning" the information, finding outliers, missing or incorrect attribute values, etc. Data mining can help to discover underlying principles, pose and prove hypotheses, and organize information by finding associations, classes, hierarchies and clusters of entities with similar characteristics.

Although challenging on its own, it is not enough to just represent the information. During system design we need to consider whether the codified knowledge will be used by human users or by other computer systems. The first approach focuses on easier comprehension while the other highlights scalability and standardized formats for data interchange.

Several approaches to scalable information management have been proposed in the area of data warehouses, organizational and business knowledge management. However, for several reasons many of these solutions are inadequate for knowledge management. Knowledge is complex and many domains in which we want to support knowledge management are characterized by complex data and information, many unknowns, lack of complete theories, and rapid knowledge evolution as a result of scientific discoveries. We must deal with not only the sheer volume of information in proteomics, but also a diversity of requirements. These requirements range from mainly sequence comparison and clustering during mass spectrometry experiments (introduced in Chapter 3), graph theory during PPI data analysis (discussed in Chapter 4), and real-time image classification in protein crystallization experiments (introduced in Chapter 5). Importantly, all of this information will then be integrated with other biological datasets and data bases (further discussed in Chapters 6 and 7) to provide a more systematic view of biological organisms.

Human experts also need to be considered. When theory is lacking, much of the reasoning process is based on experience. Experts remember positive cases for possible reuse of solutions, and can recall negative experiences for avoiding potentially unsuccessful results. Thus, storing and reasoning with experiences may facilitate efficient and effective knowledge management in continuously evolving domains [8, 281]. We will further discuss this in Chapter 5.

To support the KM process, it is beneficial to extend the notion of information systems. In general, *data* comprise values for observable, measurable, or calculable attributes. For example, individual peaks for a set of peptides constitutes our data from mass spectrometry experiments. Data in context is *information*. In mass spectrometry, identifying and characterizing a peak will provide us with information about a specific protein. *Knowledge* is validated and actionable information. Determining how the intensity of a peak differentiates, for example, healthy vs cancerous samples may give us prognostic value of that protein. In the domain of HTP proteomics we apply data analysis to interpret experimental results and generate knowledge from abundant data.

The trend is to support knowledge management and facilitate knowledge sharing by building computerized information systems with richer and more flexible representation structures. This is supplemented by new services, such as cooperative and distributed query processing [34, 302], similarity-based retrieval [177], and knowledge discovery [166, 535]. This trend also includes support for schema evolution [373, 464], integration and coexistence of unstructured, semi structured, structured and hyper-structured information. However, there is a tradeoff between expressibility of a knowledge representation scheme and the speed by which one can access and use the underlying information. We will discuss this in more detail later in the book.

Information science provides foundations for knowledge management. Traditionally, the document has been the primary piece of information. Documents used to be available in paper format only, and their content was usually individually meaningful. Although some links may have been required to interconnected documents, due to the "representation" limitation, these connections used to have little built-in semantics, were relatively sparse (i.e., a reference pointing to another document, or an untyped hypertext linking another hypertext within the World Wide Web), were quite stable, and were used by human users exclusively.

Information science has been concerned with organizing and access documents efficiently. Several powerful techniques have been developed for managing document catalogues and for retrieving documents [53, 86]. Initially, the technology targeted bibliographic information, then full-text documents, and most recently distributed, heterogeneous information sources, including multimedia biological data bases.

Managing proteomics information involves addressing several challenges, such as ensuring cross-references among diverse set of existing data bases [304] and ensuring that developed standards for experiment data representation are both "comprehensive" and "flexible" [187, 244, 424]. Defining comprehensive standards is necessary to ensure that all useful information is captured (which usually also requires some form of enforcement). In addition, flexibility is needed as introducing new instruments may lead to new data representation requirements. Cross-referencing is nontrivial because of many-to-many relationships, and because electronic information can be arbitrarily large, comprising collections of datasets or documents, or arbitrarily small, representing data (e.g., sequences, models, images, video) and text fragments.

To produce computationally useful results and support human reasoning, KM approaches have to comprise techniques for exploiting meaning. Knowledge-based decision-support systems have been designed to address this challenge. Diverse system architectures have been proposed over the years, including decision trees [452], neural networks [316], Bayesian networks [269], and case-based reasoning systems [331].

Similarly, as none of the data bases will be sufficient, none of the algorithms alone will be effective and efficient overall. There are tradeoffs and design choices that will determine which algorithm, or more usually which combination of algorithms will be most appropriate.

First, one has to decide what is the objective; focusing either on efficiency (i.e., time-performance) or accuracy (i.e., task-performance). For some problems, accurate solution may not even be an option. As we will explore in Chapter 4, comprehensive graph-theoretic analysis of complex PPI networks is intractable even now, and we need to develop powerful heuristics to enable it.

The second question relates to the form of data; we may be able to deal with numbers, complex and interconnected symbolic knowledge, or multimedia information. Clearly, some algorithms can handle only numeric data and while one can always transform any data into a numeric format, this is frequently only an approximate solution, usually ineffective and definitely inflexible.

The third issue concerns supporting information available and the type of users. Supporting information could provide cues about type of queries and analyses, which may be used to optimize data base storage. Type of users will define the needed level of sophistication for data interpretation and visualization.

Further, it is important to distinguish knowledge representation requirements for *stable* vs. *evolving* domains. In the first case, problem solving can use existing theories, principles, and generic knowledge, usually extracted from domain experts during knowledge acquisition. In the second case, the system has to rely on experiential knowledge, examples, and handle exceptions. Knowledge-based systems not only acquire different knowledge using a different process but they also need to use problem-solving algorithms that fit the process. For example, decision trees work well for well-structured information, but are less effective for categorical and continuous variables; neural networks can handle complex domains, but the process is less interactive with human experts; Bayesian networks can solve complex problems, but may not scale up to large problems; case-based reasoning is both intuitive and interactive, but may not scale to high-dimensional problems.

We will focus on case-based reasoning due to its ability to effectively manage expert knowledge, as we further discuss in Chapter 5. We enhance the system with self-organizing maps, and Bayesian networks to further improve analysis and interaction with human experts.

2.3.1 Knowledge Management in Complex Domains

There are several reasons why using conceptual modeling, knowledge management, and artificial intelligence is useful and necessary in complex domains such as biology and medicine. Below we list some of the most important characteristics:

- Domains are dynamic. As we acquire and analyze more information from experiments, our understanding of the world is continuously evolving. This progress must be supported and captured by the flexibility and evolution of the knowledge representation schemes used.

- Domains are only partially understood. Deciding what information to capture, how to represent it, and how important are individual properties is often based on a "hunch" rather then specific experience or scientific evidence. As we acquire, organize, and analyze the information, we may use machine-learning and knowledge–discovery techniques to gather evidence on how to better represent and organize information. Because domains are evolving continuously, this process must be dynamic — the feedback from biological experiments is used to improve computer analysis, which in turn improves biological understanding and design of new experiments. The process is iterative and intertwines biological experimentation with computational analysis. In addition, it may be useful to make the process interactive to further enhance the interpretation potential.

- In biology we deal with a collection of loosely coupled, federated data bases, but need to integrate existing information. This integration has to be automated, and, thus, we need a meta description of individual data bases so that generic scripts could be used to create and update various views. Individual views will combine pieces of several data bases, such that we will have the possibility to discover patterns that cross the boundaries of multiple data bases. Because of performance issues, we cannot rely on Internet speed during reasoning and analysis — but we cannot move into centralized data bases either. However, considering the evolution and need for information feedback, we need a mechanism to modify representation schemes of individual data bases based on analysis of "a centralized view of the federated data base." In addition, we need an automated mechanism to migrate data to a new schema, update programs that operate on these data and generate a request for what additional data has to be collected (if possible) for previous experiments, in order to satisfy the new schema.

Knowledge management research focuses on the development of concepts, methods, and tools supporting the management of human knowledge. Broadly speaking, this comprises support for creation, acquisition, representation, organization, usage, and evolution of knowledge in its many forms.

Supporting efficient and effective knowledge management requires the organization of computer-represented knowledge into semantically meaningful and computationally efficient structures. Conceptual information models are used to capture the meaning of information [398] and have several representation components:

1. *Primitive concepts* that describe an application, e.g., entity, activity, goal and agent;

2. *Abstraction mechanisms* that are used to organize the information, e.g., generalization, aggregation and classification [400];

3. *Operations* that can access, update and process information, also referred to as knowledge-management operations.

4. *Integrity rules* that define a set of valid information base states, or changes of states [398];

5. *Ontology* refers to the selection of appropriate concepts for modeling the world in which the system must support the operations.

In order to define a conceptual model we must make assumptions about the application we intend to model. For example, if the application consists of interrelated entities, such as protein, sequence, structure, enzyme, etc., then we need to include terms such as entity and relationship into the conceptual model. In addition, the semantics of these concepts and their relationship is used to define knowledge-management operations, such as navigation, search, retrieval, update, and reasoning. This would include sequence alignment, structure prediction, identification of structure motifs, or formulating crystallization optimization plans from initial crystallization screen results.

2.3.2 Abstraction Mechanisms

Abstraction mechanisms support organization and effective use of knowledge through suppression of irrelevant detail. The relevancy depends on the task and the use of information, and, thus, it changes with the context.

There are six main abstraction mechanisms: classification, generalization, aggregation, contextualization, materialization, and normalization [400].

Classification organizes basic representation units (e.g., entity, relationship, attribute, activity) under one or more generic units, called classes. Instances of a class share common properties. Classification has been studied in the machine-learning community to support information clustering [101, 414, 504], and problem solving [525]. Knowledge-based or conceptual clustering employs feature similarity to organize knowledge in knowledge bases [388]. In a conventional data analysis, objects are clustered into classes based on some similarity measure. The similarity of two objects is usually represented as a number, i.e., the value of a similarity function applied to symbolic descriptions of objects. Thus, this similarity measure is context-free. Such methods fail to capture the properties of a cluster as a whole and are not derivable from properties of individual entities [504]. A context-based classification helps to retrieve all entities that are relevant to a given task, which improves a system's accuracy and flexibility [279]. For example, we can define entity "protein" as shown in Figure 2.4.

Clearly, the representation we chose significantly impacts on what we can retrieve or infer. For example, we can represent information about a molecule as one-dimensional chemical formula ($C_3H_7NO_2$), which would be sufficient to infer how many atoms of carbon the molecule has. However, inferring connectivity, shape, and angle would require more complex models of protein, as a two- or three-dimensional structure.

Generalization organizes existing concepts into taxonomies, referred to as isA or generalization hierarchies [79]. The most general concept is the root of relationship Gene `codesFor` Protein

Protein `expressedIn` Tissue

entity Protein
    Name
    Species
    SwissProt_ID
    Sequence
    . . .

FIGURE 2.4: Using classification abstraction mechanism, we can define entity "protein" describing its relationships to gene and tissue, and defining its basic attributes.

the hierarchy. Inheritance inference rule states that attributes and properties of a class are also attributes and properties of its specializations.

For example, using GeneOntology [26, 234], we can define components of the hierarchy as depicted in Figure 2.5.

Aggregation views objects as aggregates of their components or parts. In this view, it represents a `partOf` relationships. Components of an object might themselves be aggregates of other simpler components. For example, the crystallization data base will contain experiments that will be aggregates of protein properties, crystallization conditions, and experiment results. In turn, protein property will comprise many measured and calculated attributes. Another example using GeneOntology [26, 234]:

Contextualization partitions the knowledge base into multiple viewpoints [89]. These viewpoints will dynamically change as a consequence of how one sees individual objects based on the task and intended use. Reasoning with information in context improves classification accuracy and system flexibility [279].

For example, knowing that two proteins interact is not sufficient if we do not have information about potential stimuli, tissue, time, and other experimental details. Such an example is presented in Figure 2.7.

Materialization relates a class to a more concrete class. The formal properties of materialization constitute a combination of those of classification and generalization [444]. This may relate to capturing information about a repeated crystallization experiment of a given molecule with the same method and conditions, or a protein may have several 3D structures available, each at different resolution.

Another example involves alternative splicing, which leads to the production of more than one related protein (or isoform) from the same gene. This regulatory mechanism varies incorporation of exons and coding regions into

Protein metabolism
    `isA` Primary metabolism
    `isA` Metabolism
    `isA` Physiological process
    `isA` Biological process

Intracellular organelle
    `isA` Membrane-bound organelle
    `isA` Organelle
    `isA` Cellular component

Protein binding
    `isA` Binding
    `isA` Molecular function

FIGURE 2.5: Generalization abstraction mechanism can be used to define components of the Gene Ontology hierarchy using `isA` relationship.

Cellular component
    & Biological process
    & Molecular function
`PartOf` Gene Ontology

FIGURE 2.6: Aggregation abstraction mechanism can be used to define `partOf` hierarchies in Gene Ontology.

mRNA (messenger RNAs). As a result, many proteins have several splice variants, such as $p73$ has 7 variants[10] (see Figure 2.8).

Similarly, transforming factor beta (TGF-β) has three isoforms: TGF-β1, TGF-β2, and TGF-β3. This results in one-to-many relationships between **genes** and **proteins**. Thus, combining classification with generalization, we can then represent TGF-β as shown in Figure 2.9.

Normalization models typical entities first and then exceptions [74]. For example, generalization may lead to over-abstraction, and, thus, there must be a systematic process to analyze and deal with the abnormal cases [73]. This is important in biology when knowledge–discovery algorithms may help us to formulate underlying principles and theories, and we need to capture information about situations when the model does not work.

Supporting knowledge management in complex and dynamic domains benefits from extending traditional approaches with automated methods. Selection

---

[10]http://www.expasy.org/cgi-bin/niceprot.pl?O15350

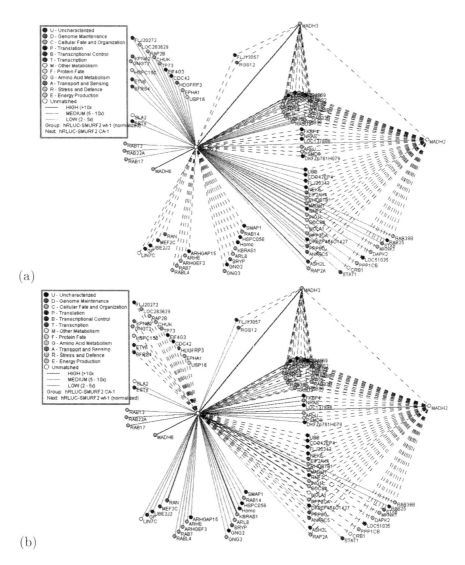

FIGURE 2.7: (See color insert following page 72.) Using an example from [48], we show the utility of HTP mammalian cell-based assays for identifying dynamic PPIs in the TGFβ pathway. Panel (a) shows SMURF2 wild-type, while panel (b) shows how PPIs change when SMURF2 is mutated (position of all proteins is unchanged between panels (a) and (b)) Dotted line represents known but inactive interaction; red line represents strong interaction; blue line represents medium strength interaction; black line represents weak interaction.

Protein name:	Tumor protein p73
Swiss-Prot ID:	O15350
Name:	Alpha
Isoform ID:	O15350-1
Name:	Beta
Isoform ID:	O15350-2
Name:	Gama
Isoform ID:	O15350-3
Name:	Delta
Isoform ID:	O15350-4
Name:	Epsilon
Isoform ID:	O15350-5
Name:	Zeta
Isoform ID:	O15350-6
Name:	Kappa
isoform ID:	O15350-7

FIGURE 2.8: Materialization-abstraction mechanism can be used to represent splice variants, such as 7 variants of tumor protein *p*73.

tgfb codesFor TGFβ

TGF-β1 isA TGFβ
TGF-β2 isA TGFβ
TGF-β3 isA TGFβ

FIGURE 2.9: Combining classification and generalization abstraction mechanisms enables us to represent many-to-many relationships between genes and proteins, by representing gene and individual splice variants.

of appropriate concepts for modeling the world in which we want to support the operations is referred to as ontology. Through this, KM can support knowledge sharing and collaboration, it can help to improve performance, i.e., competence, scalability, and flexibility of the system, and importantly KM can aid knowledge organization and new discoveries by finding and representing associations, causal effects, prediction, and evidence.

2.3.3 Ontologies

Ontology is a branch of philosophy concerned with the study of what exists. Formal ontologies have been proposed since the 18*th* century, including more recently [26, 80, 87, 95, 244, 326, 334, 544].

In computer science, ontologies have many forms, including lexicons, dictionaries, thesauri, and first order logical theories. Lexicons provide a standardized dictionary of terms that is useful during indexing and retrieval. Dictionaries can be organized according to specific relations to form hierarchies, such as taxonomies, meronomies, etc. Thesauri expand individual terms with related terms. Thus, ontologies are useful because they support terminology standardization within a given domain. When ontologies are formalized in first order logic, they can also support inference mechanisms, and, thus, they can be used to derive new facts or check for knowledge consistency. Such computational aids are clearly useful for knowledge management, especially when one is dealing with large amounts of evolving knowledge.

For example, if one is interested in protein crystallization knowledge, then intrinsic protein properties (e.g., molecular weight, isoelectric point, amino acid sequence), crystallization reagents, pH, temperature, concentration, classification of crystallization experiment outcome (i.e., clear, crystal, microcrystal, precipitate, etc.) might be among the primitive concepts that will be used to describe the application domain. These concepts and their meanings together define an ontology for protein crystallization domain. Such an ontology can be used as common knowledge that facilitates communication among crystallographers, and aid in knowledge management and discovery of principles guiding crystallization process (which we further discuss in Chapter 5).

In the broader terms, biologists have been attempting to classify and organize characteristics of all species on Earth. Although it is estimated that 10–20 million species exist [364], we have named only about 1.7 million to date. Collecting and managing this information is a challenge. One of the emerging approaches is to use DNA bar codes to aid classification of all species [238, 364]. This effort resulted in establishing the Consortium for the Barcode of Life (CBOL). The unique fingerprint is based on a small part of the mitochondrial genome — 650–750 bases of the cytochrome *c* oxidase I gene (COI). Although promising to catalog all known 1.7 million species, classifying all estimated 10–20 million species on the planet may not be directly feasible until several challenges are resolved. For example, plants or young species (e.g., orangutans) may not be tracked with this approach. To resolve these challenges, additional genes may be used.

Early work in computational ontologies includes the Cyc project [334] and the ARPA Knowledge Sharing effort [402]. The Knowledge Interchange Format (KIF) effort provides a declarative language for describing knowledge [193]. Research within artificial intelligence has formalized interesting specialized ontologies and has developed techniques for using them to represent and analyze knowledge.

As a specific example, a flexible and open specification language called MIAME (Minimum Information About Microarray Experiments) has been designed to represent information about microarray experiments [80]. Although details for particular experiments may be different, MIAME aims to define the core that is common to most experiments. MIAME is not a formal specification language, but a set of guidelines, which encourages users to provide their own qualifiers and values identifying the source of their terminology. It promotes the use of controlled vocabularies and external ontologies. An important benefit of MIAME is availability of a more complete set of information about individual experiments.

As in any other field, there is a tradeoff between rigidity and strictness, and flexibility of any specification. For example, the recent Protein Structure Initiative (PSI) ontology for PPI data bases [244] leaves many pieces of information as noncompulsory. Although this may increase the rate of acceptance initially, the resulting data sets will be limited because many important pieces of information will likely be only partially filled for individual records.

Another important consideration is the intended use of ontologies. Although most in the computational community implicitly assume that ontologies are to be used by computers to enable knowledge interchange, biology often uses ontologies to describe knowledge for humans to digest and interpret. Clearly, both approaches are valid, but since they have completely different goals, they will result in different ontologies and tools [258].

2.3.3.1   Ontology Classification

To characterize and classify current work on ontologies we have proposed four broad ontological categories, to cover static, dynamic, intentional, and social aspects [284]. Knowledge in most applications can be represented using the primitive concepts, derived from these four ontological categories. Different applications may use only a subset of defined entities.

Static ontology describes static aspects of the world, i.e., individual things, their attributes and relationships. For example, the static ontology can be used to describe properties of proteins, conditions for crystallization, etc.

A dynamic ontology describes the changing aspects of the world. The primitive concepts include state, state transition, and process. For example, transient PPIs are not present all the time, and thus describing the context of when these interactions exist and which proteins are participating is vital to understand the dynamic properties of interactomes. Dynamic ontology can also be used to describe how the crystallization outcome changes over time, and thus capture the temporal properties of the crystallization process. We can also use this ontology to describe the overall process of structure determination, starting with target selection, purification, HTP screening, diffraction-quality crystallization, actual diffraction experiment, and structure determination.

Most applications would require combined ontologies. For example, modeling a university environment may use entities and relations to model static aspects of the domain and processes to model dynamic aspects. A multiagent hospital information system may require concepts such as agent, team, goal, and social dependency to model social and intentional aspects of the application. Although agents and goals could be represented as entities, such representation would miss important properties of agency (e.g., an agent's intentionality and autonomy), which would lead to incomplete forms of inference. Representations that use elements of dynamic ontologies support simulation, a special form of inference not supported by static ontologies. Likewise, formal goal models support their own special forms of reasoning, e.g., [203] that go beyond inference mechanisms for entities and relationships.

2.3.3.2   Development and Management of Ontologies

Different applications may use all or only a subset of entities from the four ontological categories. Ubiquitously for any or all categories, information system development benefits from using tools for ontology design and integration processes.

DAML (DARPA Agent Markup Language) and OIL (Ontology Interface Layer) define a semantic markup language for web resources [117, 251]. DAML has been designed as an extension of XML (eXtended Markup Language) and RDF (Resource Description Framework). It offers a language and tools developed to facilitate a richer representation of the web, called semantic web. The semantic web extends the current web by defining meaning for information, which enables better information interchange and thus promotes cooperation among computers and people. OIL provides classification using constructs from frame-based artificial intelligence, combined with the expressiveness and reasoning power of description logics. DAML+OIL supports rich modeling primitives and provides a set of constructs for creating ontologies.

Once ontologies are defined for one or several domains, they may be organized into libraries, which enhances their reusability [240]. Such libraries can be used to build information systems by supporting requirement acquisition and design [523]. Ontolingua enables authoring and reuse of ontologies by providing tools for assembling and enhancing libraries of modular, reusable ontologies [218].

As the number and size of ontology libraries grow, tools must be used for their management. Analysis tools can help with ontology verification and validation. Verification checks whether an ontology satisfies particular constraints. For example, verification tools can check cardinality constraints for entity-relationship models or semantic consistency of rules and constraints such as "patient cannot have more embryos than she had oocytes." Validation checks the consistency of a model with respect to its application. Since the application is usually described informally, validation has to be done manually or semiautomatically.

Other supporting tools include ontology editors, viewers, and servers to assist creation, maintenance, and use of ontologies [112, 289]. For example, NEON (Networked-based Editor for ONtologies) promotes standardized radiology appropriateness criteria [289]. Individual concepts are represented in a semantic network and the system supports import and export of ontologies using SGML (Standard Generalized Markup Language). Individual entities include concept name, abbreviation, synonym, and links such as `affectedBy`, `hasPart`, `partOf` and `imagedBy`. This approach can help to not only standardize terminology but also organize existing vocabularies.

On the application side, the major effort has been developing static ontologies, such as taxonomies or controlled vocabularies [26, 206, 234, 397, 420]. The aim is to standardize terminology and create concept taxonomies in individual domains to enable information sharing and system cooperation. For example, an agent in a medical diagnosis system may use an ontology of clinical concepts, during both structured data entry and decision support. A diagnostic agent may need to cooperate with a bibliographic agent that uses an ontology for bibliographies to associate literature references with particular diseases to enable explanation and support evidence-based medicine.

2.3.3.3 Ontologies in Proteomics

Similar to the MIAME standard [80], several efforts have emerged for standardizing data from proteomics experiments. The MIAPE (Minimum Information About a Proteomics Experiment) standard defines the minimum reporting requirement for proteomic experiments in aim to support interpretation of experimental results [424]. This standard was proposed by the HUPO's (Human Proteome Organization)[11] PSI (Proteomics Standards Initiative)[12], combining early efforts from Proteome Experimental Data Repository (PEDRo) [187]. To specifically handle PPI information, a data interchange standard has been proposed [244].

The field is rapidly evolving, with new technology putting new requirements on type of data to be represented. Clearly, we cannot formalize everything, but defining comprehensive standards is needed to ensure that all useful information is captured. Ontologies must be both flexible — to handle new requirements, and enforced — so not many "voluntary" fields are present. Storing this information is context-sensitive and requires richer metadata. Notably, we are not only concerned to record information about proteomic experiments, these ontologies have to support efficient retrieval and data mining operations. Accessing information based on experimental details, mass-to-charge ratios, sequence, etc. is essential.

To further specialize generic proteomic data, the HUPO PSI General Proteomics Standards[13] (GPS) has been extended to form PSI-MS[14] (Proteomics Standards Initiative — Mass Spectrometry). The mzData standard captures peak-list data and aims to unite the large number of current formats (pkl's, dta's, mgf's, etc.) mzData is already supported by some software providers, while several other companies (e.g., Bruker Daltonics, Kratos Analytical and Matrix Science) are currently implementing the mzData standard. The mzIdent standard captures parameters and results of search engines such as Mas-

---

[11] http://www.hupo.org

[12] http://psidev.sourceforge.net/

[13] http://psidev.sourceforge.net/gps/index.html

[14] http://psidev.sourceforge.net/ontology/index.html

cot and Sequest. Although the standards have been defined and accepted, the supporting ontology has not yet been created. The goal of the PSI Mass Spectrometry work group is to to extend its coverage beyond the peak list to encompass the full chain from the mass spectrometer to the identification list.

The PSI object model data and metadata may be in different formats since they have been produced at different places. As a result, comparison and exchange is challenging. Thus, the effort is to develop a consensus object model, and XML interchange format (the XML schema — PSI-ML — will be used as a standard for data exchange). This will not only facilitate data exchange (and, thus, dissemination) but will aid the development of effective search and analysis tools. Considering that computational biology is increasingly involved with integrated analyses, the PSI ontology will eventually be merged with the MGED Ontology, under the PSI namespace.

Other GeneOntology tools include for example, The Gene Ontology Annotation (GOA) data base[15] [92], GOAnnotator[16], QuickGO[17], AmiGO[18], and I-Hop[19] [248].

2.3.3.4 Ontologies in Medicine

The most prevalent ontology category used in medicine is static, focusing on defining standard medical vocabularies, such as the International Classification of Diseases (ICD-9-CM), Systematized Nomenclature of Human and Veterinary Medicine (SNOMED), Medical Subject Headings (MeSH), Read Codes of clinical terms, Current Procedural Terminology (CPT), Unified Medical Language System (UMLS), Generalized Architecture for Languages, Encyclopedia and Nomenclatures in medicine (GALEN), etc. However, none of these standards is sufficiently comprehensive and accepted to meet the full needs of the electronic health record [328, 510]. In addition, combining and synchronizing individual versions of existing medical vocabularies remains an open problem [420].

To tackle this challenge, the National Library of Medicine has created a UMLS [257], which is a composite of about 102 sources that contain 975,354 concepts and 2.4 million concept names (UMLS 2003 AC). UMLS is available via an UMLS Knowledge Source server (UMLSKS) that provides a set of web-based interaction tools as well as an API (Application Programming Interface) to access the UMLS biomedical terminologies. The current UMLSKS release comprises three knowledge sources:

1. *UMLS Metathesaurus* (UMLS-MT) comprises information about biomedical concepts from diverse controlled vocabularies, as well as

---

[15]http://www.ebi.ac.uk/GOA

[16]http://xldb.fc.ul.pt/rebil/tools/goa/

[17]http://www.ebi.ac.uk/ego

[18]http://www.godata base.org/cgi-bin/amigo

[19]http://www.pdg.cnb.uam.es/UniPub/iHOP

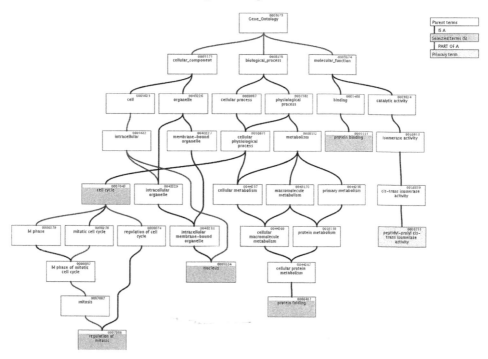

FIGURE 2.10: (See color insert following page 72.) Graphical visualization of GeneOntology hierarchy [26, 234], using EMBL-EBI QuickGO server (http://www.ebi.ac.uk/ego/DisplayGoTerm). The hierarchy covers both classification (isA relationship) and generalization (partOf relationship).

classifications used in patient records, administrative health data, bibliographic and full-text data bases, and decision-support systems;

2. *Semantic Network* (UMLS-SN) defines semantic types for consistent categorization of all concepts from UMLS-MT. The links between the semantic types represent relationships in the biomedical domain and, thus, provide the structure for the network;

3. *Specialist Lexicon* (UMLS-SL) is an English language lexicon containing syntactic, morphological, and orthographic information for biomedical concepts.

To support integrating and reusing existing terminological ontologies in medicine Steve *et al.* have designed an ontology library ON9 [505], which is written in Ontolingua [218] and Loom [355]. It includes thousands of medical concepts and organizes them into domain, generic, and metalevel theories. They use a methodology called ONIONS to aid construction of ontologies starting from existing, contextually heterogeneous terminologies. This work led to a successful integration of five medical terminology systems: the UMLS-SN (about 170 semantic types and relations, and their 890 defined combinations), SNOMED-III (about 600 most general concepts), Gabrieli Medical Nomenclature (about 700 most general concepts), ICD10 (about 250 most general concepts), and the Galen Core Model-5g (about 2,000 items).

As discussed earlier, it is important to provide tools and techniques to facilitate the design and organization of such vocabularies. Earlier models, such as ICD-9-CM, DSM, SNOMED, and Read codes Version 2 (CTV2 — Clinical Terms Version 2 of the Read Codes) use the code not only to identify a concept uniquely, but to indicate concept position within the tree-structured hierarchy. As a result, a particular concept can be associated only with one node of the hierarchy. In addition, the number of levels in the hierarchy is usually limited, since existing codes have a fixed number of alphanumeric characters and each character indicates a level.

For example, A--- specifies infectious and parasitic diseases, A1-- specifies tuberculosis, A130 specifies tuberculosis meningitis, while F00- in a different part of the hierarchy specifies bacterial meningitis, and F004 meningitis tuberculous. This results in duplication since the same concepts are represented by two different codes.

Alternatively, some systems do not use code to indicate hierarchical location, e.g., Read codes Version 3 (CTV3), the MED (Medical Entities Dictionary) and SNOMED-RT. The alphanumeric codes are a label for the concept and no longer represent the hierarchical relationship. Liu et al. [345] describes a DAG (Directed Acyclic Graph) structure enrichment of the tree-structured semantic network of the UMLS. Gu et al. proposes a methodology to partition vocabularies into contexts, where each context contains an isA tree hierarchy [220]. Liu *et al.* shows how to partition an existing MED dictionary, which comprises 48,000 concepts, over 61,000 isA links and over 71,000 additional links (e.g., categoryOf, roleOf) [344, 345]. Based on the partitioning into sets of concepts with the same sets of properties, MED has been implemented using a commercial object-oriented data base management system ONTOS.

Ontologies are also becoming an integral part of bioinformatics since they encourage common terminology for describing complex and dynamic biological knowledge [477, 506]. Ontologies can support a common access to diverse information repositories, e.g., TAMBIS (Transparent Access to Multiple Bioinformatics Information Sources) [36]. A uniform global schema and query interface in TAMBIS is achieved by using an ontology that describes diverse bioinformatics tasks and resources [37]. The TaO (TAMBIS Ontology) schema does not have materialized instances. Instead, instances are extracted from several distributed data bases.

Similarly, [77] describes the use of Doublin Core for creating a semantic medical web, to link medical web resources in a novel way to support enhanced retrieval and navigation. A step further in this direction is PharmaGKB, a frame-based ontology for pharmacogenomics [421], which was designed to support heterogeneous data integration and data acquisition.

Automating data acquisition and ontology creation further requires text mining and natural language processing [225]. The trend is moving into automating the process of ontology creation and using the existing ontologies during reasoning and simulation. This trend will result in development of new ontologies, especially dynamic and intentional. Application of reasoning systems in medicine will benefit from introducing social ontologies.

2.4 Conclusions

The KM goal is to collect and store data for interpretation. However, challenges come from the sheer volume and complexity of this information. Human genome data requires $200,000$ pages to print, which is relatively easily manageable. But we need DBs for accurate representation of maps (i.e., linkage, physical, disease loci), sequences (i.e., genomic, cDNA, proteins), and link them to each other and to biomedical bibliographic text data bases, such as PubMed with over $14,000,000$ records in 2005. Human genetic diversity puts further challenges, as any two individuals differ in about $3x10^6$ bases (0.1%). Since the population is about $5x10^9$, a catalog of all sequence differences would require $15x10^{15}$ entries.

Coping with deluge of data is no longer an option; it is a necessity of current and future biomedical research. For example, the large Hadron Collider at the EU Laboratory for Particle Physics (CERN) in Switzerland will generate 100 millions GB of data from the collider itself [21]. The US Geological Survey (USGS) is photographing the entire US from the air at high resolution, the project that generates 9 TB of uncompressed data and will grow to 12 TB [21]. In 1998 it went online with MS Terraserver, one of the largest DBs accessible on the web. NASA's Earth Observing System will produce 1 TB of satellite data every day, and a 1 PB every 3 years [21].

Bioinformatics is now expected to not only support managing data and information, but enable knowledge management as well. Some of the advanced technologies used include: cooperative query processing [110], similarity-based retrieval and browsing [282], data and text mining, knowledge discovery and visualization [104, 287, 458], text understanding [225, 462], data translation services [219], and knowledge sharing [425], to name a few. Useful support for knowledge management depends on effective embedding of the meaning in information models and ontologies.

The ontological approach with an information modeling bias derives its power from models of domain knowledge, which can be formally analyzed and processed. However, frequently there is resistance to precise formalization. Although sometimes it may be useful to have a more informal approach, true knowledge management requires a standardized approach.

For example, consider the design of a data entry form. If the content of the form fields can be arbitrary text strings, then little computational leverage can be derived from the formalization. Although such a format is highly flexible and can accommodate a broad range of inputs, not much can be done in terms of automated consistency checking. However, restricting a field content to a finite set of predefined values, which adhere to specific rules, may enable automated consistency checking, and subsequent data mining.

After the initial success of data-mining and knowledge–discovery techniques in business applications, they are now mainstream in HTP biology. However, if the data was not created having data mining and decision support in mind, how well can it serve that purpose? Data in these domains require robust approaches to deal with missing and noisy information. As Aeschylus noted: "Who knows useful things, not many things, is wise." One fallacy in dealing with HTP biological data is ascribing too much meaning to individual data points. The main reason is that protein expression profiles or PPI data contain noise that can prevent reliable conclusions for specific proteins. One promising approach is to combine orthogonal datasets, such as the same kind of information from different platforms; e.g., interaction data from phage display and 2-hybrid approaches, or from the integration of data of completely different forms, e.g., gene expression data with PPI data. We will further expand on these issues in Chapter 6.

Chapter 3

Current Status and Future Perspectives of Mass Spectrometry

Thomas Kislinger and Andrew Emili

Proteomics research is usually defined as the global detection of proteins present in biological samples. Protein sequencing has long been used to identify proteins in biochemical research. N-terminal sequencing methods using stepwise, chemical peptide degradation, such as Edman degradation [149, 150, 151], are tedious and slow. The traditional sequencing technology has proven useful for the identification of many proteins, but does not scale up to the automated high-throughput (HTP) experimentation required for global expression proteomics.

Over the last decade the field has benefited greatly from rapid improvements in protein mass spectrometry (MS). Developments in MS instrumentation and associated computing tools now permit routine protein identification and quantitation in a rapid and sensitive manner, as well as determination of sites of posttranslational modification [6, 176, 299].

In this chapter, we focus on two aspects important to biological studies. First, we describe state-of-the-art MS-based experimental procedures in widespread use in proteomics research laboratories. These techniques enable the large-scale characterization of hundreds to thousands of proteins in complex biological samples such as blood, cell and tissue extracts. Second, we review back-end bioinformatic tools that are essential for handling the flood of data emerging from global proteomic studies.

3.1 Basic Concepts of Mass Spectrometry

MS has a long tradition in organic chemistry for the detection and structural characterization of volatile, low molecular weight molecules [395]. In the mid-1980s, two mild ionization techniques were introduced, which enabled ionization of large intact macromolecules such as peptides and proteins:

- Electrospray ionization (ESI) [169];

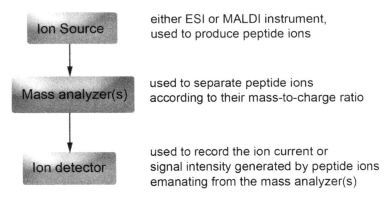

FIGURE 3.1: Essential components of MS systems.

- Matrix assisted laser desorption/ionization (MALDI) [293].

This progress transformed MS into an enabling technology in proteomics research. The tremendous impact of these technical accomplishments was recognized with the 2002 Nobel Prize in Chemistry, awarded to the ionization research pioneers J. Fenn and K. Tanaka.

Biological MS is carried out on protein and peptide analytes that have first been transferred to the gas phase, usually in a protonated form. Since lower molecular weight molecules are better suited to the mass-range limitations of most MS systems, proteins are usually cleaved chemically or digested enzymatically to produce smaller peptides prior to analysis.

Trypsin is the preferred enzyme in proteome research for three important reasons:

1. It produces peptides with a typical mass smaller than 3–4 kDa, which generally create good spectra in collision-induced dissociation fragmentation experiments (described below).

2. It cleaves C-terminal to the highly basic residues lysine and arginine, resulting in the positive charge (proton) becoming localized to the C-terminus of each generated peptide, which in turn results in more extensive and predictable peptide fragmentation patterns.

3. Peptide mixtures can be efficiently fractionated using high-performance liquid chromatography (HPLC), allowing for a sizeable reduction in sample complexity, which, in turn, can improve overall proteome detection coverage.

All MS systems comprise three essential components, integrated into a simple work-flow: an ion source, one or more mass analyzers, and an ion detector (see Figure 3.1).

A spectrum typically comprises a digitized array of the intensities of all discrete mass-to-charge (m/z) values (i.e., peptides) detected over a discrete mass range. A spectrum is usually produced after only one to two seconds of data acquisition time. Detection limits in the subnanogram (pico- to femto-molar) range are routinely achieved.

3.1.1 Electrospray Ionization

The ESI process produces ions through the application of a high potential voltage to a liquid solution containing the protein/peptide sample, typically made acidic (resulting in the peptides becoming protonated) prior to loading, that is sprayed out and nebulized from a narrow capillary tube or column under atmospheric pressure. This results in the emission of a so-called tailor cone of fine highly charged droplets (typically positively charged due to the excess of H+ ions) that is then directed into the MS inlet. Heat-assisted desolvation of the droplets, combined with the presence of a strong vacuum inside of the MS instrument, leads to the rapid generation of positively charged gas phase peptide ions.

Since macromolecules are readily transferred directly from the liquid phase into the gas phase, ESI acts as a perfect interface for coupling MS to solution-based peptide separation techniques like high performance liquid chromatography (LC) and capillary electrophoresis (CE). The peptides are generally stable under these conditions, hence the process is often referred to as a soft ionization method.

Moreover, since MS instruments are concentration-dependent detectors, reductions in sample volume and flow rate can result in a dramatic increase in detection sensitivity. The introduction of microscale columns and emitters coupled to low flow (submicroliter) commercial LC systems has, therefore, emerged as a particularly invaluable technique for protein detection. Sample flow rates in the nanoliter per minute range routinely allow for femtomole or better detection limits when combined with state-of-the-art MS instruments, which in turn enables the monitoring of lower abundance proteins in complex biological mixtures [215, 550].

3.1.2 Tandem Mass Spectrometry

Peptide "sequencing" can be performed in real-time during an LC-MS analysis using a specialized MS procedure known as *tandem mass spectrometry* (MS/MS).

In MS/MS experiments, individual peptide precursor ions are first detected using a precursor ion scan, and then electromagnetically isolated by the MS system as they elute from the column and enter into the MS system via the ESI source (see Figure 3.3). The isolated peptides are subsequently subjected to energetic collisions in order to induce peptide fragmentation. These product

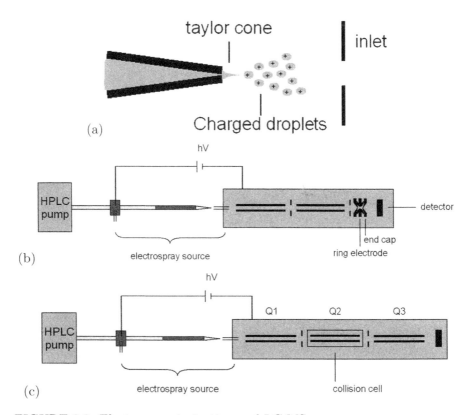

FIGURE 3.2: Electrospray ionization and LC-MS.
(a) The electrospray ionization process sprays a fine mist of ionized droplets directly from the tip (i.e., nozzle) of a microcapillary chromatography column into the inlet of a mass spectrometer. (b) Basic instrument schema of an ion trap tandem mass spectrometry coupled to a microcapillary liquid chromatography system. (c) Basic instrument setup of a triple quadrupole mass spectrometry coupled to a microcapillary column.

ions are then resolved and detected by a second round of mass analysis, which can be separated either in space or time.

Fortuitously, most protonated parent precursor ions tend to fragment along the peptide amide backbone, resulting in a predictable ladder of N- and C-terminal fragment ions (so-called *b*- and *y*-ions). In effect, this process allows for either partial or full determination of the amino acid sequence of virtually any fragmentable peptide (discussed further in detail below).

Interpretable MS/MS spectra are usually produced after only about 1–2 seconds of data acquisition time, and the entire process of peptide selection and MS/MS analysis can be fully automated (so-called data-dependent or data-directed data acquisition). This enables high-throughput, large-scale proteomic characterization of even very complex peptide mixtures in a relatively straight-forward manner.

Although the process of spectral data interpretation for the purpose of sequence determination can be performed manually by an expert in the field, computer-based data base search algorithms have been developed to both automate and vastly speed up the time of sequence analysis (discussed further in Section 3.4). Once a spectral match is made, both the peptide sequence and the corresponding cognate parent protein are in effect determined. This form of "shotgun" sequencing has proven to be a robust and effective method for protein identification since even a single reliable peptide or spectral match may often be sufficient to identify a protein candidate in a sample. These search algorithms have dramatically improved the efficiency and overall accuracy of protein expression profiling, and, therefore, serve as the backbone for a HTP informatics pipeline.

Data generated from a single LC-MS analysis can regularly be used to confidently identify upwards of several thousand peptides in a given biological sample, which in turn often map back to several hundred distinct proteins. Nevertheless, multiple technical issues can complicate such profiling studies, most of which stem from the extremes in complexity often found in complex biological samples:

1. Since the duty cycle of most MS systems is finite and rate limiting, peptide mixtures are usually "under-sampled" by the instrumentation. That is, most MS systems scan too slowly to trap and fragment all of the peptide species eluting from the chromatography system over the course of an LC-MS experiment. As a result, many peptides, especially those from lower abundance proteins, frequently go undetected and are, therefore, "lost" from any downstream data analysis.

2. Another challenge stems from the sizeable variations in the absolute abundance of different proteins in most biological mixtures. This so-called "dynamic range" consideration is further compounded by competitive ion-ion interactions at the ion source, which can result in signal suppression leading to concealment of lower intensity peptide ions from the MS detector.

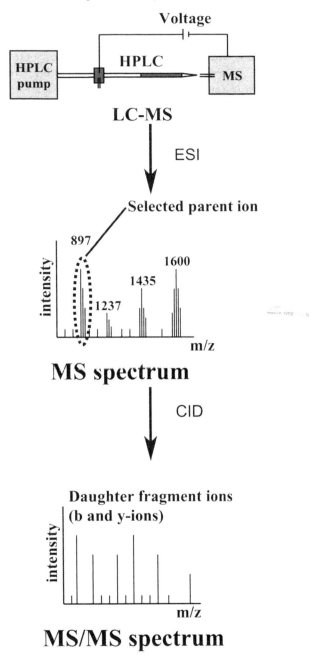

FIGURE 3.3: Collision-induced dissociation. Peptide mixtures are separated by LC. Ionized peptides are directly sprayed into the mass spectrometer. The first spectrum (full scan MS spectrum) records the m/z ratio of peptide ions as they elute from the microcapillary column. In a data-dependent process, individual peptide ions are then isolated and fragmented by collision with inert gas molecules (collision induced dissociation; CID) resulting in a tandem mass spectrum (MS/MS spectrum).

In practice, the detection efficiency of any given protein is usually dependent on its relative concentration in an organelle, cell, or tissue sample as compared to the other components. For example, peptides derived from the small handful of very high abundance proteins (e.g., serum albumin, IgGs) constituting the bulk (> 80%) of total protein concentration in human blood tend to dominate the MS/MS data acquired during an LC-MS experiment as these proteins generate high intensity ion signals that elute over large time spans during the LC separation. As a consequence, peptides from these proteins are repeatedly selected by the MS system for MS/MS fragmentation, precluding the selection and identification of peptides from lower abundance proteins.

Problems associated with limited data sampling capacity and dynamic range considerations are partly alleviated by newer generation MS instruments, such as linear ion traps. The main advantage comes from the ability of these advanced systems to perform much faster scanning and data processing on-the-fly resulting in improved duty cycles and overall data acquisition efficiencies. Alternatively, the under-sampling problem can be surmounted by repeating the LC-MS analysis several times such that more opportunity is given for a more complete set of MS/MS spectra to be generated, until saturation in protein detection is achieved (see Figure 3.4). This approach is strongly advised if comprehensive peptide detection is desired [145, 343].

As a general rule of thumb, respectable proteomic coverage and low detection limits can usually be achieved in routine practice by using an optimized sample preparation and protein/peptide fractionation procedure aimed at reducing the initial sample complexity. Subcellular fractionation, improved chromatographic fractionation (at both the intact whole protein and digested peptide level), and targeted depletion of higher abundance components are three popular methods for improving the efficiency of "shotgun" profiling studies [24, 189, 211, 231].

3.1.3 Instrumentation

The two most commonly used instrument designs used in ESI-based LC-MS shotgun sequencing studies are ion trap and triple quadrupole mass analyzers. While both are capable of sensitive MS/MS analysis of LC-fractionated peptide mixtures, each design exhibits certain advantages that are better tailored to specific proteomic applications.

3.1.3.1 Triple Quadrupole

The triple quadrupole represents a versatile and rugged family of MS systems. The design comprises three consecutive quadrupole mass analyzers, separated by consecutive focusing lenses (see Figure 3.2 a), which physically uncouple peptide ion isolation (from the stream of ion species produced by a front-end LC-ESI ion source), peptide fragmentation and mass analysis of

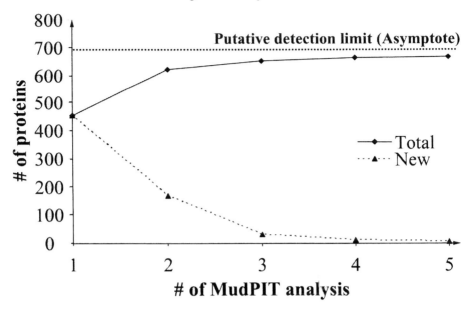

FIGURE 3.4: Saturation of detected proteins. Repeat analysis of a cytosolic fraction by MudPIT clearly results in a potential maximum of detected proteins (detection limit). Alternatively, the new number of detected proteins after each individual MudPIT run is below 10 after 4 individual injections.

the resulting daughter ions. The first quadrupole (Q1) allows for the selective filtering and retention of specific peptide ions with a desired m/z (mass-to-charge ratio) value. The isolated ions are then fragmented in the second quadrupole (Q2), typically by collision with inert gas molecules such as helium. Finally, the m/z ratio of the product ions is resolved in the third quadrupole (Q3), resulting in the generation of a tandem mass spectrum.

Triple quadrupole MS systems can be used in a number of specialized proteomic applications. For instance, they can be operated in highly selective ion-scanning modes that are optimized for monitoring precursor ions with particular features of interest, such as peptides bearing a phosphate group. Nevertheless, while long a mainstay in analytical chemistry for the analysis of small drug-like compounds, these instruments have generally been superseded in most protcomics laboratories by more versatile time-of-flight and ion trap types of mass analyzers, which we describe next.

3.1.3.2 Ion Trap

The ion trap is one of the most popular proteomics platforms, mainly because of its robustness, ease of automation, and overall cost efficiency (see Figure 3.2 b). A standard quadrupole ion trap comprises a ring electrode and two cap electrodes arranged in a manner so as to isolate and store ions prior to MS and MS/MS experiments. The ions are first "trapped" with a suitable electromagnetic field and then cooled with a neutral "bath gas" (e.g., helium) prior to analysis. In the basic MS form of mass analysis, an electromagnetic frequency is used to sequentially eject increasingly larger ion species over a particular m/z range. During MS/MS analysis, a series of waveforms is used to first isolate, then excite, and finally fragment the selected precursor ions prior to ejection from the trap.

Commercial ion trap instruments have the following two important advantages:

1. The ability to perform automated data-dependent MS/MS analysis during LC-MS experiments wherein precursor ions are selected for fragmentation on-the-fly;

2. The ability to perform drill-down MS^n experiments, wherein peptide ion fragments are sequentially retrapped and refragmented, resulting in more extensive structural information. These data can in turn be used to confirm the identity of a challenging peptide sequence or putative site of posttranslational modification, such as a phosphorylated residue [52].

Ion trap mass analyzers have several disadvantages: relatively low mass accuracy, limited resolution (i.e., the ability to discern similarly massed peptides), and narrow mass ranges. These limitations can impair peptide peak discrimination, and can result in proteins being misidentified.

These limitations are partially offset by some recent technical advances, such as the development of linear ion trap systems, which can accumulate larger ion populations and, hence, produce more reliable ion statistics, and the coupling of ion traps to ultra-sensitive Fourier transform ion cyclotron resonance mass analyzers (FT-MS), which can lead to outstanding performance in terms of markedly improved resolution and mass accuracy, albeit at a steep capital infrastructure cost.

3.1.3.3   FT-MS

FT-MS instruments use strong magnetic fields (up to 12-Tesla) under extreme vacuum to capture and resolve peptide ion species. Despite their high capital cost and the significant operational complexity of most current FT-MS systems, these instruments enjoy a rising popularity due to their exceptional mass accuracy ($< 1ppm$ with internal calibration), greatly improved mass resolution, good overall sensitivity (due to low spurious signal background) and large overall dynamic range (that is, the effective concentration range under which quantitative differences in sample composition can be reliably detected). FT-MS has been used in two distinct research settings, namely top-down as well as "shotgun" or bottom-up proteomic screening studies.

In the *top-down* approach [192], entire intact whole proteins are ionized and introduced into the FT cell, where they are then subject to extensive fragmentation using dedicated fragmentation techniques, such as electron capture dissociation (ECD) [192], which are optimized for providing sequence coverage approaching 100%. The tremendous mass accuracy of FT-MS allows ready detection and assignment of all the fragment ion peaks. If one has *a priori* information concerning the identity of the protein in question, determination of even slight deviations from expected fragment masses can enable the comprehensive mapping of posttranslational modifications [192, 417]. On the other hand, in its current implementation, FT-MS suffers from significant limitations, including a lack of user-friendly software for postacquisition data analysis, such as fully automated data base-search algorithms designed to fully exploit the enhanced mass accuracy of the acquired MS/MS spectra.

In the *bottom-up* sequencing strategy pioneered by Smith and coworkers, protein tryptic digests are reproducibly fractionated by high pressure LC-ESI and the peptide masses measured with such mass accuracy (< 1ppm) reducing ambiguity in interpretation of CID spectra, and even eliminating the need for performing MS/MS experiments in order to identify the corresponding protein. Conceptually, this method of "accurate mass tagging" offers the potential to greatly speed up comparative protein profiling studies, since it permits monitoring of large numbers of proteins based on tracking representative peptides while bypassing the need for the more tedious and lengthy process of CID "peptide sequencing." Likewise, although it remains to be determined what effect unanticipated protein processing events will have on the overall reliability of protein identification by FT-MS, the impressive mass accuracy afforded is expected to improve the overall rate of detection of posttranslational modifications in shotgun profiling studies [377].

3.1.3.4   MALDI-ToF MS

MALDI-Time-of-Flight-MS (ToF-MS) is a fundamental technology platform in proteomics research due to its tremendous versatility, reliability and sensitivity. In the basic procedure, sample peptide mixtures are mixed with an energy-absorbing matrix comprising a saturated solution of small aromatic molecules, such as α-cyano-4-hydroxybenzoic acid, sinapinic acid or 2, 5-dihydroxy benzoic acid dissolved in an acidic organic solvent. The mixtures are then spotted onto a metal target plate and allowed to dry, resulting in the formation of co-crystals of matrix and embedded peptide analytes. The target is then introduced into a time-of-flight MS instrument and analyzed under high vacuum.

Peptide ionization is induced by firing a pulsed laser directly at the sample in the presence of a high electric potential ($\sim$ 20 kV voltage) applied to the sample target. The organic matrix absorbs and transforms the photon energy in the wavelength of the used laser (usually a 337 nm nitrogen laser) into kinetic motion, resulting in sample sublimation and desorption of the analyte molecules into the gas phase (see Figure 3.5 a). The charged ions accelerate through a series of electronic lenses into a field-free drift region before reaching a detector.

Since all of the peptide ions have essentially the same kinetic energy, the time of transmission to the detector (i.e., time-of-flight) is proportional to ion mass. By calibrating the system with molecules of known mass, the exact mass of an analyte can be readily determined with a good accuracy over a broad range. In practice, MALDI-ToF systems serve as a simple and reliable analytical platform that are well suited to the analysis of purified proteins and simple peptide digests. Since each analysis can usually be performed quickly (usually in less then a minute), HTP round-the-clock MALDI-ToF-MS analysis of hundreds of gel bands can also be achieved daily using modern, fully automated instruments.

Continuous developments in MALDI-ToF-MS instrumentation has resulted in significant improvements in mass resolution and mass accuracy. In particular, two technical innovations, termed the reflectron and delayed-extraction, are now regularly used to correct for the small initial spread in the kinetic energies of the generated ions, which can impede resolution of isotopic variants (see Figure 3.5 b). The reflectron design employs two connected ToF tubes, separated by a series of ion focusing mirrors, to bring peptide ion packets into phase, resulting in increased resolution.

3.1.3.5   MALDI-ToF Mass Fingerprinting

Identification of proteins by MALDI-ToF is generally achieved after gel-based protein separation techniques are applied to a sample. To simplify complex protein mixtures prior to identification, high-resolution gel-based protein separation methods such as two-dimensional gel electrophoresis (2D-PAGE) are a common method of choice.

In 2D-PAGE, proteins are first separated using a electric potential based on differences in their isoelectric point, generally using a commercially available gel strip containing different ranges of pH. This strip is then transferred to a denaturing SDS-PAGE gel (sodium dodecylsulfate) in the second dimension, which sorts the proteins based on their molecular weight. After protein separation, the gels are usually silver stained to visualize the separated protein spots. An alternative approach is to use sensitive fluorescence stains, which support more linear quantitation/detection of separated proteins. After scanning of the gel images, individual protein spots of interest can be excised from the gel and digested in-gel with trypsin. The peptide patterns are then analyzed in order to identify the corresponding polypeptide(s).

Typically, the peptide digests are eluted from the gel pieces, mixed with matrix and spotted onto a MALDI target. The resulting experimentally acquired mass spectrum is termed a peptide mass fingerprint. Although this spectrum is simply a record of the m/z of the (detected) peptides (quite unlike MS/MS spectra, which record the fragmentation patterns of individual

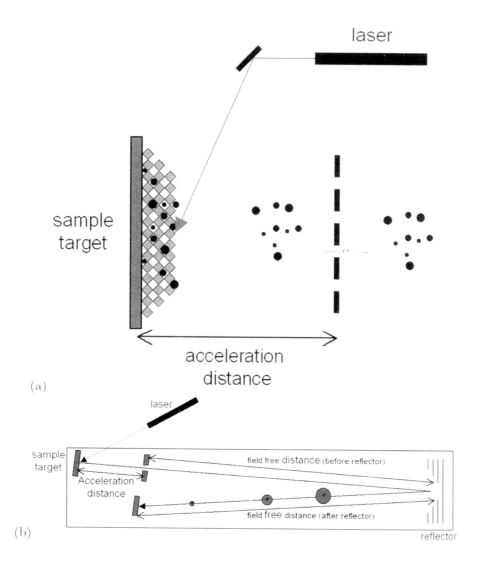

FIGURE 3.5: (a) In MALDI-ToF MS a sample analyte is spotted onto a sample target. Desorption and ionization is accomplished by illumination with a 337 nm nitrogen laser. Ionized molecules are then accelerated to a detector through the use of electric lenses. (b) Basic instrument scheme of a modern MALDI-ToF MS using a reflectron system, which focuses ions to improve resolution and sensitivity.

peptides), this pattern is usually distinctive and characteristic for the excised protein and hence can be used to identify the protein from a sequence data base (see Figure 3.6). Diverse data base search algorithms (e.g., Mascot, Protein Prospector, ProFound) have been developed for identifying candidate proteins based on these spectra.

MALDI-ToF fingerprinting is now a "workhorse" method for protein identification. Since each analysis can usually be performed quickly (typically in less then a minute), round-the-clock analysis of hundreds of gel band digests can be achieved daily using modern automated MALDI-ToF-based systems. However, since the presence of peptide signature generated from multiple polypeptides can "confuse" the downstream data base search algorithm, unambiguous detection of proteins in complex mixtures generally requires that the target proteins be highly resolved (typically by SDS-PAGE) prior to analysis.

3.1.3.6 Postsource Decay

Although MALDI is generally regarded as a "soft" ionization technique, peptide ions occasionally fragment (i.e., decay) during transit in a ToF flight tube. These metastable ion fragments have the same initial velocity as the precursor ion, and thus reach the detector simultaneously with the parent species in a linear ToF instrument. These fragments are, however, resolvable in reflectron instruments due to slight differences in kinetic energy, a process that also allows for (relatively inefficient) "postsource decay" (PSD) sequencing of individual peptides [120].

Two types of reflectors have been developed to exploit the PSD phenomena. In dual stage reflectors, the reflectron ion lense voltage is decreased in a stepwise fashion and individually recorded spectra patched together to generate a complete fragment ion series. Alternatively, more advanced curved-field reflectrons can be used to record the PSD ion series continuously [120]. Both instruments allow the gating and selection of individual precursor peptide ions.

3.1.3.7 MALDI-Q-ToF and MALDI-ToF-ToF

Further sophistication of commercial MALDI-ToF systems have resulted in the so-called hybrid systems, such as MALDI-Q-ToF (quadrupole time-of-flight) and MALDI-ToF-ToF (tandem time-of-flight) instruments. These hybrid systems enable MS/MS experiments to be performed directly. In combination with a reflectron design, MALDI-Q-ToF systems allow for low energy collision-induced dissociation of select precursor ions, which enables high accuracy sequencing of individual peptides.

MALDI-ToF-ToF instruments use high-energy fragmentation to generate more extensive structural information (e.g., single amino acid residue immonium ions) that can be used to answer specific structural questions, including discrimination of structural isomers (such as isobaric leucine and isoleucine

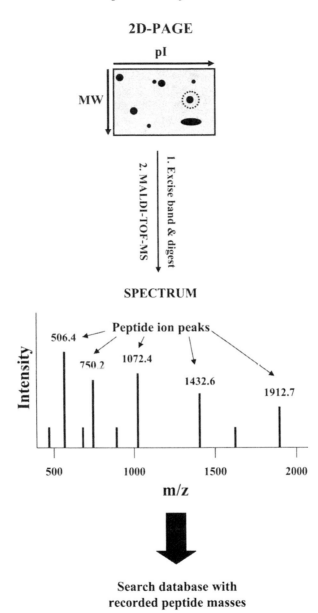

FIGURE 3.6: Peptide fingerprinting. Typically protein mixtures are separated by SDS-PAGE electrophoresis. Bands of interest are excised, in-gel digested and spotted onto a MALDI-ToF sample target. The resulting MS spectrum, termed a peptide fingerprint, contains peptides specific to the amino acid sequence of the excised protein and the protease used for enzymatic digestion. The recorded m/z values are used by data base search algorithms.

residues). These newer systems thereby overcome the modest structural information offered by the basic PSD procedure.

As mentioned above, MALDI-ToF instruments are generally used with gel-based protein separations, followed by band excision, in-gel digestion and peptide fingerprint protein identification. Another approach uses MALDI-ToF MS coupled with LC separation [214, 572] in order to simplify complex peptide mixtures prior to analysis. In this implementation, the MS instrument used must be capable of MS/MS fragmentation of individual peptides. Biological samples are proteolytically digested and the peptide mixtures fractionated by offline LC methods directly onto a MALDI sample target plate using specially designed robotic systems. Matrix is added, and the plates then analyzed using an advanced MALDI-ToF system capable of CID experiments (e.g., MALDI-q-ToF, which have an additional collision cell for fragmentation of precursor ions prior to detection). The retention time (elution) of a specific (sequenced) peptide(s) can then be correlated to a specific position on the target plate. This analysis strategy has several advantages:

1. Peptides spotted onto MALDI sample plates are generally stable and can be stored for a considerable time, allowing operators to revisit sample spots of particular interest at a later date;

2. If quantitative data is desirable, one can use a peptide labeling strategy (e.g., based on stable isotopes, for instance, as discussed below) to first measure protein expression levels or ratios (based on the MS data) prior to selecting peptide peaks for MS/MS;

3. Peaks of interest representing proteins whose levels fluctuate over a certain threshold (e.g., a two-fold increase in an experimental sample relative to a control sample) can then be targeted for follow up MS/MS CID experiments aimed at protein identification.

This process focuses the analysis on proteins of particular biological interest, bypassing the needless generation of irrelevant MS/MS spectra.

3.2 Multidimensional Chromatography

One major goal of proteome research is the simultaneous separation and detection of all the protein species expressed by an organism at any given time point. Given the multiple layers of posttranscriptional processing that occurs *in vivo*, proteomic diversity is expected to be exceedingly complex, with upwards of tens of thousands of proteins and their processed variants being produced by each cell or tissue type. Given this challenge, high performance protein/peptide separation techniques have been devised to simplify overwhelming sample mixtures prior to MS analysis.

Traditionally, complex biological mixtures have been fractionated by 2D-PAGE, a methodology that rarely allows resolution or detection of low abundance proteins, membrane proteins, or basic proteins. Hence, alternative gel-free protein profiling technologies have been developed.

A particularly promising approach developed by J. R. Yates III and colleagues, termed *multidimensional protein identification technology (MudPIT)*, has been shown to vastly outperform 2D-PAGE as a front-end to sample simplification prior to MS detection. Pilot proof-of-principle studies have shown that MudPIT can be used to identify hundreds of low-abundance proteins, even in the presence of excess higher abundance housekeeping enzymes, including important signaling molecules such membrane associated receptors, intracellular protein kinases, and their target transcription factors [174, 312, 318, 537].

In the basic MudPIT procedure, a complex protein sample (e.g., a whole cell extract or complex fraction) is enzymatically digested to completion using one or more proteases, typically endoproteinase Lys-C and trypsin which serve as a particularly efficient combination. The entire peptide mixture is then loaded manually off-line using a high pressure vessel onto a microscale (inner diameter $\sim$ 75–100 μm) capillary column containing two distinct phases of orthogonal stationary HPLC media: a plug of strong cation exchange material (SCX) followed by reverse phase RP-18 resin (see Figure 3.7). The column is then hooked up to a HPLC system and subsets of the bound peptides eluted and introduced by ESI into an online ion trap MS system programmed for data-triggered MS/MS analysis.

To maximize the coverage of peptides detected by the instrument, subsets of peptides are step-wise eluted from the SCX material by the use of increasing "salt bumps" onto the RP-18 material, and then fractionated from the second resin using a standard water/acetonitrile gradient (see Table 3.1). Tens of thousands of tandem mass spectra can routinely and automatically be captured in a multistep, multihour-long MudPIT profiling experiment.

Several notably impressive global expression proteomic studies based on the MudPIT methodology have been published to date. The first report used MudPIT to profile protein extracts isolated from three distinct rice tissues (leaf, root and seed)[318], and compared these results to 2D-PAGE-based analysis. A total of 2, 528 proteins were detected, of which 2, 363 were found by MudPIT as compared with 556 proteins by SDS-PAGE MALDI-ToF, clearly illustrating the superior performance of the gel-free shotgun technology. This publication also described many tissue-specific and previously uncharacterized proteins in this nutritionally important grain.

The second large-scale study used MudPIT to identify over 2, 400 proteins in four developmental stages of the malaria parasite Plasmodium falciparum [174]. The list comprised over 1, 200 hypothetical proteins, many of which appear to be developmentally restricted. This proteomic dataset serves as a helpful resource for understanding the complex biology of this important parasite and will hopefully lead to the identification of novel drug targets.

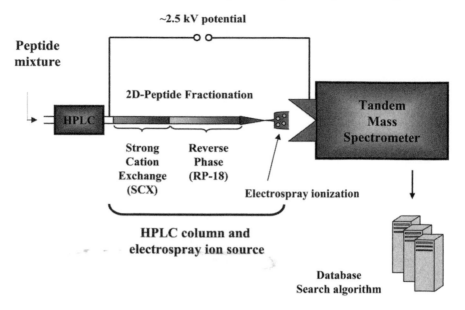

FIGURE 3.7: Multidimensional protein identification technology — Mud-PIT. Complex protein mixtures are first digested by specific proteases. The resulting peptide mixtures are then separated by two-dimensional microcapillary chromatography in a fully automated 24-hour sequence, and analyzed online by tandem mass spectrometry. The resulting tandem mass spectra are searched against a protein sequence data base using the SEQUEST algorithm.

The third study applied MudPIT towards the analysis of fractionated mouse tissue [312]. Liver and lung tissue were first separated into four distinct subcellular compartments (nuclei, mitochondria, microsomes, and cytosol) in order to improve proteomic detection coverage. This coupled strategy resulted in the identification of over 2,100 proteins, many of which displayed a tissue-specific expression pattern.

This enrichment or fractionation strategy has two major advantages as compared to the analysis of whole tissue extracts:

1. It allows for a significantly greater number of detected proteins;

2. It increases the value of the acquired biological information, as uncharacterized proteins identified only in one compartment likely have a biological function related to the isolated organelle.

This promising dual approach can be extended by combining fractionation and MudPIT-based expression profiling, to investigate proteome adaptations at defined developmental stages throughout organogenesis in mouse. Both

TABLE 3.1: Chromatographic conditions for MudPIT analysis.
A – D are different buffer solvents. Running Buffers comprise:
Buffer A 5% Acetonitrile, 0.5% acetic acid, 0.02% Heptafluorobutyric
acid
Buffer B 100% Acetonitrile
Buffer C Buffer A + 250 mM Ammonium acetate
Buffer D Buffer A + 500 mM Ammonium acetate

Strong cation exchange material			Reverse Phase					
Step	%C	%D	time (min)	A	B	C	D	ml/min
1	10	0	0	100%	0%	0%	0%	150
2	20	0	1	100%	0%	0%	0%	150
3	30	0	1.01	90%	0%	10%	0%	150
4	35	0	8	90%	0%	10%	0%	150
5	40	0	8.01	100%	0%	0%	0%	150
6	45	0	18	100%	0%	0%	0%	150
7	50	0	18.01	90%	10%	0%	0%	150
8	55	0	75	70%	30%	0%	0%	150
9	60	0	85	20%	80%	0%	0%	175
10	70	0	94	20%	80%	0%	0%	175
11	80	0	95	100%	0%	0%	0%	150
12	0	100	105	100%	0%	0%	0%	5

the subcellular (organelle-specific) location and relative expression levels (estimated based on spectral counts, an approach described further below) of $\sim 3,500$ proteins were recorded at select time-points, many of which were not previously detected in parallel comprehensive MudPIT screenings of other mouse organs, including brain, heart, kidney, liver, and placenta. Biological back-up studies of putative stage- and lung-specific targets by RNA *in situ* hybridization have confirmed the usefulness of proteomics studies to uncover previously uncharacterized gene products with potential roles in development.

MudPIT-style protein expression profiling studies of model organisms such as mouse and rat will become increasingly important for investigating the molecular basis of human disease. Importantly, the genome sequences of mouse, rat and human are now virtually complete [198, 327, 529, 538], facilitating systematic and comprehensive proteomic screening. To exploit this, we have been applying the MudPIT technology to investigate proteomic alterations associated with heart disease using several well described mouse models of heart failure, with the ultimate aim of gaining insight into the mechanistic basis of disease progression together with the discovery of clinically valuable biomarkers of early stage pathophysiology. The use of animal models of disease provides access to tissues that are likely not readily accessible in human, with the clear advantage of allowing investigators to systematically control experimental variables. MudPIT profiling of human samples (e.g., tumors or blood serum) is likewise also expected to become an important resource for monitoring clinically relevant diagnostic and prognostic indicators [453].

Despite the recent great technical progress, sample complexity and dynamic range remain two fundamentally limiting factors in all shotgun profiling studies, particularly with regard to mammalian systems. Given the tremendous complexity of human cells, even high performance LC-MS profiling technologies such as MudPIT will fail to detect all of the peptides/proteins present in a cell or tissue sample. Therefore, improved sample prefractionation procedures (albeit time consuming) must be applied in concert with shotgun profiling if the goal of complete proteome coverage is to be achieved.

We believe that additional forms of subcellular enrichment and protein fractionation using well-established biochemical methods, such as sucrose gradient ultracentrifugation or chromatographic fraction of intact proteins based on ion-exchange, size exclusion, or affinity chromatography, offer a particularly effective route for improving the scope and depth of proteomic detection coverage. These methods are highly complementary in nature to MudPIT screening, and can be readily applied to any number of biological settings.

3.3   Protein Quantitation

To compare changes or perturbations in the proteomes of distinct samples, such as diseased vs. healthy tissue, accurate quantification of large numbers of proteins has to be achieved. The most commonly used technology developed to date for this challenging task has been the separation of protein mixtures by 2D-PAGE, followed by differential staining of the resolved protein spots and visual or computer-assisted comparison of the resulting patterns to detect differences among the samples. Unfortunately, 2D-PAGE has several critical limitations that restrict the approach to the analysis of changes in high abundance proteins, which are usually of least biological interest. Therefore, alternative gel-free protein quantification methods, based on LC-MS shotgun profiling approaches, have been developed [90, 194, 214, 223, 323, 336, 491, 574].

3.3.1   Quantification Based on Isotope Labeling

Stable isotopes (e.g., C13) have long been used in analytical chemistry for the absolute quantitation of a specific analyte by MS. Stable isotopes are defined as a family of chemical elements with the same number of protons but different number of neutrons. As they have virtually the same chemical and biophysical properties as the native species, stable isotope variants can be used to address diverse quantitative analytical questions.

In traditional analytical chemistry (e.g., gas chromatography coupled to

MS or GC-MS), a known analyte can be absolutely quantified by spiking in
an isotopic derivative (that can be distinguished by MS based on differential
mass) of this same analyte to the sample of interest in a predefined concentration. As the two compounds typically co-elute from the chromatography
column, comparison of the ratio of peak areas of the internal standard and
the analyte enable quite accurate absolute quantitation. A similar strategy
has also been introduced to proteomics. The approach is called AQUA (absolute quantitation of proteins), and is based on the addition of an isotopically
labeled synthetic peptide to the digestion reaction as a internal standard or
reference [194]

3.3.1.1   Isotope Coded Affinity Tagging

The isotope coded affinity-tagging (ICAT) is by far the most widely recognized and elegant gel-free MS quantitation technology developed to date.
In its original incarnation, it was designed to evaluate changes in the relative
levels of proteins in two different cell populations, e.g., control and diseased
tissue sample. The methodology is based on selective chemical modification
of cysteine residues in peptides with a mass tag consisting of a thiol-reactive
group and a biotin moiety separated by a linker consisting of either 8 hydrogen atoms (isotopically light) or 8 deuterium atoms (isotopically heavy). The
biotin molecule is used to affinity purify labeled peptides based on its highly
specific interaction with avidin.

In a typical experiment, the proteins present in the two different experimental conditions are independently labeled with the light and heavy isotopic
versions of the ICAT reagent. The protein mixtures are then combined, digested, purified and analyzed by LC-MS. The chemically identical pairs of
sister peptides are easily resolved by MS, allowing the relative levels of the
peptides in each cell population to be compared (see Figure 3.8). In theory,
the incorporation of a built-in internal reference in each ICAT analysis can
result in more accurate quantitation of proteins based on differences in peak
areas. The potential of the basic ICAT procedure has been established in a
series of proof-of-concept LC-MS profiling experiments [223, 336, 491, 575].
This strategy has also proven to be especially useful when combined with off-
line fractionation of the isotopically-labeled peptide mixtures. These fractions
can be resolved by HPLC prior to being spotted onto a MALDI target plate,
along with a matrix solution, as they elute from the chromatography system.

A significant advantage of combining ICAT with MALDI-ToF-MS is the
possibility to repeatedly reanalyze any given sample on a target plate. For
instance, one may decide to subject peptides of interest to targeted MS/MS
sequencing only after quantitation of the precursor ions suggests the effort will
be fruitful (that is, a meaningful change in abundance is observed). Accurate
quantitation of relative peptide levels can be performed in a quick, first-pass
MS mode. Then, peaks of interest (e.g., peptides whose levels were perturbed
relative to the reference sample), are isolated and identified by more system-

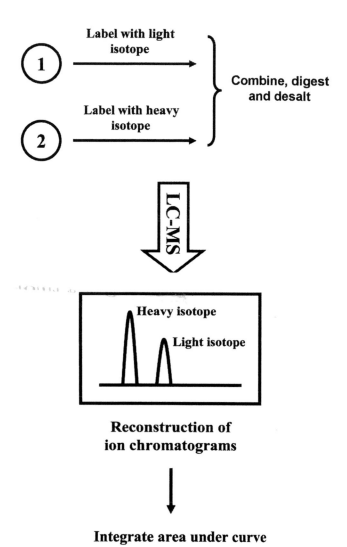

FIGURE 3.8: Peptide quantitation by ICAT: Two individual samples (e.g., (1) healthy vs. (2) disease) are chemically modified with the ICAT reagent. Specifically, one sample is modified with the isotopically light version and the other sample with the isotopically heavy version of the ICAT reagent. Samples are then combined, enzymatically digested and analyzed by LC-MS. Reconstruction and integration of co-eluting isotope peaks allows for relative quantitation of interesting proteins.

atic MS/MS sequencing [214].

Several advances to the basic ICAT methodology have been reported over the past few years. These include the use of solid phase based tagging [574], chemically cleavable reagents, and even the substitution of C12/C13 isotopic pairs for the original deuterium-based source label to improve peptide co-elution during chromatographic separations [336]. The conceptual break-through of the ICAT methodology has also stimulated development of alternative methods of isotope-based quantification, including the development of amino group-specific isotope labeling strategies, which are expected to improve peptide detection coverage as compared to cysteine-labeling, a fairly rare amino acid [476].

3.3.1.2   Metabolic Labeling and ^{18}O Labeling

An alternative enzyme-based approach has been developed for isotopic tagging of proteins *in vitro*. By digesting proteins with trypsin in the presence of ^{18}O-enriched water, two ^{18}O-atoms become incorporated at the C-terminus of virtually all generated peptides, resulting in a mass shift of 2 a.m.u. (atomic mass units) as compared to regular digests [507].

While conceptually appealing, the impact of this labeling technology has been quite restricted, in part by problems arising with incomplete labeling and exchange/loss of the ^{18}O label under the typical conditions of the LC-MS procedures [507].

To circumvent the sample handling issues associated with *in vitro* labeling, several research groups have developed methods for metabolic labeling of proteins *in vivo*, either by culturing cells in the presence of ^{14}N or ^{15}N labeled medium [536] or using medium supplemented with select isotopically-labeled amino acids [423].

Whole body metabolic labeling methods have been developed for multicellular organisms, such as worm and fly [323], These methods can potentially be adapted to free-living mouse models, for instance using ^{14}N or ^{15}N-labeled algae as the only dietary nitrogen source [558]. No matter what source is used, incorporation of heavy and light isotopic variants of amino acids in newly biosynthesized proteins results in an apparent increase in observed peptide m/z values, allowing relative abundance to be discerned directly by MS.

3.3.2   Nonisotope Based Quantitation Techniques

Stable isotopes are rare commodities, and are, therefore, expensive. To circumvent this important consideration, several less expensive chemical labeling quantitation technologies have been developed.

Mass-coded abundance tagging (MCAT) is a method based on the selective guanidination of peptide epsilon-amino lysine groups with O-methylisourea *in vitro* [90]. The addition of the mass tag results in an exact mass shift of 42 a.m.u. but does not perturb the ionization or fragmentation properties of the labeled peptides in LC–MS experiments (see Figure 3.9). Examination of the relative ratio of unmodified-to-modified forms of lysine-containing peptides enables rapid determination of the relative quantity in two different samples. Peptides exhibiting altered protein levels can then be identified by MS/MS sequencing.

It should be noted that entirely label-free protein quantification methods, based on interpretation of peptide peak intensity in MS mode across different samples, have been proposed [103, 340, 453]. These more computationally oriented approaches achieve relative quantification by measuring and integrating the area under each peptide ion peak recorded during a simple LC–MS profiling experiment, and are reasonably effective since peak area (shape and intensity) appears to correlate quite well with relative analyte abundance. That is, peptide peak area is usually proportional to the relative amount of a corresponding protein in a given sample or fraction. This is an important property, which has long been exploited in traditional analytical chemistry for determining the relative abundance of analytes in various samples, e.g., in HPLC-UV or fluorescence analysis.

Label-free profiling is generally performed in three main steps in LC–MS experiments:

1. The ion chromatogram of select MS peptide ion precursors is traced and the total ion-intensity of peaks of interest measured using appropriate algorithms and software;

2. The ratio of (matched) peak intensities of identical precursor ions is compared in different experimental LC–MS datasets to estimate relative abundance;

3. MS/MS sequence information is mapped back to peptides (and their corresponding cognate proteins) exhibiting putative differences in relative abundance.

New advances in pattern recognition and data mining enable the comparative analysis of hundreds of such peaks across many different datasets, facilitating rapid large-scale global proteomic comparisons without the burden of performing multiple isotopic controls [453]. An advantage of such an approach is the ability to incorporate statistically sound criteria for estimating the significance (as well as presumed fold-change) of putative differences in peptide (protein) abundance.

Alternative label-free quantification approaches have been introduced. Most notably, Yates and colleagues have proposed and validated the use of a simple measure based on the total cumulative spectral count (that is, the number of MS/MS spectra mapping to a given identified protein), as a semi-quantitative readout of relative protein abundance [343]. In comparative studies, the total detected spectral count acquired by LC–MS/MS for any given protein, such as a MudPIT profiling experiment, is compared across

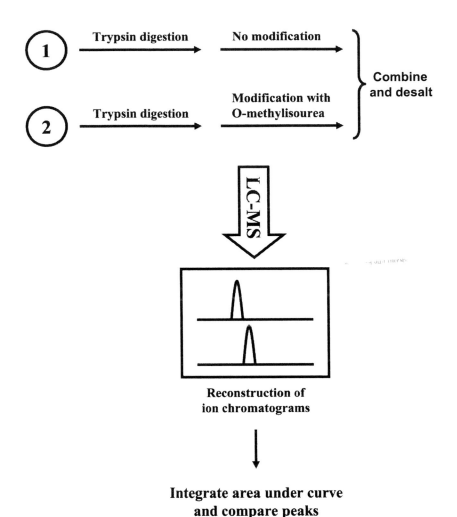

FIGURE 3.9: Peptide quantitation by MCAT: Two individual samples (e.g., healthy vs. disease) are digested by trypsin. One of the resulting peptide mixtures is chemically modified by O-methylisourea, modifying the epsilon-amino group of lysine to a guanidino group. The two samples are then combined and analyzed by LC-MS. Reconstruction and integration of the total ion chromatogram of corresponding peptides (modified and unmodified) are used for relative quantitation of particular proteins.

samples. Differential expression is indicated by substantive changes in the observed spectral count.

Based on spiking experiments of well-defined protein standards in a complex background of crude yeast cell lysate, this straight-forward method was proven to provide a faithful approximate estimate of relative protein levels over a relative large dynamic window [343]. Although there are many outstanding issues surrounding stochastic variation in spectral count due to under-sampling, a particularly appealing feature of this user-friendly accessible approach is its seamless integration with the generic MudPIT shotgun profiling method. As with ICAT, MCAT and peak integration methods of quantification, this counting approach will likewise benefit from a more formal statistical evaluation.

3.3.3   Imaging Mass Spectrometry

Several research groups have been promoting the use of MALDI-ToF-MS to directly record the proteomic patterns of intact single cells and tissue sections. This form of *imaging mass spectrometry (IMS)* [99] represents a new application area for proteome mapping.

A frozen tissue of interest is cut into thin sections ($\sim 15\mu m$), which are carefully deposited onto a chilled MALDI sample target along with a coating of matrix. A laser is then scanned across each section, and the resulting signal intensities of specific m/z values converted into a virtual digital image through specialized software programs. The patterns are then used to discriminate differences in protein accumulation or composition throughout the section. While IMS is capable of recording several hundred distinct m/z values across a 2–10 thousand a.m.u. mass range, the majority of the signals are typically $< 3,000$ a.m.u., in part because of the performance inefficiencies of MALDI-ToF instruments at higher mass ranges. Specialized matrix deposition technologies, involving spraying of matrix onto vacuum dried tissue sections using a nebulizing gas or electrospray techniques, have also had to be developed for the generation of high resolution tissue imaging.

Generally, multiple spray steps are performed to avoid extensive wetting of the tissue section, which could result in protein migration. Nevertheless, preliminary studies indicate the promise of IMS as a platform for rapid mapping of protein localization and the detection of sizeable differences in protein expression [478, 564].

Key limitations relate to inefficient protein detection, limited detection coverage, and an inability to identify the analytes of interest. Since several hundred images can be generated in a single analysis, each of which can result in gigabyte sized file depending on resolution, substantive data analysis and archiving issues also have to be dealt with.

3.4 Detection of Posttranslational Modifications

Posttranslational modifications (PTMs) are dynamic protein modifications responsible for the regulation of diverse biological properties of the modified proteins. Many, if not all, proteins are subject to posttranslational processing and modification. PTMs can be generally divided into:

1. *Enzymatic posttranslational protein modification*, such as phosphorylation, glycosylation, methylation, acetylation, ubiquitinylation, isoprenylation;

2. *Nonenzymatic posttranslational protein modifications*, including glycation, oxidation or nitrosylation.

Many PTMs can be mapped by MS, but this process remains technically challenging and often time consuming. In general, successful MS-based mapping of PTMs requires preenrichment methods tailored to the specific type of modification followed by diverse MS-based analysis methods. Here, we focus only on phosphorylation and glycosylation as representative examples of the major issues relating to experimental elucidation of biologically important PTMs.

3.4.1 Phosphorylation

In terms of biological regulation, one of the most significant PTMs is protein *phosphorylation* (and, hence, of tremendous interest to many researchers). Many examples of sequence-specific site mapping of phosphorylation serine, threonine and tyrosine residues by MS have been reported [61], including several pilot proof-of-principle global profiling studies [52, 65].

Despite these successes, significant difficulties prevent routine, comprehensive mapping of phosphorylation sites:

1. Phosphopeptides are generally substoichiometric, and often are present in relatively low yield as compared to a large molar background excess of unmodified peptide;

2. Phosphorylation adds a negative charge to the modified peptide, which is generally presumed to cause signal suppression, reducing sensitivity and impairing detection since most LC-MS studies are carried out in the positive ion mode;

3. Most established MS procedures (including MudPIT) generally only provide limited overall protein sequence coverage (that is, only a small subset of all generated peptides typically generated by a tryptic digest are subject to CID). While this is usually sufficient to unambiguously

identify a protein, the mapping of all possible phosphorylation sites necessitates virtually complete protein coverage, which is difficult, if not impossible, to achieve in practice.

Although challenges remain, promising methods for the selective enrichment of phosphopeptides have been reported. These include the use of antibodies specific for phospho-tyrosine residues [432], selective chemical replacement of the phospho-groups by a biotin moiety followed by affinity purification on avidin [418], or by immobilized metal affinity chromatography (IMAC) under conditions favoring retention of phosphopeptides [338, 445]. These methods improve the odds of successful MS detection of the modified peptide by reducing the background signal from unmodified peptides. IMAC has proven to be quite variable, but it has been shown that a significant improvement in the efficiency of MS detection can be achieved by methylation of carboxylic acid groups to methylesters prior to IMAC selection [171].

Alternatively, if a putative phosphorylation site is known (or is predicted), the AQUA strategy can be used [194]. Briefly, a synthetic isotopic variant corresponding to a representative phosphopeptide (modified on a site of interest) is added as an internal reference to the protein sample prior to tryptic digestion. As an additional control, an isotopically heavy version of the same peptide in an unphosphorylated form is likewise added to the sample as well. Although painstaking, this strategy enables the exact determination of both the exact molar amounts and overall degree of phosphorylation at a specific site within a select target protein. As proof-of-principle, this method was applied successfully to track changes in cell cycle dependent phosphorylation of serine $1,126$ of the enzyme separase , which mediates release of sister chromatids during anaphase, in HeLa cells [194].

3.4.2 Glycosylation

Protein *glycosylation* is one of the most important PTMs. Nearly all membrane proteins and secreted proteins found in body fluids (e.g., plasma) are glycosylated. While of broad interest, the relative heterogeneity of glycosylation has presented a significant challenge to routine mapping by MS.

A recent apparent breakthrough in this area has been the introduction of a method for the selective enrichment and concomitant relative quantitation of N-linked glycosylation using an ICAT-like labeling strategy [570]. This method is based on the selective oxidation of glycan sugar residues by periodate, which are then selectively coupled to hydrazide beads. The bound proteins are digested on the solid phase support by trypsin, and the nonglycosylated peptides removed by washing. If relative quantitation is desired, the N-terminal amino group can be modified by either d0 (light isotope) or d4 (heavy isotope) labeled variants of succinic anhydride after blocking the epsilon-amino groups of lysine residues by guanidination. N-glycosylated peptides are then specifically eluted from the beads by enzymatic cleavage with the glycan specific enzyme PNGase F and analyzed by LC-MS or by MALDI-Q-ToF- MS [570].

Although it is still not clear if this method will scale to global proteome settings, it has considerable promise as a generic methodology for routine sample simplication of glycosylated proteins prior to MS analysis.

3.4.3   Protein Sequencing

De novo sequencing can be regarded as the manual assignment of a peptide or partial peptide sequence of an unknown protein for the purpose of identification. Originally, *de novo* sequencing was performed by Edman degradation [149, 150, 151], but since the early 1990s and the emergence of tandem mass spectrometry, *de novo* sequencing of tandem mass spectra has become more popular.

De novo sequencing in combination with tandem mass spectra uses a peptide from full-scan mass spectra. The peptide is isolated by either a triple quadrupole or ion trap MS, and further fragmented through the collision with inert gas molecules, such as helium. The generated tandem mass spectra, which are specific for the fragmented peptide, contain the information required for manual sequence assignment.

The fragmentation of a peptide in MS can result in several theoretical breakpoints, depending on where the peptide breaks and where the charge is. The resulting fragments are called $a, b, c,$ *or* x, y, z ions, of which the y and b ions are the most important [275]:

- For the y-ions, the charge is retained on the C-terminal fragment;

- For b-ions, the charge is retained on N-terminal fragment.

However, it is generally rare that a complete peptide sequence can be identified from a single fragmented peptide precursor. Furthermore, due to complicated fragmentation rules, e.g., side-chain fragmentation, loss of neutral molecules, etc., manual interpretation of tandem mass spectra can be time consuming and complicated to interpret, and thus success is heavily dependent on the experience of the interpreter.

For this reason, *de novo* sequencing has largely been replaced by computerized data base search algorithms, which are both significantly faster and generally automate the process of spectral interpretation.

Note that *de novo* sequencing can be a valuable approach if the sequence of the protein does not exist in the data base and, therefore, automatic data base search algorithms would be expected to fail. Several methods have been developed to improve *de novo* sequencing, which mainly rely on chemical derivatization of the peptides for more rapid identification of particular ion series (y and b series) [90, 93, 241, 250, 490]. Most of these methods rely on chemical modification of the peptides and therefore enable to determine a particular ion series, e.g., *b*- or *y*-ions. These methods simplify interpretation of spectra and therefore enable easier interpretation/determination of the peptide sequence.

3.4.4 Data Base Search Algorithms

The genomics revolution and the consecutive publication of the entire DNA sequence of biologically important model organisms and human [327, 529, 538] have greatly benefited proteomics due to the generation of extensive sequence data bases (DNA and protein) used for the correlation of recorded MS/MS data. Several algorithms have been developed to correlate the obtained mass spectra to available sequence information [155, 361, 441].

In the early 1990s, M. Mann *et al.* developed an algorithm for relatively robust (error tolerant) identification of peptides in sequence data bases based on peptide mass tags [361]. This method is a combination of manual *de novo* sequencing and automated data base search based on:

1. Inference of a partial amino acid sequence of a peptide by *de novo* sequencing;

2. Screening of a sequence data base using this tag sequence together with the mass of the parent ion as a means of protein identification.

These so-called "peptide sequence tags" serve as a highly specific peptide identifier [361].

The SEQUEST algorithm developed by J. Eng and J. Yates III automatically matches tandem mass spectra data to protein data bases without recourse to sequence interpretation [155]. In a multistep process, candidate sequences are chosen from a protein sequence data base, based on the precursor mass of the isolated and fragmented parent ion. Virtual tandem mass spectra (representing predicted *b*- and *y*-ion m/z peaks that would be expected for the sequence) are then generated for the candidate sequences and compared using a cross-correlation function to the recorded mass spectrum. The final output provided by the algorithm is a list of putative peptide matches and their associated scores, which include the following components:

- Cross-correlation score (Xcorr), which is based on the spectral fit between recorded and generated spectrum;

- The normalized difference between the best match and the second highest scoring match (ΔCn);

- Preliminary ranking based on the number of matched ion peaks (RSp).

These scores, in combination with the charge of the precursor ion and the observed tryptic status of the candidate match, are usually used to evaluate the accuracy of a prediction.

Mascot[1] is a commonly used commercially available algorithm for the interpretation of mass spectrometry data [441]. It is most commonly used for the interpretation of MALDI-ToF-MS peptide mass fingerprints, under the principle that observed peptide masses are characteristic of both isolated proteins and the enzyme used for digestion.

Mascot can also be used to interrogate MS/MS spectra. In order to get appropriate results from the search, the user has to define several parameters, such as the sequence data base to be used, as well as (ideally) the identity of the source organism, mass tolerance for the measured peaks (which is highly dependent on the MS platform used), number of allowed missed-cleavages in case of an enzymatic digest, as well as knowledge of any predicted chemical modifications.

Both the Mascot and SEQUEST algorithms can be run on Unix/Linux-based parallel cluster computers, which is critical for searching the huge numbers of spectra, typically generated by MudPIT-style proteomic screening technology.

Several computer programs have been developed to rigorously assess and statistically validate the quality of the results obtained from SEQUEST and other data base search algorithms [170, 300, 312, 354].

One approach developed by our group makes use of an empirical, probabilistic scoring algorithm, called STATQUEST, to determine the likelihood of putative peptide matches. Based on empirical knowledge of a prior likelihood distribution, a confidence (p-value) can be readily calculated by STATQUEST using the observed Xcorr, ΔCn, primary ranking, charge and tryptic status of candidate peptides. The predictions can then be filtered based on a user defined quality threshold, with the aim of minimizing the rate of false positive identification without substantially reducing overall coverage (false negatives or missed identifications) [312].

3.5   Global Data Analysis, Informatics Resources, and Data Mining

Higher level analysis and interpretation of proteomics datasets is one of the main challenges facing global-expression proteome profiling studies. The main hurdle is no longer necessarily in data generation per se, but rather in extracting meaningful biological inferences from the huge amount of data commonly generated by a proteomic study.

As with researchers performing microarray-based functional genomics studies, proteomic researchers are increasingly turning to the application of more

---

[1]http://www.matrixscience.com

sophisticated, scalable bioinformatics algorithms as a means of extracting as much valuable biologically relevant information as possible. Several different computational methods can be usefully applied at various steps in the analytical pipeline.

A good first step in interpreting global proteomic datasets is to place the identified proteins into a biological context [475, 559, 557]. Hierarchical clustering and related pattern recognition algorithms are widely used to visualize the underlying patterns in large-scale gene expression dataset [153, 276]. Clustering can be used to subgroup proteins exhibiting similar expression characteristics or patterns, and can therefore reveal groups of co-expressed or co-regulated proteins, e.g., such as those that accumulate together in a particular organelle or during specific developmental time points.

Clustering of this kind offers several advantages:

1. The identified clusters provide natural structure to data organization, by grouping co-regulated proteins;

2. Co-regulated proteins might be functionally related.

There are many helpful free and commercial implementations of basic data mining tools, with the main differences in terms of scalability, ease of use and methods of visualization of the results of the clustering analysis. Many of these tools allow important biological information to be examined across entire collection of data sets. While in critical demand, easy to use publicly accessible bioinformatics tools tailored to global proteomic analysis are still in their infancy, and most of those produced to date are still found lacking in certain regards. In particular, effective and efficient data mining algorithms that support interactive and automated methods of analysis are needed.

3.5.1   Functional Prediction and Protein Annotation

Evaluating the biological function of proteins identified in large-scale proteome profiling project is an ongoing challenge. Proteomics data can, in turn, be used to predict the function of previously uncharacterized proteins. Several strategies have been developed to address this issue [64, 85, 96, 135, 237, 292, 310, 335, 485, 563] There are two main approaches:

1. Assigning function to proteins by computational and/or statistical inference using information in publicly accessible curated annotation databases;

2. Predicting protein function based on the observed pattern of subcellular localization and/or given a protein's interaction partners.

In the first approach, one examines a cluster of identified proteins for statistical enrichment in membership to a specific functional category (see Fig-

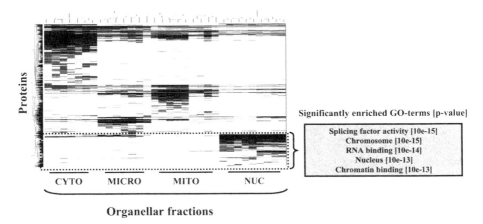

Proteins

Significantly enriched GO-terms [p-value]

Splicing factor activity [10e-15]
Chromosome [10e-15]
RNA binding [10e-14]
Nucleus [10e-13]
Chromatin binding [10e-13]

CYTO   MICRO   MITO   NUC

Organellar fractions

FIGURE 3.10: Cluster diagram of proteins identified in mouse brain. Multiple organelle fractions from brain were analyzed by MudPIT and the identified proteins were clustered using the Cluster 3.0 software package using spectral counts as a semiquantitative matrix. A nuclear unique cluster was fed into the MouseSpec software package and a small selection of significantly enriched GO-terms are displayed next to the figure.

ure 3.10). A common emerging resource is the gene ontology (GO) classification schema [26, 234]. The GO data base[2] comprises a computer-friendly schema for describing the subcellular localization, molecular functions, and biological roles of proteins. Several corollary software tools have been written to enable users to link a set of identified proteins to the GO data base, and thereby exploit the information contained therein [64, 217, 312, 563, 569]. Evidence for enrichment of one or more particular GO-terms in a sample or fraction can be highly suggestive as to the plausible roles of an unknown protein detected in that same sample. Several software tools exist for calculating the number of significantly enriched GO-terms within a defined subcluster. For example, GO-miner[3] generates a summary of GO-terms that are significantly enriched in a user inputted list of protein accession numbers as compared to a reference data base, such as SwissProt.

In a second, related approach, information concerning the organelle localization (e.g., presence in the nuclei, mitochondria, microsomes and/or cytosol) as well as the the physical (protein–protein and/or protein–nucleic acid or protein–small molecule) or functional (e.g., genetic) interactions of specificity of a protein is determined using subcellular fractionation, as well as some form of affinity enrichment, followed by analysis using MALDI-ToF MS, LC-

---

[2]http://www.geneontology.org
[3]http://discover.nci.nih.gov/gominer

MS or even exhaustive MudPIT screening. The discovery of a specific pattern of subcellular localization and select interactions, perhaps only under certain experimental conditions, can often lead to the generation of a clear hypothesis regarding biological function.

An appropriate analytical framework has to be incorporated from the start of a large-scale proteomic study so as to organize and catalog the analysis into a biologically intuitive and manageable manner. For instance, proteomic data should be formatted so as to facilitate distribution to the broader scientific community in a manner suited to user defined queries.

3.6 Conclusions

MS is a powerful, mature, and rapidly developing platform for the investigation of protein expression and function in diverse biological settings. When these data are processed with an appropriate bioinformatics framework, proteomic analyses can provide valuable insight into biological questions at hand (see Figure 3.11).

The continuous technical innovations and improvements in MS-based instruments, methods, and data processing tools combined with better sample preparation techniques, improved methodologies for large-scale protein quantification, and developments in bioinformatic algorithms and software tools are providing biologists with a powerful framework for exploration of diverse biomedical problems.

An important corollary task, which is unfortunately also still problematic, relates to the integration of proteomic datasets with the results of parallel functional genomic studies, such as DNA microarray-based readouts of global gene expression patterns (e.g., mRNA transcript levels) and large-scale screens of PPIs. Integration of diverse multivariate datasets poses a number of unresolved challenges [383, 547], some of which are specific to MS data. Nevertheless, this important emerging area will likely be solved through collaborative efforts aimed at developing an emerging standard for MS data reporting, as well as by the incorporation of data validation statistical procedures, data mining tools and machine learning algorithms developed in other fields.

At the end of the day, establishing good experimental practice together with proper controls and repeated experiments to improve the overall reliability, sensitivity and accuracy of a proteomic study can markedly improve the utility of the inferences that can be drawn. Assuming these tasks are rigorously and successfully addressed, a new era of systems biology can be expected to emerge, leading to many exiting breakthrough biological discoveries and fundamental insights concerning the mechanisms underlying complex biological processes and disease mechanisms.

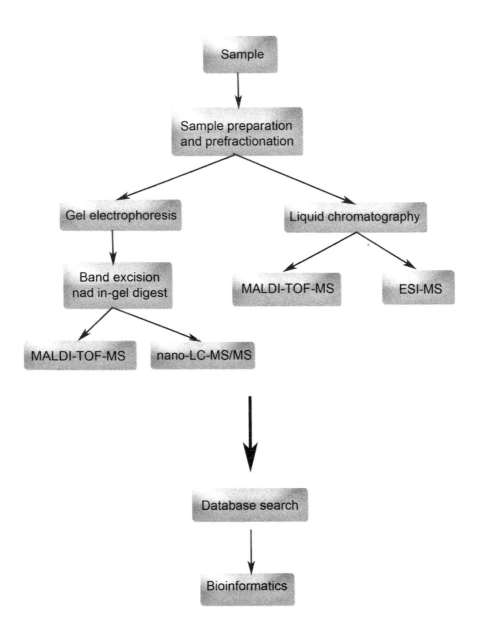

FIGURE 3.11: Flow chart overview of possible scenarios for the analysis of protein samples by mass spectrometry.

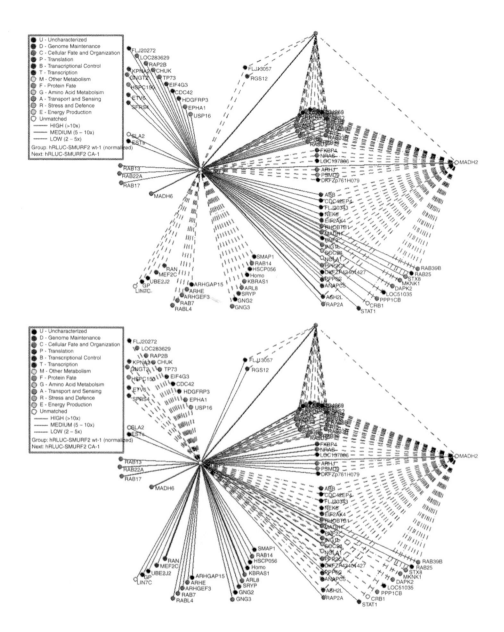

FIGURE 2.7: Using an example from [48], we show the utility of HTP mammalian cell-based assays for identifying dynamic PPIs in the TGFß pathway. Panel (a) shows SMURF2 wild-type, while panel (b) shows how PPIs change when SMURF2 is mutated (position of all proteins is unchanged between panels (a) and (b)). Dotted line represents known but inactive interaction; red line represents strong interaction; blue line represents medium strength interaction; black line represents weak interaction.

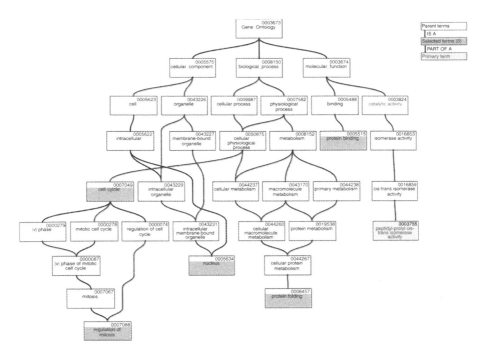

FIGURE 2.10: Graphical visualization of GeneOntology hierarchy [26, 234], using EMBL-EBI QuickGO server (http://www.ebi.ac.uk/ego/DisplayGoTerm). The hierarchy covers both classification (isA relationship) and generalization (partOf relationship).

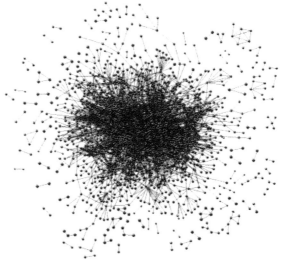

FIGURE 4.1: The PPI network constructed on 11,000 yeast interactions [531] involving 2,401 proteins. Reprinted from Przulj, N., Wigle, D., Jurisica, I. Functional topology in a network of protein interactions. *Bioinformatics*, 20(3):340–348, 2004; by permission of Oxford University Press.

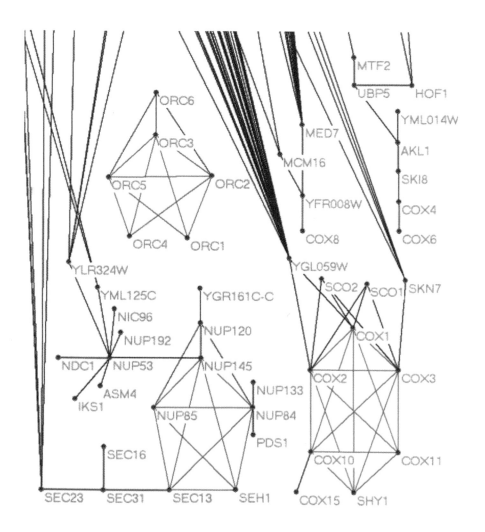

FIGURE 4.6: A subnetwork of a yeast PPI network [450] showing some of the identified complexes (green). Violet lines represent PPIs to proteins not identified as biological complex members due to stringent criteria about their connectivity in the algorithm, or due to absence of protein interactions that would connect them to the identified complex (for more details see [450]).

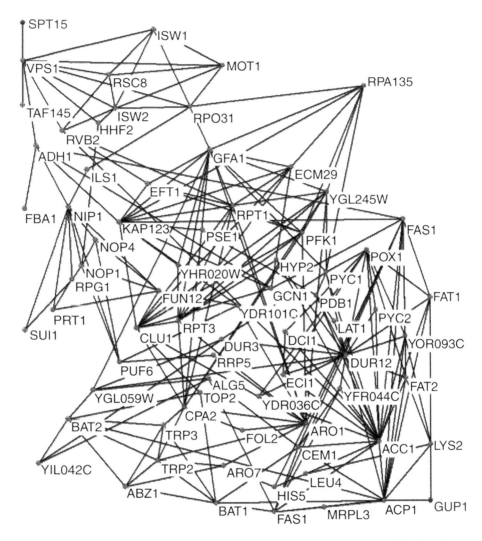

FIGURE 4.8: An example of a predicted pathway [450]. Note that this predicted pathway is presented as a subgraph of the PPI graph, and thus some of its internal nodes appear to be of low degree, even though they have many more interaction with proteins outside of this predicted pathway in the PPI graph.

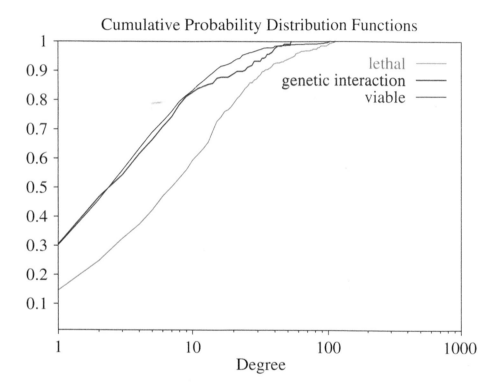

FIGURE 4.9: Cumulative distribution functions of degrees of lethal, genetic interaction, and viable protein groups in a yeast PPI network constructed on 11,000 interactions amongst 2,401 proteins [450]. Reprinted from Przulj, N., Wigle, D., Jurisica, I. Functional topology in a network of protein interactions. *Bioinformatics*, 20(3):340–348, 2004; by permission of Oxford University Press.

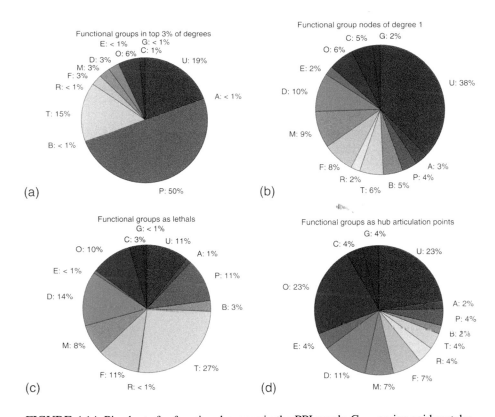

FIGURE 4.14: Pie charts for functional groups in the PPI graph: G — amino acid metabolism, C — cellular fate/organization, O — cellular organization, E — energy production, D — genome maintenance, M — other metabolism, F — protein fate, R — stress and defense, T — transcription, B — transcriptional control, P — translation, A — transport and sensing, U — uncharacterized. (a) Division of the group of nodes with degrees in the top 3% of all node degrees. (b) Division of nodes of degree 1. Compared with Figure 4.14(a), translation proteins are about 12 times less frequent, transcription about 2 times, while cellular fate/organization are 5 times more frequent, and genome maintenance, protein fate, and other metabolism are about 3 times more frequent; also, there are twice as many uncharacterized proteins. (c) Division of lethal nodes. (d) Division of articular points which are hubs. Reprinted from Przulj, N., Wigle, D., Jurisica, I. Functional topology in a network of protein interactions. *Bioinformatics*, 20(3):340–348, 2004; by permission of Oxford University Press.

FIGURE 5.2: (a) – (c) Crystallization experiments in micropipettes (crystal, precipitate, clear). (d) Subset of a 1,536-well plate with highlights of recognized drop and area of interest for further image analysis.

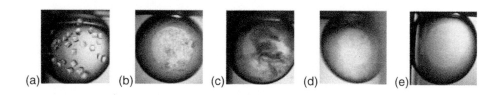

FIGURE 5.9: The set of images ranging from crystal, through precipitates to clear drop show crystallization results from a large drop volume experiment. Although useful for manual inspection, possibly x-ray diffraction experiment, and definitely computational image analysis, these examples will likely not be used in a screen due to prohibitive large volume of protein required.

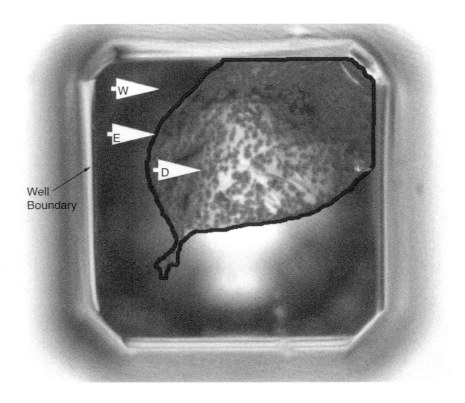

FIGURE 5.15: Drop segmentation eliminates the edges of the drop before feature extraction and image classification steps. The algorithm divides drop into the empty well — W, the inside of a drop — D, and the edge of a drop — E.

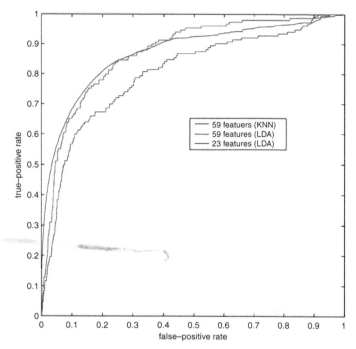

FIGURE 5.32: The ROC curves of image retrieval exercises for the three classifiers — different set of image features, and comparing LDA with k-nearest neighbor algorithm (with $k = 50$). The evaluation is conducted on 127 human-classified protein screens containing 5,600 crystal images and 189,472 noncrystal images.

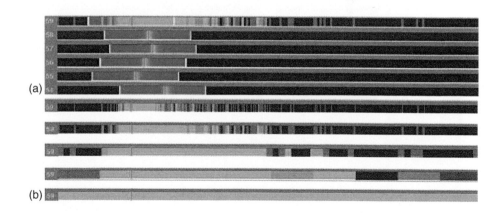

FIGURE 5.37: One of the methods for conext relaxation is generalization. Iterative generalization results in extending interval of values that are considered a match (a). (b) depicts a hierarchy of one relaxed attribute that was obtained by iteratively generalizing the attribute values.

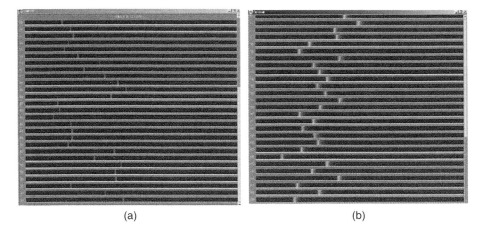

(a)   (b)

FIGURE 5.38: An input case is used to specify the initial context (a). If the initial retrieval does not return any similar cases, the algorithm relaxes value constraints for individual attributes, resulting in context relaxation (b).

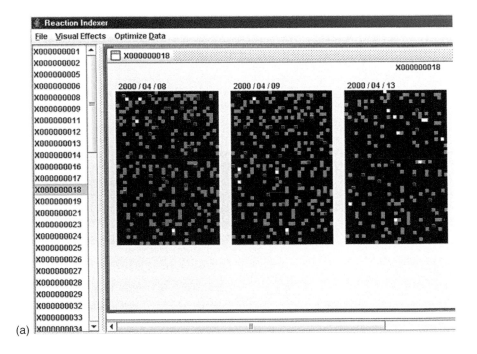

FIGURE 5.39: Visualization and optimization of the initial crystallization plan. White squares denote crystals, green squares show precipitates, black squares represent clear drops, and red square are unknowns. (a) depicts full view, (b) shows precipitates and crystal hits, and (c) highlights only crystals.

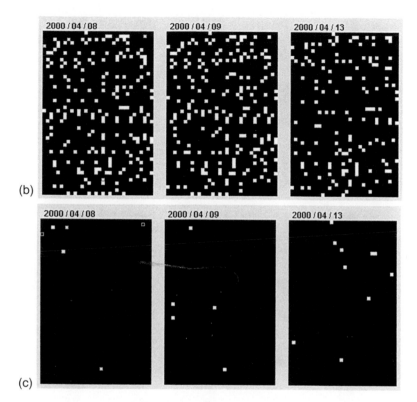

(b)

(c)

FIGURE 5.39 (continued) : Visualization and optimization of the initial crystallization plan. White squares denote crystals, green squares show precipitates, black squares represent clear drops, and red square are unknowns. (a) depicts full view, (b) shows precipitates and crystal hits, and (c) highlights only crystals.

	Reactions	Well #	Cocktail #	Chemical Additive	Chemical Formula	Concentration	Buffer Type	Buffer Concentration	pH
■ ■ □ ▦ ▦		13	2_C0013	Ammonium chloride	NH4Cl	2.5M	TAPS	0.1 M	9
□ □ ■ ▦ ▦		72	2_C0072	Lithium bromide	LiBr	1.87M	TAPS	0.1 M	9
■ ■ □ ▦ ▦		272	2_C0272	Ammonium thiocyanate	NH4SCN	2.7M	CAPS	0.1 M	10

FIGURE 5.40: The panel displays time course data and crystallization conditions for selected wells.

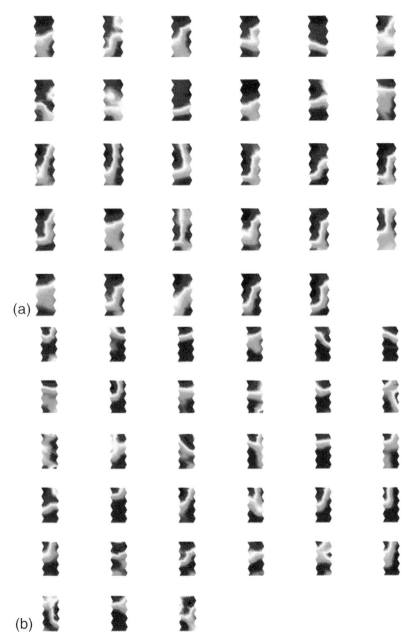

(a)

(b)

FIGURE 5.44: Two clusters of proteins ((a) and (b)) show different profiles across clusters, but homogeneity within clusters. Self-organizing maps are used to render the similarity among the proteins. A specific area within the component plane represents the same cluster of cocktails across all proteins in the study. The color renders individual outcomes, ranging from the blue that represents clear, through yellow that corresponds to precipitate, to red that corresponds to crystals.

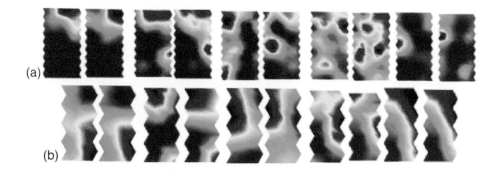

FIGURE 5.45: Clustering cocktails identifies both similarities and differences among individual conditions (a). Analogously, one can cluster proteins based on their propensity to crystallize under the same conditions (b).

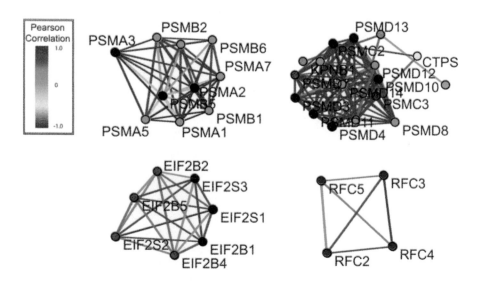

FIGURE 6.3: Protein–protein interactions identified through high-throughput screens (yeast high confidence [533]), mapped to human proteins, and integrated with the GeneAtlas mRNA expression data [515]. Nodes in graphs are proteins, while edges indicate interactions between proteins. The darker blue edges indicate a higher correlation in the gene expression data, suggestive of higher confidence in the individual protein interactions.

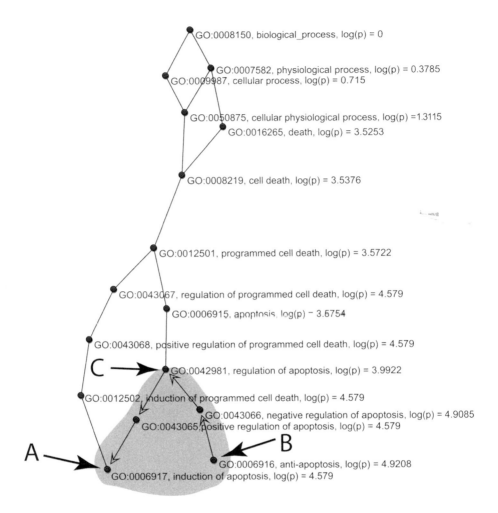

FIGURE 6.4: The GO biological process tree is illustrated, where each node represents an individual GO term, and the labels include GO identifiers, description, and semantic similarity score (log(p-value)) for each node. The maximum semantic similarity for ICE3 (SP:P42574) and BIR4 (SP:P98170) are found between the GO terms "induction of apoptosis" (A) and "anti-apoptosis" (B), through the common parent term "regulation of apoptosis" (C). This provides a final semantic similarity of 3.992 for ICE3 and BIR4.

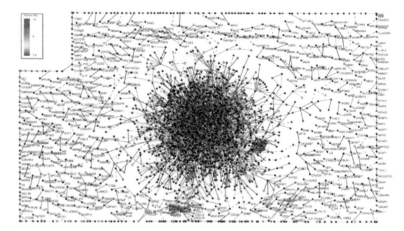

FIGURE 6.5: The protein interaction data from OPHID was combined with the gene expression data from GeneAtlas. The interactions were filtered to include only those with transcripts on the microarray, leaving 30, 495 PPIs. Shown here are 4,674 high confidence interactions, including data from DIP, MINT, *C. elegans, D. melanogaster, S. cerevisiae,* and *M. musculus.* Each node represents a protein, edges indicate interactions, and the color indicates the degree of correlation between the gene transcripts.

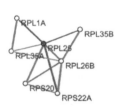

Party Hub: Rpl25
RPL25: ribosomal protein L23a.e
RPL1A: ribosomal protein L1
RPL35A: 60S large subunit ribosomal protein
RPL35B: 60S large subunit ribosomal protein
RPL26B: 60S large subunit ribosomal protein
RPS20: 40S ribosomal protein S20
RPS22A: 40S ribosomal protein S15a.e.c10

Date Hub: Skp1
SKP1: kinetochore protein complex CBF3, subunit D
SGT1: subunit of SCF Ub ligase complex/ essential
     component of kinetochore complex
CDC53: controls G1/S transition; non-catalytic scaffolding
     protein for Cdc34p to Skp/F-box proteins
CTF13: kinetochore protein complex, CBF3, 58kD subunit
     (chromosome segregation and movement of
     centromeres along microtubules)
MET30: cell cycle progress; involved in degradation of the
     Cdk-inhibitory kinase
RUB1: ubiquitin-like protein
UFO1: involved in degradation of Ho protein
HRT3: putative nuclear ubiquitin ligase
YN51: putative member of Ub ligase complex Swe1p
YNL311C: putative member of Ub ligase complex
GRR1: req'd for glucose repression and glucose and cation
     transport

FIGURE 6.6: A party hub, such as RPL25, is defined as a hub (a highly connected node) with a high average Pearson correlation (PCC). These proteins are assumed to interact with their partners in the same time and space. They are also more likely to represent stable protein interactions, such as those found in large protein complexes. Alternatively, the date hub is defined as a hub with a low average PCC. These proteins interact with one or few partners at a time, such as proteins that are involved in distinct cellular pathways. These are more likely to represent transient protein interactions.

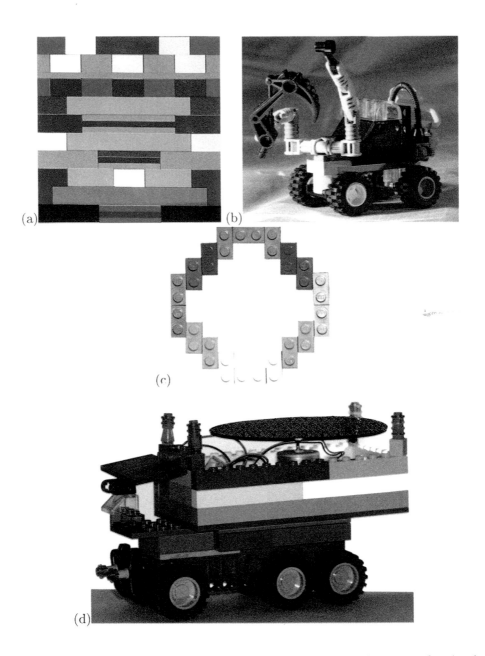

FIGURE 7.3: Generic block can be used to construct simple wall (a) or more functional and diverse toys (b), more intriguing structures, which create forms different from the basic components (circular shape from rectangular pieces) (c) or when using specialized parts even highly mobile and controllable devices (d). Biological systems too share many basic components, and higher organisms achieve complex behavior by not only new "parts," but also by more complex use of individual components.

Chapter 4

Graph Theory Analysis of Protein–Protein Interactions

Natasa Przulj

Understanding protein–protein interactions (PPIs) is an important problem in proteomics. It is widely believed that studying networks of these interactions will provide valuable insight about the inner working of cells, and will lead to important insights into complex diseases. The deluge of experimental PPI data available has made graph theory approaches an important part of computational biology and the knowledge–discovery process.

PPI networks are commonly represented as graphs, with nodes corresponding to proteins and edges representing PPIs. An example of a PPI network constructed in this way is presented in Figure 4.1. In general, these networks have directed edges and varying length; however, most of current PPI networks are undirected and represent only binary interactions.

Using high-throughput (HTP) techniques such as mass spectrometry (described in Chapter 3), and yeast 2-hybrid screening, a large volume of experimental PPI data has been generated. For example, the yeast *S. cerevisiae* contains over 6, 000 proteins, and over 78, 000 PPIs have now been identified. The analogous networks for mammals are expected to be much larger. For example, humans are expected to have around 120, 000 proteins, and thus approximately 10^6 PPIs. However, some of the largest public data sets of human PPIs currently available, such as the Data base of Interacting Proteins [562], and A Molecular INTeraction data base [568], contain less than 2, 500 interactions. Several manually created human PPI data sets have become available [440], comprising about 13, 000 interactions; however, this is in a format not amenable to further automated data analysis (at the time of writing). Another new data set comprising predicted protein interactions called HPID [228] is also available only for manual and visual browsing, and thus not directly suitable for knowledge–discovery approaches. A different approach to generate large numbers of putative human protein interactions is to use model organism data, and map to human orthologues using sequence homology. This enables us to generate a much larger human PPI data set, currently almost 50, 000 interactions [84]. An important aspect of OPHID [84] is effective access to data to facilitate retrieval, visualization, and analysis. Thus, multiple output formats (including PSI [244]) are supported. Without

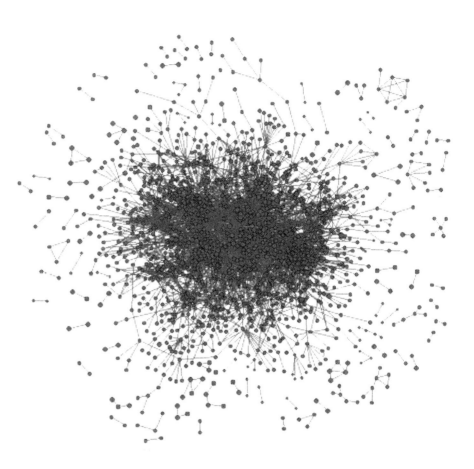

FIGURE 4.1: (See color insert following page 72.) The PPI network constructed on 11,000 yeast interactions [531] involving 2,401 proteins. Reprinted from Przulj, N., Wigle, D., Jurisica, I. Functional topology in a network of protein interactions. *Bioinformatics*, 20(3):340–348, 2004; by permission of Oxford University Press.

such flexibility, bioinformatic use of PPI data for knowledge discovery would be much harder.

PPI data sets provide both an opportunity and a challenge. Analyzing these networks may provide useful clues about the function of individual proteins, protein complexes, signaling pathways, and larger subnetworks. However, the data volume and noise within it renders many algorithms for its analysis intractable.

One of the goals of systems biology is to explain relationships between structure, function, and regulation of molecular networks by combining theoretical and experimental approaches, as described in Chapter 7. Graph theory is an integral part of this process, as it enables us to analyze structural properties of PPI networks, and link them to other information, such as function. Using this analysis leads to building predictive models for hypothesis generation, and, thus, more efficient and effective experiment planning.

After describing graph theoretic and biological terminology used in the PPI literature, we will introduce three large research areas necessary for understanding the issues arising in studying PPI networks:

1. Mathematical models of large networks and the most important properties of these models;

2. PPI identification methods, publicly available PPI data sets, some of the biological structures embedded in the PPI networks and methods used for their detection, and the mathematical properties of the currently available PPI networks;

3. Graph theoretic algorithms that have successfully been used in biological applications, and which may be used to identify biological structures in PPI networks.

4.1   Graph Theoretic Terminology

A *graph* is a collection of points with lines connecting pairs of points [543]. The points are called *nodes* or *vertices*, and the lines are called *edges*. A graph is usually denoted by G, or by $G(V, E)$, where V is the set of nodes and $E \subseteq V \times V$ is the set of edges of G. We often use n to represent number of nodes, $|V|$, and m to represent number of edges, $|E|$. We also use $V(G)$ to represent the set of nodes of a graph G, and $E(G)$ to represent the set of edges of a graph G.

A graph is *undirected* if its edges (node pairs) are undirected, otherwise it is *directed*. A graph is *weighted* if there is a weight function associated with its edges, or nodes. A graph is *complete* if it has an edge between every pair of nodes. Such a graph is also called a *clique*. A complete graph on n nodes is commonly denoted by K_n. A graph G is *bipartite* if its node set can be partitioned into two sets, A and B, such that every edge of G has one node in A and the other in B.

Nodes joined by an edge are called *adjacent*. A *neighbor* of a node v is a node adjacent to v. We denote by $N(v)$ the set of neighbors of node v (called the *neighborhood* of v), and by $N[v]$ the *closed neighborhood* of v, which is defined as $N[v] = N(v) \cup \{v\}$. The *degree* of a node is the number of edges incident with the node. In directed graphs, an *in-degree* of a node is the number of edges ending at the node, and the *out-degree* is the number of edges originating at the node.

A *path* in a graph is a sequence of nodes and edges, such that a node belongs to the edges before and after it, and no nodes are repeated. A path with k nodes is commonly denoted by P_k. The path *length* is the number of edges in the path. The shortest path length between nodes u and v is commonly denoted by $d(u, v)$. The *diameter* of a graph is the maximum of $d(u, v)$ over all nodes u and v. If a graph is disconnected, we assume that its diameter is equal to the maximum of the diameters of its connected components.

A *subgraph* of G is a graph whose nodes and edges belong to G. An *induced subgraph* H of G, denoted by $H \lhd G$, is a subgraph of G on $V(H)$ nodes, such that $E(H)$ consists of all edges of G that connect nodes of $V(H)$. The *minimum-edge cut* of a graph G is the set of edges S, such that $|S|$ is of minimum size over all sets of edges that disconnect the graph upon removal. The minimum number of edges whose deletion disconnects G is called *edge connectivity*. A graph is k-edge-connected if its edge connectivity is $\geq k$. In a weighted graph, the minimum weight edge cut of a graph can be defined.

4.2 Biological Terminology

Proteins are key components of cellular machinery. They play multiple roles — transferring signals, controlling the function of enzymes, and regulating production and activities in the cell. To do this, they interact with other proteins, DNA, and other molecules. Some of the protein interactions are permanent, while others are transient and happen only during certain cellular processes. Groups of proteins that together perform a certain cellular task are called *protein complexes*. There is some evidence to suggest that protein complexes correspond to complete or "nearly complete" subgraphs of PPI networks (see Section 4.4.3.1 and [450]).

A *domain* is part of a protein that has its own unique binding properties or function. The combination of domains in a protein determines its overall function. Examples of protein function include cell growth and maintenance, signal transduction, transcription, translation, metabolism, and others. These are systematically described in Gene Ontology [119]. Many domains mediate protein interactions with other biomolecules. Returning to knowledge management terminology from Chapter 2, there is a many-to-many relationship between proteins and domains: a protein may have several different domains and the same domain may be found in different proteins.

A *molecular pathway* is a directed sequence of molecular reactions involved in cellular processes. Modeling them in PPI networks will lead to adding directionality and causality, and, thus, will be necessary for simulations. Shortest paths in PPI networks have been used to model pathways (see Section 4.4.3.2 and [450]).

Homology is a relationship between two biological features (here we consider genes, or proteins) that have a common ancestor. The two subclasses of homology are *orthology* and *paralogy*. Two genes are *orthologous* if they have evolved from a common ancestor by speciation; they often have the same function, taken over from the precursor gene in the species of origin. Orthologous gene products are believed to be responsible for essential cellular activities. In contrast, *paralogous* proteins have evolved by gene duplication; they either diverge functionally, or all but one of the versions is lost.

4.3 Large Network Models

Describing real-world phenomena by a network may improve our understanding of the phenomena, and allow for simulations and predictions. Diverse complex systems can effectively be described by large networks. For example:

- The cell, where we model genes/proteins/metabolites by nodes and their interactions by edges;

- The Internet, which is a complex network of routers and computers connected by various physical or wireless links;

- The World Wide Web, which is a virtual network of Web pages connected by hyper-links;

- Networks of infection spread, which represent spread of biological or computer viruses;

- Electronic circuits, which represent connection among microelectronic components;

- Food chain webs, comprising networks of food-dependent linkages among animals;

- Human collaboration networks, which link scientists based on collaboration or actors based on appearance in the same movies, etc.

Several possible network models can be created for given real-world phenomena. However, the choice of a model is not arbitrary; the model must resemble properties of a true network. This leads to two challenges: (1) objectively describing characteristic properties of complex, real-world networks, and (2) defining network models that maximize overlap of characteristic network properties. As a result, it is likely that these models will change, as we generate more data and improve our understanding of given phenomena.

Despite progress, the field is still in its early phase. The emergence of the Internet, the World Wide Web, and cellular function data has made a significant impact on the modeling of large networks. Network theory has consequently become an important area of research on its own. Several articles give good surveys of large network models [11, 410, 411, 511].

4.3.1   Properties of Large Networks

One can study *global* and *local* properties of networks. While global properties provide an overall view of a given network, they fail to describe intricate differences among networks. Local properties measure small, local substructures or patterns, called *motifs* [387, 489, 566] or *graphlets* [447]. The main advantage of the local properties is evident when we study networks with incomplete node and edge sets. The reason is that while the local structures of these networks are more likely to be complete, the global properties are highly biased. For example, PPI networks are still under-studied and, thus, global properties cannot truly describe these networks, when, for example, instead of millions of interactions for human, we only have tens of thousands available now.

So far, the greatest research focus and progress has been made on studying global properties, such as: diameter [12], clustering [236], and degree distribution [413].

As mentioned in Section 4.1, the *diameter* is a maximum of shortest path lengths between any two nodes in the network. Despite their large sizes, most real-world networks have small diameters. This property is often referred to as the *small-world* property [541].

A network shows *clustering* (or *network transitivity*) if the probability of a pair of nodes being adjacent is higher when the two nodes have a common neighbor. *Clustering coefficient C* is defined as the average probability that two neighbors of a given node are adjacent [541]. More formally, if a node v in the network has d_v neighbors, the ratio between the number of edges E_v between the neighbors of v, and the largest possible number of edges between them, $\frac{d_v(d_v-1)}{2}$, is called the clustering coefficient of node v, and is denoted by C_v:

$$C_v = \frac{2E_v}{d_v(d_v - 1)}.$$

The clustering coefficient C of the whole network is the average of C_vs for all nodes v in the network. Complex, real-world networks exhibit a large degree of clustering, i.e., their clustering coefficient is large.

The *degree distribution* characterizes the distribution of degrees in a network. Let us denote by $P(k)$ the probability that a randomly selected node of a network has degree k. Most large real-world networks have nonPoisson degree distributions. For example, a large number of these networks has the degree distribution with a power-law tail, $P(k) \approx k^{-\gamma}$. Such networks are called *scale-free* [40].

Measuring global properties of real-world, complex networks led to proposing multiple models of these networks. Proposed in late 1950s [159], *random graphs* represent the simplest model of a complex network, yet are still an active research area (see below). The *small-world model* [541] was motivated by clustering, and it interpolates between the highly clustered regular ring lattices (defined below) and random graphs. The *scale-free* model [40] was motivated by the discovery of the power-law degree distribution. The *geometric random graph model* [200, 224, 438] has been used to model real-world networks, such as wireless communication [63, 113, 222], electrical power-grid [386], protein structure [386], and PPI networks [447].

Measuring local properties follows a bottom-up approach by focusing on finding small, over-represented patterns in a network [264, 386, 387, 489]. In this approach, *motifs* of a network are identified as small subgraphs of a large network that appear significantly more frequently than in the randomized network. Not surprisingly, different types of real-world networks have different motifs [387]. Furthermore, different real-world evolved and designed networks have been grouped into superfamilies according to their local structural properties [386].

A slightly different approach to measuring local network structure has been proposed. *Graphlets* have been defined as small, induced subgraphs of a large network [447]. They do not need to be over-represented in a real-world network and this, along with being induced, distinguishes them from motifs. The distribution of graphlet frequencies in real-world networks can be close to those of model networks pointing to limitations of previous models and suggesting new ways to model real-world networks [386, 447].

4.3.2 Random Graphs

Random graphs are based on the principle that the probability that there is an edge between any pair of nodes (denoted by p) is distributed uniformly at random. Thus, a random graph on n nodes has approximately $\frac{n(n-1)}{2}p$ edges, distributed uniformly at random.

Random graphs represent the earliest model of a complex network. Since the 1950s, large networks with no apparent design have been modeled by random graphs. The pioneering work of Erdös and Rényi [159, 160, 161] led to this field becoming a significant research area (for details see survey in [69]).

Erdös and Rényi defined several versions of the model, out of which the most commonly studied one is denoted by $G_{n,p}$, where each possible edge in the graph on n nodes is present with probability p and absent with probability $1-p$. The properties of $G_{n,p}$ are often expressed in terms of the average degree z of a node. The average number of edges in the graph $G_{n,p}$ is $\frac{n(n-1)}{2}p$, each edge contains two nodes, and, thus, the average degree of a node is:

$$z = \frac{n(n-1)p}{n} = (n-1)p,$$

which is approximately equal to np for large n.

These graphs have many properties that can be calculated exactly in the limit of large n, which makes them appealing as models of real networks. Thus, the following terminology is commonly used in the literature on random graphs. It is said that *almost all* random graphs (or *almost every* random graph) on n nodes have a property X, if the probability $Pr(X)$ that a graph has the property X satisfies $\lim_{n \to \infty} Pr(X) = 1$. A graph on n nodes *almost always*, or *almost surely*, satisfies a property X, if $\lim_{n \to \infty} Pr(X) = 1$.

Examples of properties that can be calculated exactly in the limit of large n include the following. When the number of edges (m) is small, the graph is likely to be fragmented into many small connected components having node sets of size at most $O(\log n)$. As m increases, the components grow at first by linking to isolated nodes, and later by fusing with other components. A transition happens at $m = \frac{n}{2}$, when many clusters cross-link spontaneously to form a unique largest component called the *giant component*, whose node set size is much larger than the node set sizes of any other component. The giant component contains $O(n)$ nodes, while the second largest component contains $O(\log n)$ nodes. Furthermore, the shortest path length between pairs of nodes in the giant component grows with $\log n$ (more details are given later), and, thus, these graphs are small worlds. This result is typical for random-graph theory whose main goal usually is to determine at what probability p a certain graph property is most likely to appear. Erdös and Rényi's greatest discovery was that many important properties, such as the emergence of the giant component, appear quite suddenly, i.e., at a given probability either almost all graphs have some property, or almost no graphs have it.

The probability $P(k)$ of a given node in a random graph on n nodes having degree k is given by the binomial distribution:

$$P(k) = \binom{n-1}{k} p^k (1-p)^{n-1-k},$$

which in the limit where $n \gg kz$ becomes the Poisson distribution:

$$P(k) = \frac{z^k e^{-z}}{k!}.$$

Both of these distributions are strongly peaked around the mean z and have a tail that decays rapidly as $1/k!$. Minimum and maximum degrees of random graphs are finite for a large range of p. For instance, if $p \approx n^{-1-1/k}$, almost no random graph has nodes with degree higher than k. However, if:

$$p = \frac{\ln n + k \ln(\ln n) + c}{n},$$

almost every random graph has a minimum degree of at least k. If $pn/\ln n \to \infty$, the maximum degree of almost all random graphs has the same order of magnitude as the average degree. Thus, a typical random graph has rather homogeneous degrees.

Random graphs tend to have small diameters. Random graphs on n nodes and probability p have a narrow range of diameters, usually concentrated around $\frac{\ln n}{\ln np} = \frac{\ln n}{\ln z}$ [111]. However, for $z = np < 1$, a typical random graph is composed of isolated trees and its diameter equals the diameter of a tree. If $z > 1$, the giant component emerges, and the diameter of the graph is equal to the diameter of the giant component if $z > 3.5$, and is proportional to $\frac{\ln n}{\ln k}$. If $z \geq \ln n$, almost every random graph is totally connected and the diameters of these graphs on n nodes and with the same z are concentrated on a few values around $\frac{\ln n}{\ln z}$. The average path length also scales with the number of nodes as $\frac{\ln n}{\ln z}$, which is a reasonable estimate for average path lengths of many real-world networks [410].

Random graphs have served as idealized models of gene networks [297], ecosystems [372], and the spread of infectious diseases [296] and computer viruses [303]. Although, random graph models reasonably approximate the corresponding properties of these real-world networks, they still differ from them in two fundamental ways:

1. **Degree distribution:** Real-world networks appear to have power-law degree distributions [10, 12, 40, 163, 221], i.e., a small but not negligible faction of their nodes has a very large degree. These degree distributions differ from the rapidly decaying Poisson degree distribution, and they have profound effects on the behavior of the network. Examples of degree distributions of real-world networks are presented in Figure 4.2.

2. **Clustering properties:** While real-world networks have strong clustering, the Erdös and Rényi model does not [540, 541]. The probabilities of pairs of nodes being adjacent in Erdös-Rényi random graphs are by definition independent, i.e., the probability of two nodes being adjacent is the same regardless of whether they have a common neighbor. Thus, the clustering coefficient for a random graph is $C = p$. Table 4.1 illustrates this by comparing clustering coefficients of real-world and random networks [410].

TABLE 4.1:   Clustering coefficients, C, for a number of different networks; n is the number of node, z is the mean degree. Taken from [410].

Network	n	z	C measured	C for random graph
Internet [433]	6,374	3.8	0.24	0.00060
World Wide Web (sites) [4]	153,127	35.2	0.11	0.00023
power grid [541]	4,941	2.7	0.080	0.00054
collaborations:				
biology [404]	1,520,251	15.5	0.081	0.000010
mathematics [405]	253,339	3.9	0.15	0.000015
film actor [413]	449,913	113.4	0.20	0.00025
company directors [413]	7,673	14.4	0.59	0.0019
word co-occurrence [97]	460,902	70.1	0.44	0.00015
neural network [541]	282	14.0	0.28	0.049
metabolic network [167]	315	28.3	0.59	0.090
food web [391]	134	8.7	0.22	0.065

Since random graphs do not provide an adequate model for real-world networks with respect to degree distributions and network clustering properties, we will review other network models, which fit the real-world networks better.

4.3.3   Generalized Random Graphs

This model captures power-law degree distribution in a graph, while leaving all other aspects as in the random graph model. That is, the edges are randomly chosen with the constraint that the degree distribution is restricted to a power law. A systematic analysis of these scale-free random networks showed that there is a threshold value of γ in the degree distribution $P(k) \approx k^{-\gamma}$, at which the properties of these networks suddenly change.

Generating a random graph with a nonPoisson degree distribution is relatively simple and has been discussed in a number of papers starting with [55]. Given a degree distribution (as a degree sequence), one can generate a random graph by assigning to a node i a degree k_i from the given degree sequence, and then choosing pairs of nodes uniformly at random to make edges so that the assigned degrees remain preserved. When all degrees have been used up to make edges, the resulting graph is a random member of the set of graphs with the desired degree distribution. Naturally, the sum of degrees has to be even to successfully complete the above algorithm. Note that this method does not allow the clustering coefficient to be specified, which is one of the crucial properties of these graphs that makes it possible to solve exactly for many of their properties in the limit of large n. For example, if we want to find the mean number of second neighbors of a randomly chosen node in a graph with clustering, we have to account for the fact that many of the second neighbors of a node are also its first neighbors as well. However, in a random

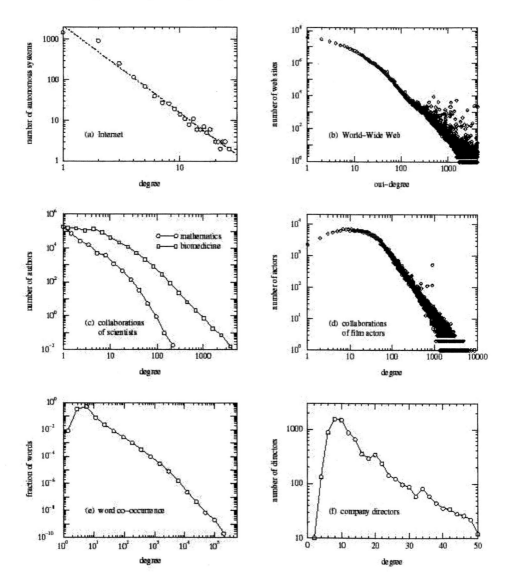

FIGURE 4.2: Degree distributions for different networks. (a) Physical connections between autonomous systems on the Internet in 1997 [163], (b) a 200 million page subset of the World Wide Web in 1999 [83], (c) collaborations between biomedical scientists and between mathematicians [404, 405], (d) collaborations of film actors [17], (e) co-occurrence of words in English [97], (f) board membership of directors of Fortune 1000 companies for year 1999 [413]. Taken from [410]. Reprinted from Random graphs as models of networks, in *Handbook of Graphs and Networks*, S. Bornholdt and H. G. Schuster (eds.), Wiley-VCH, Berlin (2003), with permission from Wiley-VCH, Berlin.

graph without clustering, the probability that a second neighbor of a node is also its first neighbor behaves as $\frac{1}{n}$, regardless of the degree distribution, and, thus, can be ignored in the limit of large n [410].

It has been proven that almost all random graphs with a fixed degree distribution, and no nodes of degree smaller than 2, have a unique giant component [350]. There is a simple condition for the birth of the giant component, as well as an implicit formula for its size [389, 390]. More specifically, for $n \gg 1$ and $P(k) = \frac{d_k}{n}$, Q is defined as:

$$Q = \sum_{k=1}^{\infty} P(k)k(k-2),$$

and it was shown that if $Q < 0$ the graph almost always consists of many small components, the average component size almost always diverges as $Q \to 0^-$, and a giant component almost surely emerges for $Q > 0$, under the condition that the maximum degree is less than $n^{\frac{1}{4}}$.

One can apply these results to a random graph model for scale-free networks. Aiello, Chung and Lu showed that for a power-law $P(k)$, the condition on Q implies that a giant component exists if and only if $\gamma < 3.47875... = \gamma_0$ [9]. They also observed several interesting properties for different values of γ:

- When $\gamma > \gamma_0$, the random graph is disconnected and made of independent finite clusters.

- When $\gamma < \gamma_0$, there is almost surely a unique infinite cluster.

- When $2 \leq \gamma < \gamma_0$, the second largest component almost surely has a size of the order of $\ln n$.

- When $1 < \gamma < 2$, every node with degree greater than $\ln n$ almost surely belongs to the infinite cluster, and the size of second largest component does not increase as the size of the graph goes to infinity. Thus, the fraction of nodes in the infinite cluster approaches 1 as the system size increases meaning that the graph becomes totally connected in the limit of infinite system size.

- When $0 < \gamma < 1$, the graph is almost surely connected.

Using the mathematics of generating functions [548], one can calculate exactly many statistical properties of these graphs in the limit of large n [413], such as the emergence of the giant component, the size of the giant component, the average distribution of the sizes of the other components, the average numbers of nodes at a certain distance from a given node, the clustering coefficient, the typical distance between a pair of nodes in a graph, etc. As described in [413], one can define the generating function

$$G_0(x) = \sum_{k=0}^{\infty} P(k)x^k$$

FIGURE 4.3: A regular ring lattice for $k = 2$. Reprinted from M. E. J. Newman, Random graphs as models of networks, *Handbook of Graphs and Networks*, S. Bornholdt and H. G. Schuster (eds.), Wiley-VCH, Berlin, 2003; with permission from Wiley-VCH, Berlin.

for the probability distribution of node degrees k, where the distribution $P(k)$ is assumed to be normalized so that $G_0(1) = 1$. Then, the condition for the emergence of the giant component is $\sum_k k(k - 2)P(k) = 0$ [389, 413] (a positive sum leads to the appearance of a giant cluster). The size of the giant component is $S = 1 - G_0(u)$, where u is the smallest nonnegative real solution of the equation $u = G_1(u)$ [390, 413].

This theory has been applied to the modeling of collaboration graphs, which are bipartite, and the World Wide Web, which is directed [413]. It has been shown that this theory gives good order of magnitude estimates of the properties of known collaboration graphs of business people, scientists, and movie actors, although there are measurable differences between theory and data that point to the presence of interesting effects, such as sociological effects in collaboration networks [413].

4.3.4   Small-World Networks

Networks of many biological, social, and artificial systems often exhibit *small-world* topology, i.e., a small-world character along with unusually large clustering coefficients independent of network size. Watts and Strogatz proposed this one-parameter model of networks in order to interpolate between an ordered finite-dimensional lattice and a random graph [541]. They start from a ring lattice with n nodes and m edges in which every node is adjacent to "its first k neighbors" on the ring (an illustration is presented in Figure 4.3), and "rewire" each edge at random with probability p, not allowing for self-loops and multiple edges. This process introduces $\frac{pnk}{2}$ "long-range" edges. Thus, the graph can be "tuned" between regularity ($p = 0$) and disorder ($p = 1$).

Structural properties of these graphs have been quantified by their *characteristic path length* $L(p)$ (the shortest path length between two nodes averaged over all pairs of nodes), and the clustering coefficient $C(p)$ as functions of rewiring probability p [541]. Remember that if:

$$C_v = \frac{|E(N(v))|}{\frac{1}{2}|N(v)|(|N(v)| - 1)},$$

then C is the average of C_v over all nodes v.

Watts and Strogatz established that the regular lattice at $p = 0$ is a highly clustered "large world" in which L grows linearly with n, since:

$$L(0) \approx \frac{n}{2k} \gg 1 \text{ and } C(0) \approx \frac{3}{2}.$$

On the other hand, as $p \to 1$ the model converges to a random graph, which is a poorly clustered "small world," where L grows logarithmically with n, since:

$$L(1) \approx \frac{\ln n}{\ln k}, \quad and \quad C(1) \approx \frac{k}{n}.$$

Note that these limiting cases do not imply that large C is always associated with large L, and small C with small L. On the contrary, even the small amount of rewiring transforms the network into a small-world with short paths between any two nodes, like in the giant component of a random graph, but at the same time such a network is much more clustered than a random graph.

This is in agreement with the characteristics of real-world networks. For example, the collaboration graph of actors in feature films, the neural network of the nematode worm *C. elegans*, and the electrical power grid of the western United States all have a small-world topology [541]. Thus, this model is generic for many large, sparse networks found in nature [541].

This pioneering work led to developing a large area of research, and many additional empirical examples of small-world networks have been documented [17, 40, 274, 497, 503, 532]. Graphs associated with many different search problems have a small-world topology, and the cost of solving them can have a heavy-tailed distribution [534]. This is due to the fact that local decisions in a small-world topology quickly propagate globally. One can use randomization and restarts to eliminate these heavy tails [534]. Finding a short chain of acquaintances linking oneself to a random person using only local information is solvable only for certain kinds of small worlds [315] (this problem was originally posed by Milgram's sociological experiment [384]).

Significant research in small-world networks exists outside computer science as well. Epidemiologists have studied how local clustering and global contacts together influence the spread of infectious disease, trying to make vaccination strategies and understand evolution of virulence [38, 72, 298, 533]. Neurobiologists have asked about the possible evolutionary significance of small-world neural topology. They have argued that small-world topology combines fast signal processing with coherent oscillations [330], and thus was selected by adaptation to rich sensory environments and motor demands [49].

One of the areas most active in research of small-world networks is within statistical physics (a good overview can be found in [408]). A variant of the Watts-Strogatz model was proposed by Newman and Watts [406, 407] in which no edges are removed from the regular lattice and new edges are added between randomly chosen pairs of nodes. This model is easier to analyze, since it does not lead to the formation of isolated components, which could happen in the original model. The formula for a characteristic path length in these networks has been derived to be [412]:

$$L(p) = \frac{n}{k} f(nkp),$$

where

$$f(x) \approx \frac{1}{2\sqrt{x^2 + 2x}} \tanh^{-1} \frac{x}{\sqrt{x^2 + 2x}}.$$

This solution is asymptotically exact in the limits of large n and when either $nkp \to \infty$, or $nkp \to 0$ (large or small number of shortcuts). This result can be improved by finding a rigorous distributional approximation for $L(p)$, together with a bound on the error [45].

Small-world networks have a relatively high clustering coefficient. In a regular lattice ($p = 0$), the clustering coefficient does not depend on the lattice size, but only on its topology. It remains close to $C(0)$ up to relatively large values of p as the network gets randomized. A slightly different, but equivalent definition of C, $C'(p)$, is defined as the fraction between the mean number of edges between the neighbors of a node and the mean number of possible edges between those neighbors [47]. It was used to derive a formula for $C(p)$ [47]. Starting with a regular lattice with a clustering coefficient $C(0)$, and observing that for $p > 0$ two neighbors of a node v that were connected at $p = 0$ are still neighbors of v and connected by an edge with probability $(1 - p)^3$, since there are three edges that need to remain intact, it follows that $C'(p) \approx C(0)(1 - p)^3$. The deviation of $C(p)$ from this expression is small and goes to zero as $n \to \infty$ [47]. The corresponding expression for the Newman-Watts model [409] is:

$$C'(p) = \frac{3k(k - 1)}{2k(2k - 1) + 8pk^2 + 4p^2k^2}.$$

The degree distribution of small-world networks is similar to that of a random graph. In the Watts-Strogatz model for $p = 0$, each node has the same degree k. A nonzero p introduces disorder in the network and widens the degree distribution while still maintaining the average degree equal to k. Since only one end of an edge gets rewired, $\frac{pnk}{2}$ edges in total, each node has degree at least $\frac{k}{2}$ after rewiring. Thus, for $k > 2$ there are no isolated nodes. For $p > 0$, the degree k_v of a node v can be expressed as $k_v = \frac{k}{2} + c_v$ [47], where c_v is divided into two parts, $c_v = c_v^1 + c_v^2$, so that $c_v^1 \leq \frac{k}{2}$ edges have been left in place with probability $1 - p$, and c_v^2 edges have been rewired towards v, each with probability $\frac{1}{n}$. For large n the probability distributions for c_v^1 and c_v^2 are:

$$P_1(c_v^1) = \binom{\frac{k}{2}}{c_v^1}(1 - p^{c_v^1})p^{\frac{k}{2} - c_v^1}$$

and

$$P_2(c_v^2) = \binom{\frac{pnk}{2}}{c_v^2}\left(\frac{1}{n}\right)^{c_v^2}\left(1 - \frac{1}{n}\right)^{\frac{pnk}{2}-c_v^2} \approx \frac{(pk/2)^{c_v^2}}{c_v^2!}e^{-pk/2}.$$

Combining these two factors, the degree distribution is:

$$P_p(c) = \sum_{n=0}^{min(c-k/2,k/2)} \binom{k/2}{n}(1-p)^n p^{k/2-n} \times \frac{(pk/2)^{c-k/2-n}}{(c-k/2-n)!}e^{-pk/2},$$

for $c \geq \frac{k}{2}$. As p grows, the distribution becomes broader, but it stays strongly peaked at the average degree with an exponentially decaying tail.

4.3.5   Scale-Free Networks

In many real networks connectivity of some nodes is significantly higher than for the other nodes. For example, the degree distributions of the Internet backbone [163], metabolic reaction networks [274], the telephone call graph [1], and the World Wide Web [83] decay as a power law $P(k) \approx k^{-\gamma}$, with the exponent $\gamma \approx 2.1 - 2.4$. This form of heavy-tailed distribution would imply an infinite variance, but in reality there are only a few nodes with many links, such as search engines for the World Wide Web.

The earliest work on the theory of scale-free networks dates back to 1955 [492], but it has been rediscovered [40, 41, 76]. A heavy-tailed degree distribution in these networks emerges automatically from a stochastic growth model, in which new nodes are added continuously and they preferentially attach to existing nodes with probability proportional to the degree of the target node [41]. That is, high-degree nodes become of even higher degree with time and the resulting degree distribution is $P(k) \approx k^{-3}$. Further, if either the growth, or the preferential attachment is eliminated, the resulting network does not exhibit scale-free properties [41]. Thus, both the growth and preferential attachment are needed simultaneously to produce the power-law distribution observed in real networks.

The average path length in the Barabasi-Albert network [41] is smaller than in a random graph, indicating that a heterogeneous scale-free topology is more efficient in bringing nodes close together than the homogeneous random graph topology. Analytical results show that the average path length, ℓ, satisfies $\ell \approx \frac{\ln n}{\ln \ln n}$ [70]. Interestingly, while in random graph models with arbitrary degree distribution the node degrees are uncorrelated [9, 413], nontrivial correlations develop spontaneously between the degrees of connected nodes in the Barabasi-Albert model [320]. There has been no analytical prediction for the clustering coefficient of the Barabasi-Albert model.

It has been observed that the clustering coefficient of a scale-free network is about five times higher than that of a random graph, and that this factor slowly increases with the number of nodes [11]. However, the clustering coefficient of the Barabasi-Albert model decreases with the network size approximately as $C \approx n^{-0.75}$, which is a slower decay than the $C = \frac{<k>}{N}$ for random graphs, where $< k >$ denotes the average degree, but is still different from the small-world models in which C is independent of n.

The Barabasi-Albert model is a minimal model that captures the mechanisms responsible for the power-law degree distributions observed in real networks. There are discrepancies between this model and real networks. While the exponent of the predicted power-law distribution for the model is fixed, real networks have measured exponents varying between 1 and 3.

This discrepancy has led to an increased interest in addressing network evolution questions. The theory of evolving networks offers insights into network topology and its evolution. More sophisticated models including the effects of adding or rewiring edges, allowing nodes to age so that they can no longer accept new edges, or varying the form of preferential attachment have been developed [10, 140, 321].

In addition to scale-free degree distributions, these generalized models also predict exponential and truncated power-law degree distribution in some parameter regimes. Scale-free networks are resistant to random failures due to a few high-degree "hubs" dominating their topology: any node that fails probably has a small degree, and, thus, does not severely affect the rest of the network [13]. However, such networks are vulnerable to deliberate attacks on the hubs. These intuitive ideas have been confirmed numerically [13, 83] and analytically [91, 115] by examining how the average path length and size of the giant component depend on the number and degree of the removed nodes. Implications have been made for the resilience of the Internet [76], the design of therapeutic drugs [274], and the evolution of metabolic networks [274, 532].

To generate networks with scale-free topologies in a deterministic, rather than stochastic way, Barabasi, Ravasz, and Vicsek have introduced a simple model, which they solved exactly showing that the tail of the degree distribution of the model follows a power law [44]. The first steps of the construction are presented in Figure 4.4. The construction can be viewed as follows. The starting point is a P_3. In the next iteration, add two more copies of a P_3 and connect the mid-point of the initial P_3 with the outer nodes of the two new P_3s. In the next step, make two copies of the 9-node module constructed in the previous step, and connect "end" nodes of the two new copies to the "middle" node of the old module (as presented in Figure 4.4). This process can continue indefinitely. The degree distribution of such a graph behaves as $P(k) \approx k^{\frac{\ln 3}{\ln 2}}$. An additional property that these networks have is the hierarchical combination of smaller modules into larger ones. Thus, these networks are called "hierarchical".

Another deterministic graph construction, called "pseudo-fractal", has been proposed to model evolving scale-free networks [139]. The scheme of the growth of the scale-free pseudo-fractal graph is presented in Figure 4.5. The degree distribution of this graph can be characterized by a power law with exponent $\gamma = 1 + \ln 3/\ln 2 \approx 2.585$, which is close to the distribution of real growing scale-free networks. All main characteristics of the graph were determined both exactly and numerically [139]. For example, the shortest

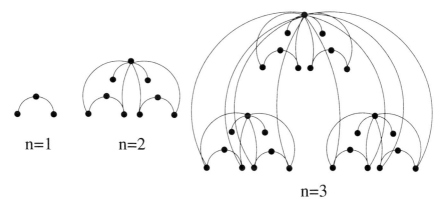

FIGURE 4.4: Scheme of the growth of a scale-free deterministic hierarchical graph. Adapted from [44].

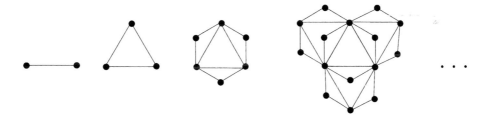

FIGURE 4.5: Scheme of the growth of the scale-free pseudofractal graph. The graph at time step $t + 1$ can be constructed by "connecting together" three graphs from step t. Adapted from [139].

path length distribution follows a Gaussian of width $\approx \sqrt{\ln n}$ centered at $\bar{l} \approx \ln n$ (for $\ln n \gg 1$), clustering coefficient of a degree k node follows $C(k) = 2/k$, and the eigenvalue spectrum of the adjacency matrix of the graph follows a power-law with the exponent $2 + \gamma$.

4.4 Protein Interaction Networks

With DNA sequence becoming available for an increasing number of organisms, there is a growing interest in correlating the genome with the proteome to explain biological function and to develop new, effective protein-targeting drugs. One of the key questions in proteomics today is identifying interactions of proteins and other molecules. The hope is to exploit this information to understand both the disease and healthy states, and to use this information for developing new therapeutic approaches.

Different methods have been used to identify protein interactions, including diverse biochemical and computational approaches. A survey of biochemical methods used to identify proteins and PPIs can be found in [431]. Unfortunately, most of these methods are lab-intensive, have strong biases, and low accuracy. Despite the resulting low confidence in the identified interactions, maps of PPIs are being constructed and their analysis is increasingly attracting attention. Many laboratories throughout the world are contributing to one of the ultimate goals of the biological sciences — the creation and understanding of a full molecular map of a cell. The computational focus is to analyze currently available networks of PPIs.

Despite many shortcomings of representing a PPI network using the standard mathematical representation of a network, with proteins being represented by nodes and interactions being represented by edges, this has been the only mathematical model used so far to represent and analyze these networks. This model does not address the following important properties of PPI data sets:

1. There is a large percentage of false-positive interactions in these data sets. For example, a common class of false-positive PPIs happens when in reality proteins indirectly interact, i.e., through mediation of one or more other molecules, but an experimental method detects this as a direct physical interaction. This may be a reason why very dense subnetworks are being detected inside PPI networks.

2. False-negative interactions are also present in these networks resulting from imperfect experimental techniques used to identify interactions. Since different biochemical methods have different biases, they detect different subsets of false-negatives. Thus, finding "high confidence" PPIs as overlap of multiple data sets may discard real interactions.

3. The following PPI properties are not captured by this model: spatial and temporal information, information about experimental conditions, strength of the interactions, number of experiments confirming the interactions, etc. Without this information, no true modeling and simulation can be performed. This has partially been addressed by von Mering *et al.* who classified PPIs into groups depending on the number of experiments that detected a specific PPI; they call this a *level of confidence* that a given PPI is a true interaction [531].

Next, we give an overview of several PPI identification methods, currently available PPI data sets and their repositories, biological structures contained in the PPI networks (such as protein complexes and pathways), and computational methods used to identify them in PPI networks. Then we focus on surveying the literature on PPI network properties and structure. We point to open problems and future research directions in the area of PPI networks in Section 4.6.

4.4.1   PPI Identification Methods

The lists of genes and encoded proteins are becoming available for an increasing number of organisms. Data bases, such as Ensembl [253] and GenBank [56] (described in the next section) contain publicly available DNA sequences for more than 105,000 organisms, including whole genomes of many organisms in all three domains of life, bacteria, archea, and eukaryota, as well as their protein data. In parallel to the increasing number of genomes becoming available, high-throughput PPI detection methods have been producing a huge amount of interaction data. Such methods include yeast 2-hybrid systems [261, 262, 526], protein complex purification methods using mass spectrometry [190, 247], correlated messenger RNA (m-RNA) expression profiles [255], genetic interactions [381], and *in silico* interaction predictions derived from gene fusion [157], gene neighborhood [128], and gene co-occurrences or phylogenetic profiles [259]. None of these PPI detection methods is perfect and the rates of false positives and false negatives vary from method to method. A brief summary describing these methods can be found in [531]. Next we outline the main characteristics of each of these methods.

Yeast 2-hybrid assay is an *in vivo* technique involving fusing one protein to a DNA-binding domain and the other to a transcriptional activator domain. An interaction between the two proteins is detected by the formation of a functional transcription factor. This technique detects even transient and unstable interactions. However, it is not related to the physiological setting. Also, the method will not detect interactions where three or more proteins need to be involved.

Mass spectrometry of purified complexes involves tagging individual proteins which are used as hooks to biochemically purify whole protein complexes. The complexes are separated and their components identified by mass spectrometry. There are two protocols, tandem affinity purification (TAP) [190, 460], and high-throughput mass-spectrometric protein complex identification (HMS-PCI) [247, 360]. This technique detects real complexes in their physiological settings and enables a consistency check by tagging several members of a complex at the same time. However, its drawbacks are that it might miss some complexes that are not present under the given conditions, tagging can disturb complex formation, and loosely associated components can be washed off during purification. More details are presented in Chapter 3.

LUMIER (LUinescence-based Mammalian IntERactome Mapping[1] [48]. In LUMIER, protein interactions in mammalian cells are measured after immunopurification and enzymatic detection of the partner protein. Briefly, flag epitope-tagged versions of various proteins are expressed together with a panel of partner proteins engineered to contain a Renilla luciferase tag (RLuc). Rluc-tagged proteins bound to flag-tagged partners can thus be quantitatively measured in immunoprecipitates without the need for western blotting. This

---

[1]http://ophid.utoronto.ca/LUMIER

method has proven to be remarkably robust, very sensitive and particularly amenable to analyzing PPIs that are subject to dynamic regulation [48, 428].

Interactions between membrane proteins and intracellular effectors is less amenable to analysis by Y2H, as membrane proteins may need to exist in their native environment for effective structure. Therefore, we predict that interactions between membrane proteins and intracellular interaction networks will be underrepresented in the Y2H dataset.

Correlated m-RNA expression (synexpression) involves measuring m-RNA levels under many different cellular conditions and grouping genes that show a similar transcriptional response to these conditions. The groups that encode physically interacting proteins were shown to frequently exhibit this behavior [191]. This is an indirect *in vivo* technique, which has a much broader coverage of cellular conditions than other techniques. However, it is very sensitive to parameter choices and clustering methods used during the analysis, and thus is not very accurate for predicting direct physical interactions.

Genetic interactions is an indirect *in vivo* technique that involves detecting interactions by observing the phenotypic results of gene mutations. An example of a genetic interaction is *synthetic lethality*, which involves detecting pairs of genes that cause lethality when mutated at the same time. These genes are frequently functionally associated, and, thus, their encoded proteins may physically interact. However, synthetically lethal genes may also represent alternative cellular wiring, and not a protein interaction.

In silico predictions using genome analysis involve screening whole genomes for the following types of interaction evidence: (1) finding conserved operons in prokaryotic genomes which often encode interacting proteins [128]; (2) finding similar phylogenetic profiles, since interacting proteins often have similar phylogenetic profile, i.e., they are either both present, or both absent from a fully sequenced genome [259]; (3) finding proteins that are found fused into one polypeptide chain [157]; (4) finding structural and sequence motifs within the protein–protein interfaces of known interactions that allow the construction of general rules for protein interaction interfaces [277, 278].

In silico methods are fast and relatively inexpensive techniques whose coverage expands with more and more organisms becoming fully sequenced. However, they require orthology between proteins and fail when orthology relationships are not clear. Also, these methods generally favor high sensitivity at the cost of low specificity, and, thus, they produce a large number of false positives.

4.4.2 Public Data Sets

Vast amounts of biological data that are being generated by many HTP methods are deposited in numerous data bases. Different PPI data bases contain PPIs from different single experiments, HTP experiments, and literature sources. PPIs resulting from the most recent studies are usually only available on the journal web sites where the corresponding papers appeared. Standardization efforts to represent and organize PPI data bases [244] will eventually improve public data sets. However, due to relatively high flexibility in the data format proposed, data curation and integration will remain challenging.

Here we briefly introduce the main data bases, including nucleotide sequence, protein sequence, and PPI data bases. Nucleotide and protein sequence data bases do not suffer from the lack of standardization that is present in PPI data bases. A comprehensive list of major molecular biology data bases can be found in [51].

The largest nucleotide sequence data bases are EMBL[2] [509], GenBank[3] [56], and DDBJ[4] [516]. They contain sequences from the literature as well as those submitted directly by individual laboratories. These data bases store information in a general manner for all organisms. Organism specific data bases exist for many organisms. For example, the complete genome of bakers yeast and related yeast strains can be found in Saccharomyces Genome Data base (SGD)[5] [146]. FlyBase[6] [25] contains the complete genome of the fruit fly *Drosophila melanogaster*. It is one of the earliest model organism data bases. Ensembl[7] [253] contains the draft human genome sequence along with its gene prediction and large scale annotation. It currently contains over 4,300 megabases and 29,000 predicted genes, as well as information about predicted genes and proteins, protein families, domains etc. Ensembl is not only free, but is also open source.

SwissProt[8] [35] and Protein Information Resource (PIR)[9] [374] are two major protein sequence data bases. They are both manually curated and contain literature links. They exhibit a large degree of overlap, but still contain many sequences that can be found in only one of them. SwissProt maintains a high level of annotation for each protein including its function, domain structure, and posttranslational modification information. It contains over 101,000 curated protein sequences. Computationally derived translations of EMBL nucleotide coding sequences that have not yet been integrated into the SwissProt resource can be found in Trembl[10]. The NonRedundant Data base (NRDB)[11] merges two sequences into a representative sequence if they exhibit a large degree of similarity. This is useful when a large scale computational analysis needs to be performed.

---

[2]http://www.ebi.ac.uk/embl/

[3]http://www.ncbi.nlm.nih.gov/Genbank/

[4]http://www.ddbj.nig.ac.jp/

[5]http://genome-www.stanford.edu/Saccharomyces/

[6]http://flybase.bio.indiana.edu/

[7]http://www.ensembl.org/

[8]http://www.ebi.ac.uk/swissprot/

[9]http://pir.georgetown.edu/

[10]http://www.ebi.ac.uk/trembl/

[11]http://www.ebi.ac.uk/ holm/nrdb90

There is a growing number of public data bases comprising PPI data for one or multiple organisms, including the following major resources:

- The Munich Information Center for Protein Sequences (MIPS)[12] provides high quality curated genome related information, such as PPIs, protein complexes, pathways etc., for several organisms [381].

- Yeast Proteomics Database (YPD)[13] is a curated data base, comprising bakers yeast, *S. cerevisiae*, protein information, including their sequence and genetic information, related proteins, PPIs, complexes, literature links, etc. [121].

- The Database of Interacting Proteins (DIP)[14] is a curated data base containing information about experimentally determined PPIs. It catalogs around 11,000 unique interactions between 5,900 proteins from over 80 organisms including yeast and human [562]. However, it still maintains only 988 human PPIs, compared to the 15,358 interactions in *S. cerevisiae*, 13,178 for *D. melanogaster*, and 5,500 for *C. elegans*.

- Human Protein Reference Database (HPRD)[15] is a large-scale effort to manually catalog many known human PPIs related to diseases in a data base. Currently, HPRD comprises 12,641 human protein interactions derived from literature sources [440]. However, it is currently in a form not suitable for large-scale analysis, although the trend is to provide export in PSI format [244].

- The Biomolecular Interaction Network Database (BIND)[16] archives biomolecular interaction, complex, and pathway information [28]. BIND stores interactions of diverse biological objects: a protein, RNA, DNA, molecular complex, small molecule, photon, or gene. This data base includes Pajek [50] as a network visualization tool. It includes a network clustering tool as well, called the Molecular Complex Detection (MCODE) algorithm [31] (described in Section 4.4.3), and a functional alignment search tool (FAST) [28] (details of FAST are not yet available in the literature), which displays the domain annotation for a group of functionally related proteins.

- The General Repository for Interaction Datasets (GRID)[17] stores genetic and physical interactions. It contains interactions from several genome and proteome wide studies, as well as the interactions from

---

[12]http://mips.gsf.de
[13]http://www.incyte.com/sequence/proteome/databases/YPD.shtml
[14]http://dip.doe-mbi.ucla.edu/
[15]http://www.hprd.org/
[16]http://www.binddb.org/
[17]http://biodata.mshri.on.ca/grid/

MIPS and BIND data bases [81]. It also provides a powerful network visualization tool called Osprey [82].

- A Molecular INTeraction data base (MINT)[18] contains about 2, 500 interactions curated manually from the literature [568].

- Online Predicted Human Interaction Database (OPHID)[19], is a web-based data base of predicted and known human PPIs [84]. In January 2005, OPHID comprised 16, 034 known human PPIs obtained from BIND, DIP, MINT and HPRD, and additional 23, 889 predicted interactions from model organisms [84]. Predictions are made from *S. cerevisiae*, *C. elegans*, *D. melanogaster*, and *M. musculus*. Individual predictions have been extensively investigated to provide biological evidence for their support. The OPHID data base can be queried using single IDs or in a batch mode, and results can be displayed as text, HTML, or visualized using our custom visualization tool. In addition, the entire data base is available in tab-delimited text or PSI-compliant XML format [244].

As experimentally derived PPIs include both false positives and negatives, and individual detection methods have different sensitivity and specificity, it is important to assess the quality of these datasets.

Von Mering *et al.* have performed a systematic synthesis and evaluation of PPIs detected by major high-throughput PPI identification methods for yeast *S. cerevisiae* [531], a model organism relevant to human biology [466]. They integrated 78, 390 interactions between 5, 321 yeast proteins, out of which only 2, 455 are supported by more than one PPI detection method. This low overlap between the methods may be due to a high rate of false positives, or to difficulties in detecting certain types of interactions by specific methods. Research bias is another potential explanation of the low overlap; research groups usually have interest focused on finding interactions between certain types of proteins. Further, each PPI identification technique produces a unique distribution of interactions with respect to functional categories of interacting proteins [531]. Assessing the quality of interaction data produced a list of 78, 390 yeast PPIs ordered by the level of confidence (high, medium, and low) with the highest confidence being assigned to interactions confirmed by multiple methods [531]. The resulting list of PPIs currently represents the largest publicly available collection of PPIs for *S. cerevisiae*, and also the largest PPI collection for any organism.

There are other approaches to improve the quality of PPI data sets, such as correlating transcriptome and interactome data [216, 191], computational use of statistical and topological descriptors combined with transcriptional

---

[18]mint.bio.uniroma2.it/mint/

[19]http://ophid.utoronto.ca

information and other annotations [32], probabilistic modeling of interactions based on the available evidence [27], combining information about evolutionary conserved and essential proteins [560], integrating large screens with information about chromosomal proximity, and phylogenetic profiling and domain fusion [379].

Clearly, false negatives will dominate error in PPI datasets. Due to diverse biases, nonoverlap among available data bases does not always imply false positives. Careful integration of multiple information sources [106] and hypothesis-driven design of HTP experiments, similar to [329], will enable effective increases in PPI data bases, while improving data quality.

4.4.3   Biological Structures within PPI Networks and Their Extraction

Biochemical studies used to identify biological structures, such as complexes and pathways, are expensive, time consuming, and of low accuracy. One approach to reduce the time and cost, and increase accuracy of these studies is to computationally detect biological structures from PPI networks. The hope is that with the emergence of high confidence PPI networks, such as the one constructed by Mering et al. [531], computational approaches will become inexpensive and reliable tools for extraction of known and prediction of still unknown members of these structures.

Despite a large body of literature involving purely theoretical aspects of networks, such as finding clusters in graphs (see Section 4.5), only a few such methods have been developed specifically for biological applications and applied to PPI networks.

4.4.3.1   Protein Complexes

Cellular processes are usually carried out by groups of proteins acting together to perform a certain function. These groups of proteins are called *protein complexes.* They are not necessarily of invariable composition, i.e., a complex may have several core proteins, which are always present in the complex, as well as more dynamic, perhaps regulatory proteins, which are only present in a complex from time to time. Also, the same protein may participate in several different complexes during different cellular activities. One of the most challenging aspects of PPI data analysis is determining which of the myriad of interactions in a PPI network comprise true protein complexes [27, 31, 152, 190, 247, 309, 519].

Several mass spectrometry studies have been used to identify protein complexes in yeast *S. cerevisiae*, and report improved data quality compared to yeast 2-hybrid method. Ho *et al.* used HMS-PCI to extract complexes from the *S. cerevisiae* proteome [247]. They reported an approximately three-fold higher success rate in detection of known complexes when compared to large-scale yeast 2-hybrid studies [261, 526]. Gavin *et al.* have performed a mass-spec analysis of the *S. cerevisiae* proteome to identify protein complexes [190], with about 70% probability of detecting the same protein in two different purifications. Amongst $1,739$ yeast genes, including $1,143$ human orthologues, they purified 589 protein assemblies, out of which 98 corresponded to protein complexes in the Yeast Protein Database (YPD), 134 were new complexes, and the remaining ones showed no detectable association with other proteins. This led to proposing a new cellular function for 344 proteins, including 231 proteins with no previous functional annotation. They attempted investigating relationships between complexes in order to understand the integration and coordination of cellular functions by representing each complex as a node and having an edge between two nodes if the corresponding complexes share proteins. By color-coding complexes according to their cellular roles they noticed grouping of the same colored complexes, suggesting that sharing of components reflects functional relationships. No graph theoretic analysis of this protein complex network has been done so far. Comparing human and yeast complexes showed that orthologous proteins preferentially interact with complexes enriched with other orthologues, supporting the existence of an "orthologous proteome," which may represent core functions for all eukaryotic cells [190]. This leads to stronger evolution-based interaction conservation [179, 512, 560], which will increasingly play a role as we move to more complex organisms [75].

Diverse computational approaches have been proposed to identify protein complexes from PPI networks. They have involved measuring connectedness (e.g., k-core concept [30]), node neighborhood "cliquishness" [541] (e.g., MCODE method [31]), partitioning the network's node set into clusters based on a cost function that is assigned to each partitioning [309], or the reliance on reciprocal bait-hit interactions as a measure of complex involvement. The challenge in this analysis is complexity and scalability of the algorithm, and most importantly its specificity and sensitivity. Evaluating these performance characteristics is not trivial, due to the following reasons:

- Since existing data bases of protein complexes are incomplete, computationally identified complexes may incorrectly appear as false positives. A few biological validations also does not conclusively prove quality of the algorithm, as only a small fraction of many tests may succeed overall. Thus, multiple computational, comparative, and experimental combinations have to be used for performance evaluation.

- Increasing sensitivity usually decreases specificity, and, thus, multilevel algorithms that use additional filters to remove potential false positives are necessary. However, depending on the overall goal of the analysis one can change the tradeoff between sensitivity and specificity, and a given algorithm can be tuned for a specific task. Thus, when comparing algorithms one must consider this selection.

The Molecular Complex Detection (MCODE) algorithm [31] exploits the notion of a clustering coefficient [541] (described in Section 4.3). Bader and Hogue used the notion of a *k-core*, a graph of minimum degree k, and a notion of the "highest k-core of a graph", the most densely connected k-core of a graph, to weight PPI network nodes in the following way. A core-clustering coefficient of a node v is defined as a density of the highest k-core of $N[v]$, where density is the number of edges of a graph divided by the maximum possible number of edges of the graph. The weight of a node v is the product of the node core-clustering coefficient and the highest k-core level, k_{max}, of the $N(v)$. A complex is seeded in the weighted graph with the highest weighted node, and nodes with weights above a given threshold are recursively included in the complex. The process is repeated for the next highest weighted unexplored node. This generation phase is followed by postprocessing, which discards complexes that do not contain a k-core with $k \geq 2$.

The following two options are also included in the algorithm: an option to "fluff" the complexes by adding to them their neighbors unexplored by the algorithm of weight bigger than the "fluff" parameter, and an option to "haircut" the complex by removing nodes of degree 1 from the complex. The resulting complexes are scored according to the product of the complex node set size and the complex density, and they are ranked according to the scoring function. The MCODE algorithm also offers an option to specify a seed node.

MCODE was evaluated against known complexes in the MIPS data base [381] and 221 complexes from [190]. In an attempt to maximize the overlap between predicted and known complexes, all combinations of the parameters (true/false for haircut and fluff, and node weight percentage in 0.05 increments) have been varied. However, only 88 out of the 221 complexes from [190] matched, and only 52 of 166 predicted complexes matched MIPS complexes. MCODE identified complexes of high density, which were highly likely to match real complexes. Thus, high sensitivity was achieved at the cost of low specificity.

These results suggest that a different approach of finding efficient graph clustering algorithms to identify highly connected subgraphs should be used to identify protein complexes in PPI networks. We explored this approach and used Hartuv and Shamir's Highly Connected Subgraph (HCS) algorithm [235, 236] (described in Section 4.5) to identify protein complexes from a yeast PPI network with 11,000 interactions amongst 2,401 proteins [450]. The algorithm detected a number of known protein complexes (an illustration is presented in Figure 4.6). Also, 27 out of 31 clusters identified in this way had high overlaps with protein complexes documented in MIPS data base. The remaining 4 clusters that did not overlap MIPS contained a functionally homogeneous 6-protein cluster Rib 1-5 and a cluster Vps20, Vps25, Vps36, which are likely to correspond to protein complexes. In addition, the clusters identified in this way had a statistically significant functional homogeneity.

A similar approach explored three different methods for identification of highly connected subgraphs in a PPI network constructed on the MIPS data base [496]. The first method involves identifying complete subgraphs of the

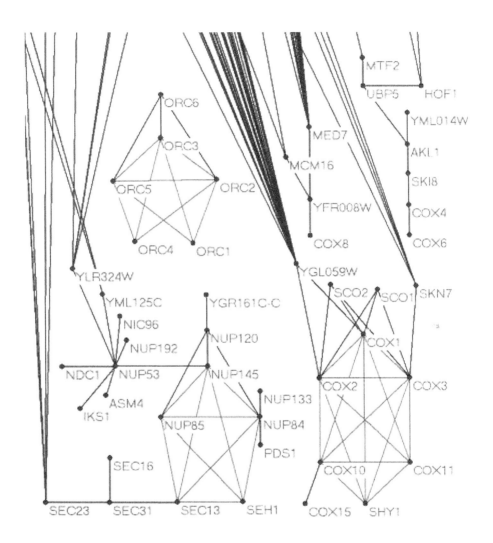

FIGURE 4.6: (See color insert following page 72.) A subnetwork of a yeast PPI network [450] showing some of the identified complexes (green). Violet lines represent PPIs to proteins not identified as biological complex members due to stringent criteria about their connectivity in the algorithm, or due to absence of protein interactions that would connect them to the identified complex (for more details see [450]).

PPI graph. The second method is the Super-Paramagnetic Clustering algorithm [66] to cluster objects in a nonmetric space of an arbitrary dimension. The third method comprises a novel Monte-Carlo optimization technique to identify highly connected subgraphs in a network (the details of this algorithm have not been published yet). Most of the dense subnetworks that were identified in this way had consistent functional annotation revealing the function of the whole complex [496]. Also, dense subgraphs had a good agreement with the known protein complexes from MIPS, BIND, and the data from [247]. Also, several novel complexes and pathways were predicted, but no further details about these are given [496].

A similar approach has been used to predict functions of uncharacterized proteins [85]. Spectral graph theory methods, previously used for analyzing the World Wide Web [199, 314], have been applied to the yeast PPI network constructed on high and medium confidence interactions from [531]. This method identified "quasi-cliques" and "quasi-bipartites" in the PPI network. Since proteins participating in quasi-cliques usually shared common functions, this method was used to assign function to 76 uncharacterized proteins.

4.4.3.2 Molecular Pathways

Molecular *pathways* are chains of cascading molecular reactions involved in maintaining life. Different processes involve different pathways. Some examples include metabolic, apoptosis, and signaling pathways for cellular responses. An example of a signaling pathway transmitting information from the cell surface to the nucleus where it causes transcriptional changes, is presented in Figure 4.7.

Disruption in a pathway function may cause severe diseases, such as cancer. Thus, understanding molecular pathways is an important step in understanding cellular processes and the effects of drugs on cellular processes. As a consequence, modeling and computational pathway prediction from PPI networks has become an active research area.

The Biopathways Consortium[20] was founded to catalyze the emergence and development of computational pathways biology. One of their main goals is to coordinate the development and use of open technologies, standards, and resources for representing, handling, accessing, and analyzing pathway information. Numerous papers addressing these topics have been presented at the Biopathways Consortium Meetings. Many of them used classical graph algorithms in order to integrate genome-wide data on regulatory proteins and their targets with PPI data in yeast [565], reconstruct microbial metabolic pathways [376], determine parts of structure and evolution of metabolic networks [353], etc.

Using simple graph theory and a yeast 2-hybrid data from [261, 479, 526], a model of a *S. cerevisiae* signal transduction network has been constructed

---

[20]http://www.biopathways.org/

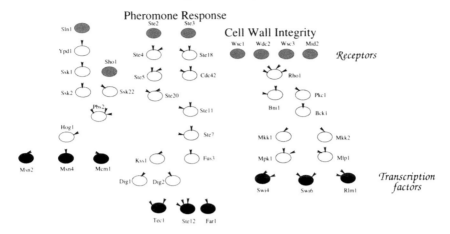

FIGURE 4.7: Examples of MAPK (mitogen-activated protein kinases) signal transduction pathways in yeast. Gray nodes represent membrane proteins, black nodes represent transcription factors, and white nodes represent intermediate proteins. Reprinted from Przulj, N., Wigle, D., Jurisica, I. Functional topology in a network of protein interactions. *Bioinformatics*, 20(3):340–348, 2004; by permission of Oxford University Press.

[501]. In order to reduce the number of candidate signaling pathways from around 17 million to around 4.4 million, the most highly connected nodes have been first deleted, and then shortest paths of length at most eight between every membrane protein and every DNA-binding protein has been identified in the modified network. To maximize high-scoring pathways in the real PPI network and minimize those in the randomized networks, several parameters have been optimized: the number of clusters in which genes were grouped, the microarray expression datasets used in clustering, the maximum path length of their pathways, and the scoring metric. The resulting putative pathways have been compared with the ones obtained in the same way in three randomized PPI networks. Only paths with lengths of 8 at most have been identified, because the average shortest path length between any two proteins in the given PPI graph was 7.4, and because a fraction of pathways with high microarray clustering ratios over various shortest path lengths peaked at 8. This method reproduced many essential elements of the pheromone response, cell wall integrity, and filamentation MAPK pathways, but it failed to model the High Osmolarity (HOG) MAPK pathway due to missing interactions (false-negatives) in the PPI network.

Although in general, pathways are complex, one can identify and model linear components of some pathways. Using this model, it is possible to predict novel linear pathways [450]. We focused on finding and exploiting the basic structure that these pathways have in PPI networks. We used MAPK as our model pathway, because they are among the most thoroughly studied networks in yeast and because they exhibit linearity in structure [463]. Initial analysis showed that these pathways have source and sink nodes of low degree and internal nodes of high degree in the yeast PPI network. This information was used to create a linear pathway model, and to extract such putative pathways from a PPI network. The approach is based on the following steps [450]:

- Construct a shortest path from a transmembrane or sensing protein of low degree to a transcription factor protein of low degree, such that the internal nodes on the shortest path are of high degree.

- Increase sensitivity by including high degree first and second neighbors of internal nodes of such a shortest path into these predicted pathways (for more details see [450]).

Using this approach, we extracted 399 putative pathways (an example of a predicted pathway is presented in Figure 4.8).

Other theoretical approaches have been proposed to model pathways. They involve system stoichiometry, thermodynamics, etc. (for example, see [474]). Also, methods for extraction of pathway information from on-line literature are being developed [181, 322, 415].

4.4.4   Properties of PPI Networks

Systematically analyzing PPI network properties may be used to assess functional meaning of individual proteins and subgraphs within these networks. In addition, comparing properties of multiple PPI networks can be used to evaluate evolutionary characteristics, robustness of the organism, and quality of a given data set.

4.4.4.1   Scale-Free Network Topology

One of the first approaches to modeling biological networks focused on metabolic pathway networks of different organisms [272, 274] from the WIT data base [426]. This data base contains predicted pathways based on the annotated genome of an organism and data established from the biochemical literature. This analysis showed that metabolic networks of 43 organisms from the WIT data base, containing 6 archaea, 32 bacteria, and 5 eukaryota, all have scale-free topology with $P(k) \approx k^{-2.2}$ for both in- and out-degrees [274]. The diameter of the metabolic networks was the same for all 43 organisms, indicating that with increasing organism complexity, nodes are increasingly connected. A few hubs dominated these networks, and upon the sequential removal of the most connected nodes, the diameter of the network rose sharply. Only around 4% of the nodes were present in all species, and these were the ones that were most highly connected in any individual organism; species-specific differences among organisms emerged for less connected nodes. A

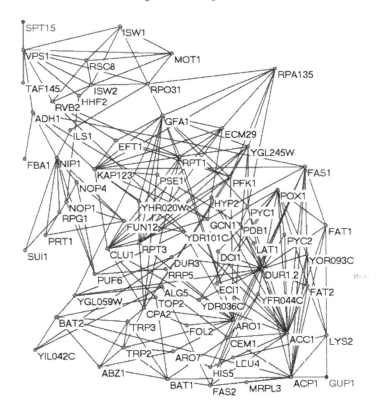

FIGURE 4.8: (See color insert following page 72.) An example of a predicted pathway [450]. Note that this predicted pathway is presented as a subgraph of the PPI graph, and thus some of its internal nodes appear to be of low degree, even though they have many more interactions with proteins outside of this predicted pathway in the PPI graph.

potential bias introduced by a high interest, research conducted on some, and a lack of interest and research conducted on other proteins may also have contributed to this effect. In addition, randomly removing nodes from these networks, the average shortest path lengths did not change, indicating insensitivity to random errors in these networks.

Subjecting the *S. ccrevisiae* PPI network constructed on 1,870 proteins and 2,240 interactions derived from the yeast 2-hybrid study [526] and the DIP data base [562] to the same analysis established that the yeast PPI network and the PPI network of the human gastric pathogen *Helicobacter pylori* [454] also have heterogeneous scale-free network topology with a few highly connected proteins and numerous less connected proteins. To study robustness of a PPI network, one can correlate removal of a protein of a certain degree and lethality. Interestingly, the same tolerance to random errors, coupled with fragility against the removal of high-degree nodes was observed as in the metabolic networks:

- Even though about 93% of proteins had degree at most 5, only about 21% of them were essential;

- Only 0.7% of the yeast proteins with known phenotypic profiles had degree at least 15, but 62% of them were essential.

These results suggest that there is evolutionary selection of a common large-scale structure of biological networks and that future systematic PPI network studies in other organisms should uncover an identical PPI network topology [272]. Our results on a larger yeast PPI network confirm this hypothesis [450].

A *genetic regulatory network* of a cell is formed by all pairs of proteins in which one protein directly regulates the abundance of the other. In most of these networks regulation happens at the transcriptional level, where a transcription factor regulates the RNA transcription of the controlled protein. These networks are naturally directed. The analysis of a regulatory network from the YPD data base with 682 proteins and 1,289 edges [121] as well as of the PPI network from 2-hybrid screens with 2,378 proteins and 4,549 interactions [261] revealed that both networks had a small number of high-degree nodes (hubs) [366]. Both of these networks had edges between hubs systematically suppressed, while those between a hub and a low-connected protein were favored [366]. Furthermore, hubs tended to share few neighbors with other hubs. This led to the hypothesis that these effects decrease the likelihood of "cross talk" between different functional modules of the cell and increase the overall robustness of a network by localizing effects of harmful perturbations [366]. This may explain why the correlation between the connectivity of a given protein and the lethality of the mutant cell lacking this protein was not strong [272].

However, an alternative explanation of this phenomenon suggests that hubs whose removal disconnects the PPI graph are likely to cause lethality [450]. Analyzing the top 11,000 interactions among 2,401 proteins from [531], which utilizes high confidence interactions detected by diverse experimental methods [450], we confirmed the previously noted result on smaller networks [272], demonstrating that:

- *Viable* proteins, whose disruption is nonlethal, have a degree that is half that of *lethal* proteins, whose mutation causes lethality (see Figure 4.9);

- Proteins participating in *genetic interaction pairs* in the PPI network, i.e., combinations of nonlethal mutations which together lead to lethality or dosage lethality, appeared to have a degree closer to that of viable proteins;

- *Lethal* proteins are more frequent in the top 3% of high degree nodes compared to viable ones, while viable mutations were more frequent amongst the nodes of degree 1.

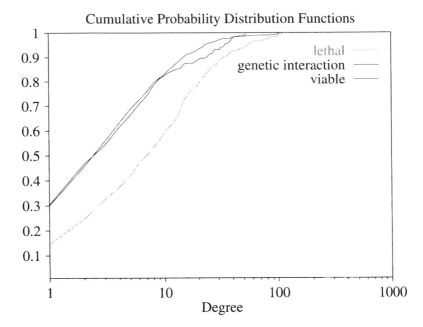

FIGURE 4.9: (See color insert following page 72.) Cumulative distribution functions of degrees of lethal, genetic interaction, and viable protein groups in a yeast PPI network constructed on 11, 000 interactions amongst 2, 401 proteins [450]. Reprinted from Przulj, N., Wigle, D., Jurisica, I. Functional topology in a network of protein interactions. *Bioinformatics*, 20(3):340–348, 2004; by permission of Oxford University Press.

Interestingly, lethal mutations were not only highly connected nodes within the network, but were nodes whose removal caused a disruption in network structure — it disconnected one part of the network from the other [450].

The obvious interpretation of these observations in the context of cellular wiring is that lethality can be conceptualized as a point of disconnection in the PPI network. A contrasting property to hubs, which are at the same time points of disconnection, is the existence of alternative connections, called *siblings*, i.e., nodes in a graph with the same neighborhood. The analysis shows that viable mutations have an increased frequency in the group of proteins that could be described as siblings within the network, compared to lethal mutations or genetic interactions [450]. This suggests the existence of alternate paths bypassing viable nodes in PPI networks, and offers an explanation why null mutation of these proteins is less likely to be lethal [450].

However, it is possible that the reason why lethal mutations in current PPI networks have large degrees is due to the extensiveness of research conducted in the neighborhood of these nodes. It is possible that the observed low-degrees of viable nodes are due to the high range of false negatives present in

PPI networks, i.e., the lack of interest and research conducted in these parts of PPI networks. This is further supported by the observed weak correspondence between mutational robustness and the topology of PPI networks [122].

4.4.4.2 Hierarchical Network Topology

Further analysis of metabolic networks of 43 organisms from the WIT data base [426] presented a dichotomy between the two phenomena found in metabolic networks [457]:

1. These networks are scale-free with the observed power law degree distribution [40, 274, 532];

2. These networks include hubs, which integrate all nodes into a single, integrated network;

3. These networks have high clustering coefficients [532], which imply modular topologies.

Determining the average clustering coefficients of metabolic networks of 43 different organisms established that all were an order of magnitude larger than expected for a scale-free network of similar size [457]. This suggested high modularity of these networks. Also, the clustering coefficients of metabolic networks were independent of their sizes, contrasting the scale-free model, for which the clustering coefficient decreases as $n^{-0.75}$. It is possible to integrate the seemingly contradicting phenomena of modularity and integration using a heuristic model of metabolic organization, called a "hierarchical" network [457]. Thus, metabolic networks appear to be organized into many small, highly connected modules, which are combined in a hierarchical manner into larger units [457].

The "hierarchical" network construction is similar to the one described in [44] (see Section 4.3 for the discussion), but a starting point in this network is a K_4 as a hypothetical module (rather than a P_3 [44]). Nodes of this starting module are connected with nodes of three additional copies of K_4 so that the "central node" of the initial K_4 is connected to the three "external nodes" of new K_4s, as presented in Figure 4.10 (b), which generates a 16-node module. This process is repeated by making three additional copies of this 16-node module and connecting the "peripheral nodes" of the three new 16-node modules with the "central node" of the initial 16-node module (see Figure 4.10 (c)). This process can be repeated indefinitely. The architecture of this network has the following characteristics:

- It integrates a scale-free topology with a modular structure;

- It has a power law degree distribution with $P(k) = k^{-2.26}$, which is in agreement with the observed $P(k) \approx k^{-2.2}$ [274];

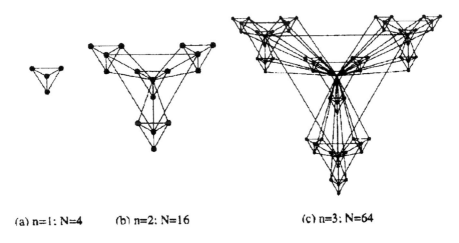

(a) n=1: N=4   (b) n=2: N=16   (c) n=3: N=64

FIGURE 4.10: Three steps in the construction of a hierarchical model network. Reprinted with permission from E. Ravasz, A. L. Somera, D. A. Mongru, Z. N. Oltvai, and A.-L. Barabasi, Hierarchical organization of modularity in metabolic networks, *Science* 279, Number 5586, Issue of 30 August, pp. 1551–1555, (Figure S4 is on page 5 of the Supporting Online Material); Copyright 2002 AAAS.

- It has a clustering coefficient $C \approx 0.6$, which is comparable to coefficients observed for metabolic networks;

- It has a clustering coefficient independent of the network size.

The hierarchical structure of this model is a feature that is not shared by the scale-free, or modular network models. It was also demonstrated that for each of the 43 organisms, the clustering coefficient $C(k)$ of a degree k node is well approximated by $C(k) \approx k^{-1}$. This is in agreement with the theoretical result establishing that the clustering coefficient of a degree k node of a scale-free network follows the scaling law $C(k) \approx k^{-1}$ [139] (further discussed in Section 4.3.5). Thus, this hierarchical network model includes all the observed properties of metabolic networks: the scale-free topology, the high, system size independent clustering coefficient, and the power law scaling of $C(k)$.

To inspect whether this model reflects the true functional organization of cellular metabolism, the focus was turned on the extensively studied metabolic network of *E. coli* [457]. It was established that the model closely overlaps with *E. coli*'s known metabolic network. It was hypothesized that this network architecture may be generic to cellular organization [457]. The existence of small, highly frequent subgraphs in these networks, called "network motifs" [387, 489] makes this hypothesis even more plausible.

These results have been further extended [271] by applying the same analysis on the complete biochemical reaction network of the 43 organisms from the WIT data base [426]. The networks were constructed by combining all pathways deposited in the WIT data base for each organism into a single network. These networks are naturally directed, which made it possible to examine their in- and out-degree distributions. All of the 43 networks obtained in this way exhibited a power-law distribution for both in- and out-degrees, which suggested that scale-free topology is a generic structural organization of the total biochemical reaction networks in all organisms in all three domains of life [271].

However, the largest portion of the WIT data base contains data on core metabolism pathways, followed by the data on information transfer pathways. Thus, these results may have largely been influenced by the domination of metabolic pathways.

To resolve this issue, the same analysis has been performed on the information transfer pathways alone, since apart from the metabolic pathways, these were the only ones present in high enough quantities for doing statistical analyses. The analysis of the information transfer pathways of 39 organisms (four of the 43 organisms had their information pathways of too small size for doing statistics) revealed the same power-law degree distribution both for in- and out-degree as seen for metabolic and complete biochemical reaction networks [271]. Similarly, it was confirmed that the network diameter (which they defined as the average of shortest path lengths between each pair of nodes) remained constant and around 3 for biochemical reaction networks, metabolic networks, and information transfer networks of all 43 organisms, irrespective of the network sizes. Thus, in these networks, the average degree of a node increases with the network size. This is contrary to the results on real nonbiological networks, in which the average degree of a node is fixed, so the diameter of the network increases logarithmically with the network size [40, 49, 541]. Further, only about 5% of all nodes were common to the biochemical reaction networks of all 43 species, and these were the highest degree nodes. The same result was observed when repeated analysis was conducted for metabolic and information transfer networks alone.

Applying the same approach, with minor variation of the hierarchical network model (presented in Figure 4.11) to four independent yeast PPI networks derived from the DIP data base [562], data from [261], the MIPS data base [381], and data from [526], showed that all networks had hierarchical structures with $C(k)$ scaling as k^{-1} [42].

4.4.4.3 Network Motifs

"Network motifs" can be defined as patterns of interconnections that recur in many different parts of a network at frequencies much higher than those found in randomized networks [489]. The *transcriptional regulation network* can be represented as a directed graph in which each node represents an

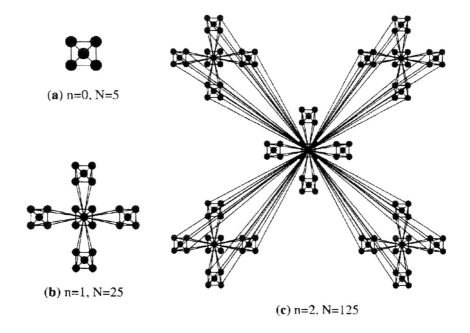

(a) n=0, N=5

(b) n=1, N=25

(c) n=2, N=125

FIGURE 4.11: Three steps in the construction of a hierarchical model network. Reprinted with permission from Laszlo Barabasi.

operon, a group of contiguous genes that are transcribed into a single m-RNA molecule, and each edge is directed from an operon that encodes a transcription factor to an operon that is regulated by that transcription factor [489]. Using network motifs to analyze the transcriptional regulation network of *Escherichia coli* showed that much of the network is composed of repeated appearances of three highly significant motifs, each of which had a specific function in determining gene expression [489]:

1. Feed-forward loop;

2. Single-input module (SIM);

3. Dense overlapping regulons (DOR) (a *regulon* stands for a group of coordinately regulated operons).

An illustration of these three motifs is presented in Figure 4.12. Network motifs on 3 and 4 nodes and SIMs have been detected using straightforward adjacency matrix manipulation algorithms. DORs have been identified applying a standard average-linkage clustering algorithm [144], and using a non metric distance measure between operons [489].

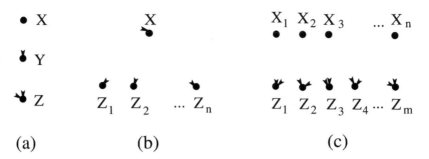

FIGURE 4.12: Motifs from [489]: (a) feed-forward loop, (b) single input module (SIM), (c) dense overlapping regulons (DOR).

Individual motifs work differently. Feed-forward loops can act as circuits that reject transient activation signals and respond only to persistent signals, while allowing a rapid system shutdown: X and Y act in an AND-gate-like manner to control operon Z [489]. SIMs allow temporal ordering of activation of different genes with different activation thresholds, which is useful for processes that require several stages to complete, such as amino-acid biosynthesis processes.

In addition to explaining functionality, motifs can be used to represent the transcriptional network in a compact, modular form [489]. This suggests that performing similar analyses would enable us to identify diverse motifs describing different functional protein groups in large, undirected PPI networks. This in turn would provide mathematical description of cell processes, predict new processes, and aid in determining function of uncharacterized proteins (see Section 4.6).

Further network motif analysis of multiple large networks (*E. coli* and *S. cerevisiae* gene regulation networks (transcription), the neuron connectivity network of *C. elegans*, seven food web networks, the ISCAS'89 benchmark set of sequential logic electronic circuits, and a network of directed hyper-links between World Wide Web pages within a single domain) showed that different networks have different motifs [387]. All possible 3- and 4-node directed subnetworks in these real large networks were found and the frequencies of occurrences of each of these small subnetworks in a real network were compared with the frequencies of their occurrences in randomized networks that have the same connectivity properties and the same number of $(n-1)$-node subgraphs as the real networks, where n is the size of the motif being detected.

The challenge of this analysis is to generate appropriate random networks. During the analysis, one must account for patterns that appear only because of the single-node characteristics of the network, such as the presence of nodes with a large degree, and also to ensure that a high significance is not assigned to a pattern only because it has a highly significant subpattern. Following this principle, "network motifs" are defined as those patterns for which the probability of appearing in a randomized network an equal or greater number of times than in the real network is lower than 0.01 [387]. This again brings the tradeoff of higher specificity for lower sensitivity, as this approach could miss functionally important but not statistically significant patterns.

It was observed that the number of appearances of each motif in the real networks appears to grow linearly with the system size, while it drops in the corresponding random networks [387]. This drop is in accordance with an exact result on Erdös-Rényi random graphs in which the concentration C of a subgraph with n nodes and m edges (i.e., the fraction of times a given n-node subgraph occurs among the total number of occurrences of all possible n-node subgraphs) scales with network size S, as $C \approx S^{n-m-1}$ [69], which in the study of Milo *et al.* is equal to $\frac{1}{S}$, since all but one of their motifs have $n = m$.

In addition, it was established that the identified motifs were insensitive to data errors, since they do not change after addition, deletion, or rearrangement of 20% of the edges at random.

This approach was also applied to an undirected yeast PPI network on $1,843$ nodes and $2,203$ edges [272], which identified one 3-node motif, one 4-node motif, and two 4-node "anti-motifs." Anti-motifs are defined as patterns whose probability of appearing in randomized networks fewer times than in the real network is less than 0.01, and $N_{rand} - N_{real} > 0.1 N_{rand}$, where N_{rand} and N_{real} are the number or subgraph appearances in a real and in randomized networks respectively [387].

An approach to study similarity in the local structure of networks was proposed [386]. A real network was compared with a set of randomized networks with the same degree sequence in the following way. For each 3- and 4-node subgraph i, the statistical significance was expressed by a Z score, $Z_i = (Nreal_i - <Nrand_i>)/std(Nrand_i)$, where $Nreal_i$ is the number of times (i.e., the frequency) the subgraph appears in the network, and $<Nrand_i>$ and $std(Nrand_i)$ are the mean and standard deviation of its appearances in the randomized networks. Then the *significance profile (SP)* for the network is the vector of normalized Z scores, $SP_i = Z_i/(\sum Z_i^2)^{1/2}$. Super-families of previously unrelated networks were found based on the similarity of their SPs [386]. For example, protein signaling, developmental genetic networks, and neuronal wiring formed a distinct superfamily. Also, power grids, protein-structure networks and geometric networks formed a network superfamily. Structure and function of some network motifs in transcription networks has been suggested [358, 359].

4.4.4.4 Geometric Network Topology

We introduced another bottom-up approach of measuring local network structure and showed that the local structure of PPI networks corresponds to the local structure of geometric random graphs [147]. In the *geometric random*

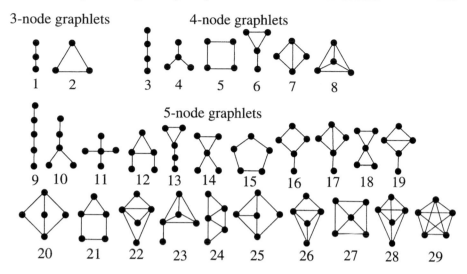

3-node graphlets   **4-node graphlets**

1   2   3   4   5   6   7   8

5-node graphlets

9   10   11   12   13   14   15   16   17   18   19

20   21   22   23   24   25   26   27   28   29

FIGURE 4.13:   All 3-node, 4-node, and 5-node connected networks (graphlets), ordered within groups from the least to the most dense with respect to the number of edges when compared to the maximum possible number of edges in the graphlet; they are numbered from 1 to 29. Reprinted from Przulj, N., Corneil, D., Jurisica, I. Modeling interactome: Scale-free or geometric?, *Bioinformatics*, 20(18):3508–3515, 2004; by permission of Oxford University Press.

graph model, nodes correspond to independently and uniformly randomly distributed points in a metric space, and two nodes are linked by an edge if the distance between them is smaller than or equal to some radius r, where distance is an arbitrary distance norm in the metric space (more details about geometric random graphs can be found in [438]).

We use the term *graphlet* to denote a connected network with a small number of nodes. All of the 29 different 3-, 4-, and 5-node graphlets are presented in Figure 4.13. We exhaustively searched and found all occurrences of every one of these 29 graphlets in four PPI networks and the corresponding Erdös-Rényi random graphs, generalized random graphs, scale-free networks, and geometric random graphs.

Graphlet counts quantify the local structural properties of a network. Currently, our knowledge of the connections in PPI networks is incomplete. The edges we *do* know are dominated by experiments focused around proteins that are currently considered "important." However, we hypothesize that the local structural properties of the full PPI network, once all connections are made, are similar to the local structural properties of the currently known, highly studied parts of the network. Thus, we expect that the *relative* frequency of graphlets among the currently known connections is similar to the relative frequency of graphlets in the full PPI network, which is as yet unknown. Therefore, we use the *relative frequency of graphlets* $N_i(G)/T(G)$ to characterize PPI networks and the networks we use to model them, where $N_i(G)$ is the number of graphlets of type i ($i \in \{1, \ldots, 29\}$) in a network G, and $T(G) = \sum_{i=1}^{29} N_i(G)$ is the total number of graphlets of G. In this model, then, the "similarity" between two graphs should be independent of the total number of nodes or edges, and should depend only upon the differences between relative frequencies of graphlets. Thus, we define the *relative graphlet frequency distance* $D(G, H)$, or *distance* for brevity, between two graphs G and H as

$$D(G, H) = \sum_{i=1}^{29} |F_i(G) - F_i(H)|,$$

where $F_i(G) = -\log(N_i(G)/T(G))$.

Using this distance measure, the graphlet frequency distributions of high-confidence PPI networks were closest to the graphlet frequency distributions of geometric random networks [447]. In addition, all global network parameters of high-confidence PPI networks, except for the degree distribution, were closest to the parameters of geometric random networks. However, with added noise in PPI networks, their graphlet distribution and global network parameters become closer to these parameters of scale-free networks. Thus, we expect that the structure of noise-free PPI networks is close to the structure of geometric random networks.

Exhaustively searching for all instances of a graphlet in a large network is computationally intensive. Thus, heuristic techniques for finding approximate frequencies of small subgraphs in large networks have been developed [294]. Analytical solutions for the numbers of 3- and 4-node subgraphs in directed and undirected geometric networks have been obtained [263]. Also, topological generalizations of network motifs have been suggested [295].

4.4.4.5 Function–Structure Relationship in PPI Networks

As discussed above, network organization is not random, and network properties characterize its function. It has been established that complex networks comprise simple building blocks [387, 489], and that distinct functional classes of proteins have differing network properties [450].

Since different building blocks and modules have different properties, it can be expected that they serve different functions. Assuming the functional classifications in the MIPS data base [381], it is possible to statistically determine simple graph properties for each functional group [450]. This analysis shows that proteins involved in translation appear to have the highest average degree, while transport and sensing proteins have the lowest average degree. Figures 4.14(a) and (b) support this result as half of the nodes with degrees in the top 3% of all node degrees are translation proteins, while none belong to amino-acid metabolism, energy production, stress and defense, transcriptional control, or transport and sensing proteins. This is further supported by the observation that metabolic networks across 43 organisms tested have an average degree of < 4 [274].

Intersecting each of the lethal, genetic interaction, and viable protein sets with each of the functional groups shows that amino-acid metabolism, energy production, stress and defense, transport and sensing proteins are less likely to be lethal mutations (see Figure 4.14(c)). Of all functional groups, transcription proteins have the largest presence in the set of lethal nodes on the PPI graph; approximately 27% of lethals on the PPI graph are transcription proteins, as illustrated in Figure 4.14(c). Notably, amongst all functional groups, cellular organization proteins have the largest presence in hub nodes whose removal disconnects the network (the nodes whose removal disconnects the network we called *articulation points*; see Figure 4.14(d)).

This strong network structure–function relationship can be exploited for computational prediction. For example, we constructed a simple model for predicting new genetic interaction pairs in the yeast PPI network [450]. The predictive model is based on the distribution of shortest path lengths between known genetic interaction pairs in the PPI network.

New approaches integrating different HTP methods in order to describe known and to predict new biological phenomena continue to appear. Since all HTP techniques contain noise, but often noise caused by different phenomena, resulting data sets not only complement each other, but their integration may also reduce the noise. One approach in this direction is integrating graph–theoretic PPI analysis with the results of microarray experiments (for example, see [273]). We discuss some novel approaches in Chapter 6.

4.5 Detection of Dense Subnetworks

There is growing evidence that although currently available PPI networks contain a high degree of false positives and false negatives, they do have structure. One of the main goals is to discover more PPI network structure and ultimately exploit it for designing efficient, robust, and reliable algorithms for extracting graph substructures embedded in these networks that have biological meaning and represent biological processes.

Our previous discussion suggests that one example of such substructures may be dense subgraphs of these networks, representing core proteins of protein complexes. Thus, next we describe some useful graph–theoretic techniques that can be used as a first step towards addressing extraction of these dense subgraphs in PPI graphs. However, it is possible that protein complexes (and pathways, too) have a distinct graph–theoretic structure requiring novel

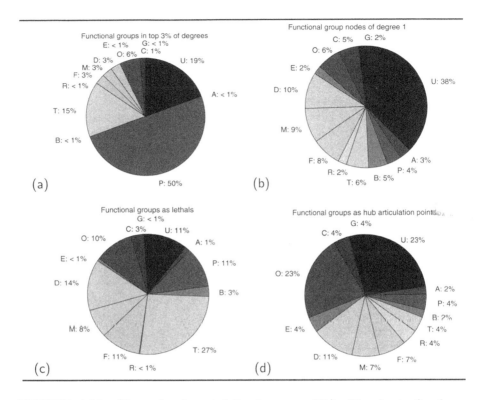

FIGURE 4.14: (See color insert following page 72.) Pie charts for functional groups in the PPI graph: G — amino acid metabolism, C — cellular fate/organization, O — cellular organization, E — energy production, D — genome maintenance, M — other metabolism, F — protein fate, R — stress and defense, T — transcription, B — transcriptional control, P — translation, A — transport and sensing, U — uncharacterized. (a) Division of the group of nodes with degrees in the top 3% of all node degrees. (b) Division of nodes of degree 1. Compared with Figure 4.14(a), translation proteins are about 12 times less frequent, transcription about 2-times, while cellular fate/organization are 5 times more frequent, and genome maintenance, protein fate, and other metabolism are about 3 times more frequent; also, there are twice as many uncharacterized proteins. (c) Division of lethal nodes. (d) Division of articulation points which are hubs. Reprinted from Przulj, N., Wigle, D., Jurisica, I. Functional topology in a network of protein interactions. *Bioinformatics*, 20(3):340–348, 2004; by permission of Oxford University Press.

graph–theoretic approaches for their detection in PPI networks.

Clustering is an important problem in many disciplines, including computational biology. Its goal is to partition a set of elements into subsets called *clusters*, so that the elements of the same cluster are similar to each other (this property is called *homogeneity*) and elements from different clusters are not similar to each other (this property is called *separation*). Homogeneity and separation can be defined in many different ways leading to different optimization problems. Elements belonging to the same cluster are usually called *mates* and the elements belonging to different clusters are called *nonmates*.

Clustering problems and algorithms can be expressed in graph–theoretic terms. For example, a *similarity graph* can be constructed so that nodes represent elements and edge weights represent similarity values of the corresponding elements. In the analysis of PPI networks, weights on edges have not yet been incorporated in the model, but it may be useful to incorporate them to represent the confidence that the two proteins interact (as in [531]), the strength of the interaction, or possibly the timing of an interaction to distinguish those that are transient.

Several graph–theoretic techniques have been developed to cluster microarray gene expression profiles. An overview of these and other gene expression clustering methods can be found in [486]. Identifying protein complexes using graph–theoretic methods requires recognizing "dense" subgraphs of PPI networks. Existing algorithms can be characterized as exact algorithms when they have proven properties in terms of solution quality and time complexity, and approximate when heuristics are used to make them more efficient [236, 488]. Some of the algorithms have a probabilistic nature [54, 156, 528].

The early work on graph–theoretic clustering determined that highly connected regions of similarity graphs are useful in cluster analysis [367, 368, 369, 370]. This work introduced the *cohesiveness function* for every node and edge in a graph as the maximum edge-connectivity of any subgraph containing that node/edge. By deleting all elements of a graph of cohesiveness less than k, a maximal k-edge-connected subgraphs of the graph are obtained. First, clusters are identified by using a constant k [368]. This approach was modified to obtain, for any k, clusters which are maximal k-edge-connected subgraphs that do not contain a subgraph with higher connectivity [369]. Several important graph cluster concepts have been introduced in [370]:

- k-bond — a maximal subgraph S such that every node in S has degree at least k in S;

- k-component — a maximal subgraph S such that every pair of nodes in S is joined by k edge-disjoint paths in S;

- k-block — a maximal subgraph S such that every pair of nodes in S is joined by k node-disjoint paths in S.

These notions imply cluster methods with successive refinements going from bond to component to block. These algorithms require solving a minimum cut network flow problem and their time complexities are at least cubic in the input graph node set size for connected graphs. To cope with the complexity, several faster exact graph-clustering algorithms have been proposed, some using heuristics for further speed up [235, 236, 488]. A good survey of other graph–theoretic-clustering techniques, including the probabilistic ones can be found in [14].

The Highly Connected Subgraph (HCS) [235, 236] and CLuster Identification via Connectivity Kernels (CLICK) [488] algorithms operate on a similar principle. The input is a similarity graph, and the algorithm first considers if the graph satisfies a stopping criterion, in which case it is declared a "kernel." Otherwise, the graph is partitioned into two subgraphs, separated by a minimum weight edge cut, and the algorithm recursively proceeds on the two subgraphs, outputting in the end a list of kernels that represent a basis for the possible clusters. The overview of this general algorithm scheme is presented in Algorithm 1 (adapted from [486]). HCS and CLICK have several distinguishing properties, they construct similarity graphs differently and have different stopping criteria.

Algorithm 1: FORM-KERNELS(G)

---

if $V(G) = \{v\}$ **then**
  | move v to the singleton set
end
else
  | **if** G *is a kernel* **then**
  | | output $V(G)$
  | **end**
end
else
  | $(H, \overline{H}) \leftarrow MinWeightEdgeCut(G)$;
  | Form-Kernels(H);
  | Form-Kernels($\overline{H}$);
end

The input into the HCS is an unweighted similarity graph G. A *highly connected subgraph (HCS)* is defined to be an induced subgraph H of G, such that the number of edges in a minimum-edge cut of H is bigger than $\frac{|V(H)|}{2}$. Thus, if any $\lfloor \frac{|V(H)|}{2} \rfloor$ of edges of H are removed, H remains connected. The algorithm uses these highly connected subgraphs as kernels. Clusters satisfy the two properties [235]:

- Clusters are homogeneous, since the diameter of each cluster is at most 2 and each cluster is at least half as dense as a clique;

- Clusters are well separated, since any nontrivial split by the algorithm happens on subgraphs that are likely to be of diameter at least 3.

The running time of the HCS algorithm is bounded by $2N \times f(n, m)$, where N is the number of clusters found (often $N \ll n$) and $f(n, m)$ is the time complexity of computing a minimum-edge cut of a graph with n nodes and m edges. Currently, the fastest deterministic minimum-edge cut algorithms for unweighted graphs are of time complexity $O(nm)$ [371, 401]. The fastest simple deterministic minimum-edge cut algorithm for weighted graphs is of time complexity $O(nm + n^2 \log n)$ [508] (it is implemented by Mehlhorn and is part of the Leda library [378]).

Several heuristics can be used to speed up the HCS algorithm. Since HCS arbitrarily chooses a minimum-edge cut when the same minimum cut value is obtained by several different cuts, this process will often break small clusters into singletons. To avoid this, *Iterated HCS* runs several iterations of HCS, until no new cluster is found. Theoretically, this would add another $O(n)$ factor to the running time, but in practice only between 1 and 5 iterations are usually needed. *Singletons Adoption* heuristics is based on the principle that singleton nodes get "adopted" by clusters based on their similarity to the clusters. For each singleton node x, the number of neighbors of x in each cluster as well as in the set of all singletons, $\mathcal{S}$, is computed, and x is added to a cluster (never to $\mathcal{S}$) with the maximum number of neighbors, $\mathcal{N}$, if $\mathcal{N}$ is sufficiently large. This process is repeated a specified number of times to account for changes in clusters resulting from previous adoptions. The last HCS algorithm heuristic is based on *removing low-degree nodes*. If the input graph contains many low-degree nodes, one iteration of the minimum edge cut algorithm may only separate a low-degree node from the rest of the graph contributing to increased computational cost at a low informative value in terms of clustering. This is especially expensive for large graphs with many low-degree nodes. For example, around 28% of the nodes of the PPI graph constructed on the top $11,000$ interactions (and $2,401$ proteins) using data from [531], and around 13% of the nodes of the PPI graph constructed on all $\approx 78K$ of the yeast PPIs (and $5,321$ proteins) [531] are of degree 1. Thus, this heuristic may significantly speed up the HCS algorithm applied to these data sets.

We implemented the HCS algorithm without any heuristics and applied it to several PPI graphs constructed on the data set from [531], as described in Section 4.4.3.1. Our results show that clusters identified this way have high overlap with known MIPS protein complexes and a much higher functional group homogeneity than expected at random [450]. Thus, high specificity is favored by this method of protein complex identification. In contrast, Bader and Hogue's approach improves sensitivity at the expense of specificity [31].

The CLICK algorithm builds on the HCS algorithm [235] by incorporating statistical techniques to identify kernels [488]. Similar to HCS, a weighted similarity input graph is recursively partitioned into components using minimum weight edge cut computations. The edge weights and the stopping criterion of the recursion have probabilistic meaning. Pairwise similarity values between mates are assumed to be normally distributed with mean μ_T and variance σ_T, and pairwise similarity values between nonmates are assumed to be normally distributed with mean μ_F and variance σ_F, where $\mu_T > \mu_F$ (this is observed on real data and can also be theoretically justified [488]). The probability p_{mates} of two randomly chosen elements being mates is taken into account when computing edge weights.

If the input similarity matrix between elements is denoted by $S = (S_{ij})$, the weight of an edge (i, j) in the similarity graph is computed as:

$$w_{ij} = \ln \frac{Prob(i, j \text{ are mates}|S_{ij})}{Prob(i, j \text{ are nonmates}|S_{ij})} = \ln \frac{p_{mates} f(S_{ij}|i, j \text{ are mates})}{(1 - p_{mates}) f(S_{ij}|i, j \text{ are nonmates})},$$

where

$$f(S_{ij}|i, j \text{ are mates}) = \frac{1}{\sqrt{2\pi}\sigma_T} e^{-\frac{(s_{ij} - \mu_T)^2}{2\sigma_T^2}}$$

and

$$f(S_{ij}|i, j \text{ are nonmates}) = \frac{1}{\sqrt{2\pi}\sigma_F} e^{-\frac{(s_{ij} - \mu_F)^2}{2\sigma_F^2}},$$

and, thus,

$$w_{ij} = \ln \frac{p_{mates}\sigma_F}{(1 - p_{mates})\sigma_T} + \frac{(S_{ij} - \mu_F)^2}{2\sigma_F^2} - \frac{(S_{ij} - \mu_T)^2}{2\sigma_T^2}.$$

To increase efficiency, edges whose weight is below a predefined nonnegative threshold are removed from a graph. A connected subgraph G is called *pure*, if $V(G)$ contains only elements of some cluster. For each cut C of a connected graph, the following hypotheses are tested:

- H_0^C : C contains only edges between nonmates.

- H_1^C : C contains only edges between mates.

G is called a kernel if it is pure, which it is if and only if H_1^C is accepted for every cut C of the graph G. If G is not a kernel, then it is partitioned along a cut C for which the ratio $Pr(H_1^C|C)/Pr(H_0^C|C)$ is minimum. Kernels obtained this way are expanded to obtain clusters, first by singletons adoptions, then by merging "similar" clusters, and finally by performing another round of singletons adoptions. For more details, see Algorithm 2 and [488].

Algorithm 2: CLICK(G)

Singletons $\mathcal{S} \leftarrow$ complete set of elements N;
while *some change occurs* **do**

> Execute FORM-KERNELS($G(\mathcal{S})$);
> Let $\mathcal{K}$ be the list of produced kernels;
> Let $\mathcal{S}$ be the set of singletons produced;
> Adoption($\mathcal{K}, \mathcal{S}$)

end
Merge($\mathcal{K}$);
Adoption($\mathcal{K}, \mathcal{S}$)

CLICK performance can be improved by using the following two heuristics. Similar to removing low-degree nodes for HCS, low-weight nodes[21] are filtered from large components in the following way. The average node weight W is computed for the component and multiplied by a factor proportional to the logarithm of the component size; the result is denoted by W^*. Nodes with weight less than W^* are removed repeatedly, updating the weight of the remaining nodes each time a node is removed, until the updated weight of all remaining nodes is greater than W^*. The removed nodes are added to the singletons set and handled later. Although CLICK uses the fastest minimum weight edge cut algorithm [102], with the complexity $O(n^2\sqrt{m})$ [232], the second heuristic exploits a different approach to computing a minimum-edge cut. Instead of finding computationally expensive minimum weight edge cuts, they computed a minimum $s-t$ cut of the underlying unweighted graph using $O(nm^{2/3})$ time algorithm [162], with s and t chosen to be nodes that achieve the diameter d of the graph, when $d \geq 4$ (the $O(n+m)$ time breadth first search algorithm is used to find the diameter of the graph).

A polynomial time algorithm for finding the clustering with high probability under the stochastic model of the data has been introduced in [54]. It assumes that the correct structure of the input graph is a disjoint union of cliques (cliques represent clusters), but that errors were introduced to it by independently adding or removing edges with probability $\alpha < \frac{1}{2}$. This heuristic Cluster Affinity Search Technique (CAST) algorithm is built on the theoretical Parallel Classification with Cores (PCC) algorithm, which solves the problem to a desired accuracy with high probability in time $O(n^2(\log n)^c)$ [54].

The input to CAST is the similarity matrix S. CAST uses the notion of the *affinity* of an element v to a putative cluster C, $a(v) = \sum_{i \in C} S(i, v)$, and the affinity threshold parameter t. It generates clusters sequentially by starting with a single element and adding or removing elements from a cluster if their

---

[21]The weight of a node v is the sum of weights of the edges incident on v.

affinity is larger or lower than t, respectively. This process is repeated until it stabilizes. The details are shown in Algorithm 3. In the end, an additional heuristic tries to ensure that each element has the affinity to its assigned cluster higher than to any other cluster by moving elements until the process converges, or some maximum number of iterations is completed.

Algorithm 3: CAST(S)

---

while *there are unclustered elements* **do**

> Pick an unclustered element to start a new cluster C;
>
> **repeat**
>
> > add an unclustered element v with maximum affinity to C, if $a(v) > t|C|$;
> >
> > remove an element u from C with minimum affinity, if $a(u) \leq t|C|$;
>
> **until** *no changes occur*;
>
> Add C to the list of final clusters;

end

A Markov Cluster (MCL) algorithm has been introduced to cluster undirected unweighted and weighted graphs [528]. MCL simulates flow within a graph, promoting flow where the current is strong and demoting flow where the current is weak until the current across borders between different groups of nodes withers away, revealing a cluster structure of the graph (an illustration is presented in Figure 4.15).

The MCL algorithm deterministically computes the probabilities of random walks through the graph and uses two operators, expansion and inflation, to transform one set of probabilities into another. It uses stochastic matrices (also called Markov matrices) that capture the mathematical concept of random walks on a graph.

Following the notation from [528], for a weighted directed graph $G = (V, E)$, with $|V| = n$, its *associated matrix* M_G is an $n \times n$ matrix with entries $(M_G)_{pq}(1 \leq p, q \leq n)$ being equal to weights of edges between nodes p and q (clearly, weights of all edges of an unweighted graph are equal to 1). Similarly, every square matrix M can be assigned an *associated graph* G_M. For a graph G on n nodes and its associated matrix $M = M_G$, the *Markov matrix* associated with G, denoted by $\mathcal{T}_G$, is obtained by normalizing each column of M so that it sums to 1, i.e., if D is a diagonal matrix with $D_{kk} = \sum_i M_{ik}$ and $D_{ij} = 0$ for $i \neq j$, then $\mathcal{T}_G$ is defined as $\mathcal{T}_G = M_G D^{-1}$. A column j of $\mathcal{T}_G$ corresponds with node j of the stochastic graph associated with $\mathcal{T}_G$, and the matrix entry $(\mathcal{T}_G)_{ij}$ corresponds to the probability of going from node j to node i.

Given such a matrix $\mathcal{T}_G \in \mathbb{R}^{n \times n}, \mathcal{T}_G \geq 0$, and a real number $r > 1$, let

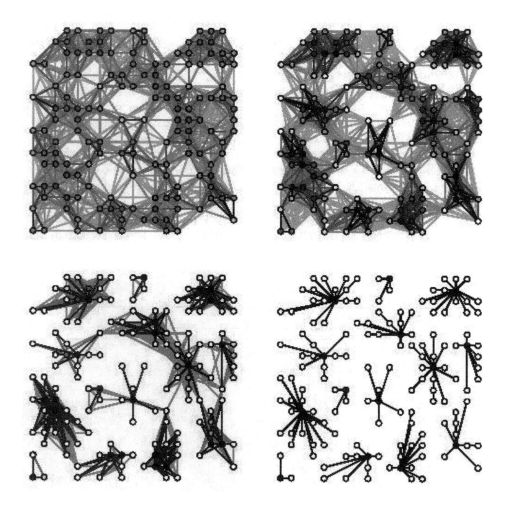

FIGURE 4.15: Stages of flow simulation by the MCL process [528]. Reprinted with permission from Stijn van Dongen.

$\Gamma_r : \mathbb{R}^{k \times k} \to \mathbb{R}^{k \times k}$ be defined as:

$$(\Gamma_r \mathcal{T}_G)_{pq} = ((\mathcal{T}_G)_{pq})^r / \sum_{i=1}^{n} ((\mathcal{T}_G)_{iq})^r.$$

Γ_r is called the *inflation* operator with power coefficient r and the Markov matrix resulting from inflating each of the columns of $\mathcal{T}_G$ with power coefficient r is written as $\Gamma_r \mathcal{T}_G$. For $r > 1$, inflation changes the probabilities associated with the collection of random walks departing from a node (corresponding to a matrix column) by favoring more probable walks over less probable ones. Inflation can be altered by changing r: larger r makes inflation stronger and produces "tighter" clusters.

Expansion corresponds to computing "longer" random walks. It associates new probabilities with all pairs of nodes with one node being the point of departure and the other being the destination. It relies on the observation that longer paths are more common within clusters than between different clusters, and, thus, the probabilities associated with node pairs, which are within the same cluster will, in general, be relatively large, because there are many ways of going from one node to the other. Expansion is achieved via matrix multiplication. The MCL algorithm iterates the process of expanding information flow via matrix multiplication and contracting it via inflation. The basics of the MCL algorithm are presented in Algorithm 4.

Algorithm 4: MCL$(G, \Delta, e_{(i)}, r_{(i)})$

\# G is a graph with every node of degree ≥ 1;
\# $e_{(i)}$ is a sequence of $e_i \in \mathbb{N}, e_i > 1, i = 1, \dots$;
\# $r_{(i)}$ is a sequence of $r_i \in \mathbb{R}, r_i > 0, i = 1, \dots$;
$G = G + \Delta$; \# Possibly add (weighted) self-loops;
$T_1 = \mathcal{T}_G$;
$k = 0$;
repeat
$\quad | \quad k = k + 1$;
$\quad | \quad T_{2k} = (T_{2k-1})^{e_k}$; \# Expansion;
$\quad | \quad T_{2k+1} = \Gamma_{r_k}(T_{2k})$; \# Inflation;
until T_{2k+1} *is (near-) idempotent*;
Interpret T_{2k+1} as a clustering.

Iterating expansion and inflation results in the matrix that is idempotent under both matrix multiplication and the inflation (such a matrix is called *doubly idempotent*), that is, an equilibrium state is reached when a matrix does not change with further expansion and inflation. The graph associated with such a matrix consists of different directed connected star-like components with an attractor in the center (see bottom right picture in Figure 4.15). Each of these components is interpreted as a cluster.

Theoretically, there may be nodes connected to different stars, which is interpreted as cluster overlap [528]. The algorithm iterants converge nearly always to the doubly idempotent matrix. In practice they start noticeably converging after 3 to 10 iterations. Van Dongen proved quadratic convergence of the MCL process in the neighborhood of doubly idempotent matrices [528]. The row of expansion powers, $e_{(i)}$, and the row of inflation powers, $r_{(i)}$, in Algorithm 4 influence the granularity of the resulting clustering.

MCL was successfully applied to cluster protein sequences into families [156, 158]. For this purpose, nodes of the graph represent proteins, edges represent sequence similarities between the corresponding proteins, and edge weights corresponded to sequence similarity scores obtained from an algorithm such as BLAST [15, 16]. The overview of this algorithm, called Tribe-MCL, is presented in Algorithm 5. Tribe-MCL allowed hundreds of thousands of sequences to be accurately classified in a matter of minutes [156].

Algorithm 5: TRIBE-MCL(SET OF PROTEIN SEQUENCES S)

All versus all BLAST(S);

Parse results and symmetrify similarity scores;

Produce similarity matrix M;

Transform M to normalize similarity scores $(-\log(\text{E-value}))$ and generate transition probabilities;

MCL(G_M);

Postprocess and correct domains of resulting protein clusters (families).

The Restricted Neighborhood Search Clustering Algorithm (RNSC) has been used to extract identify protein complexes in PPI networks [309]. This algorithm partitions the node set of a network $G(V, E)$ by searching the space of partitions of V, each of which has an assigned cost, for a clustering with low cost. The initial clustering is random, or user-input. The RNSC searches for a low-cost clustering starting from an initial random clustering, by iteratively moving one node from one cluster to another in a randomized fashion to improve the cost of the clustering. A general move is one that reduced the clustering cost by a near-optimal amount. To avoid the tendency to settle in poor local minima, diversification moves shuffle the clustering by occasionally dispersing the contents of a cluster at random. Also, RNSC maintains a list of tabu (forbidden) moves to prevent cycling back to the previously explored partitioning.

Since the RNSC is randomized, different runs on the same input data will result in different clusterings. Thus, to achieve high accuracy in predicting true protein complexes in PPI networks, the RNSC output is filtered for functionally homogeneous, dense, and large clusters. This resulted in an accurate and scalable method for detecting and predicting protein complexes within a PPI network [309].

4.6 Conclusions

There is growing evidence that the analysis of PPI networks is useful, but multidisciplinary approaches need to be taken. Existing algorithms provide encouraging results, but novel methods have to be designed in order to improve both the accuracy and scalability of these algorithms, and biological relevance of the models and hypotheses they generate. We emphasize here those open problems that we consider the most interesting and most pressing.

Understanding interactions between proteins in a cell may benefit from a better model of a PPI network. A full description of protein interaction networks requires a model that would encompass the undirected physical PPIs, other types of interactions, interaction confidence level, source (or method) and multiplicity of an interaction, directional pathway information, temporal information on the presence or absence of a PPI, information on the strength of the interactions, and possibly protein complex information. This may be achieved by designing a weighting function and assigning weights to nodes and edges of a PPI network to incorporate temporal and other interaction specific information, adding directionality to the network subgraphs, and building a hypergraph [22] structure on the top of the network to incorporate information about complexes, or pathways in which proteins take part.

Another interesting research topic is finding an efficient and robust graph clustering algorithm that would reliably identify protein complexes, separate stable from transient complexes [267], or detect pathways despite the noise present in PPI networks. Identifying graph–theoretic structural properties that are common to protein complexes or certain pathway types in PPI networks may be crucial to designing such an algorithm. Similarly, modeling signaling pathways and finding efficient algorithms for their identification in PPI networks is another interesting topic. These algorithms would likely have to be stochastic, massively parallel, and use local search techniques, due to the presence of noise and large network sizes.

The existence of a "core proteome" has been hypothesized. It has been proposed that approximately 40% of yeast proteins are conserved through eukaryotic evolution [107]. We are approaching the moment when enough information would be available to verify the existence of such a proteome and

---

[22] A generalization of graph in which edges may be any subset of the nodes

discover its structural properties within PPI networks. It is already possible to take the first steps towards this goal with the currently available data. We propose to construct putative PPI networks for a number of eukaryotic organisms with mapped genomes by combining protein sequence similarities between different organisms with the known PPI networks of model organisms. With the set of putative PPI networks constructed in this way, it may be possible to do PPI network structural comparisons over different organisms. Preferential attachment to high degree nodes in real world networks has been suggested, implying that the core proteome would consist of high-degree nodes in PPI networks. It is interesting to observe the discrepancy between the high degree of a supposed "core proteome" proteins (hubs) and the separation of hubs by low-degree nodes [366]. Exploring the structural properties of this discrepancy may give an insight not only in the processes of evolution, but also in the properties that a better PPI network model should have. Research in this direction may result in construction of a stochastic, or deterministic large network model (similar to the model in [457] and the model in [271]), which would provide a better framework for understanding PPI networks. Also, the evolutionary mechanisms that produce the observed geometric properties of PPI networks [447] need to be explored.

Other interesting topics for future research include distinguishing different graph–theoretic properties of proteins belonging to different functional groups. Our results suggest that such differences exist [450]. One way to approach this problem would be to identify different network motifs [489] in the neighborhood of proteins belonging to the same functional group, and compare the enrichment of these motifs over the functionally different sets of proteins. Along the same lines, it may be interesting to compare graph structures of the "same-function modules" over ZZ putative (or real, when they become available) PPI networks of different species and possibly infer common and differing elements in the structures of these modules. This could lead to construction of new models, which could be used for identification of false positives and false negatives in PPI networks.

Integration of microarray data with PPI data may be beneficial for finding solutions to many of the above mentioned open problems, as we further discuss in Chapter 6. Also, combining PPI data and orthology information based on protein sequence comparison can be used to find conserved protein complexes [487].

Complex biological and artificial networks show graph–theoretic properties that reflect the function these networks carry [205, 387, 148, 502, 524, 567, 549]. A similar analysis could be applied to call graphs of large software [448, 449]. A comparison between PPI networks, software call graphs, and other biological and artificial networks may give further insight into the properties of large, evolving networks. Comparing these networks to intentionally designed and optimized networks, such as circuit designs, could provide further information about the true nature of biological networks — as being truly scale-free, or following other, perhaps geometric properties. This hypothesis is supported by the strong correlation between missing data and scale-free properties, suggesting that it is the noise that is scale free [447]. Further, bias in biological experiments amplifies this problem by artificially decreasing connectedness of less studied proteins. Computational and modeling techniques, coupled with experimental validations will help identify such biases, improve experimental design and subsequently, improve the quality of data.

Chapter 5

HTP Protein Crystallization Approaches

Christian A. Cumbaa and Igor Jurisica

5.1 Protein Crystallization

Proteins are involved in every biochemical process that maintains life. Understanding protein structure will help us to unravel functions of these important molecules. Correct protein function depends on their three-dimensional structure. Thus, the key to understanding gene function is to understand protein structure. Three-dimensional motifs are more conserved than amino acid sequences. Similar motifs may serve similar functions in different proteins, e.g., Bcl-Xl and Diphtheria toxin have different sequences but similar structure and function.

Despite the progress in experimental techniques, it remains challenging to crystallize novel proteins, and we still do not know most of the laws by which proteins adopt their three-dimensional structure. Thus, understanding these laws is one of the primary challenges in modern molecular biology.

There are two main approaches for structure determination: *in silico* and experimental. *In silico* approaches can be categorized into three main methods:

1. *Comparative modeling*, which takes advantage of evolutionarily related proteins with similar sequences since they often have similar structures [202];

2. *Threading* compares a target sequence against a library of structural templates, producing a list of ranked scores [494];

3. *Ab initio prediction* is based on modeling all energetics involved in the process of folding, and finding the structure with the lowest free energy (e.g., IBM's BlueGene project) [527].

Experimental methods include x-ray crystallography [451] and NMR (nuclear magnetic resonance) [446]. A crystallography experiment begins with a well-formed crystal that ideally diffracts x-rays to high resolution (see Chapter 5 for further discussion). Currently, the most powerful method for protein structure determination is single crystal x-ray diffraction, although new breakthroughs in NMR approaches [306] are growing in their importance. X-ray crystallography and NMR are complementary approaches to structure determination [473].

In spite of new breakthroughs in NMR (nuclear magnetic resonance, [306]) and *in silico* [88] approaches, single crystal x-ray diffraction remains the most powerful method for protein structure determination.

A crystallography experiment begins with a well-formed crystal that ideally diffracts x-rays to high resolution. Often, it is difficult to grow protein crystals suitable for diffraction. Unfortunately, we frequently encounter an inverse relationship between protein "interestingness" and its propensity to crystallize; the more interesting the protein is the more difficult is to crystallize it.

Crystallization is a multiparametric process with three classical steps: nucleation, growth and cessation of growth. Technical difficulties in protein crystallization are due to mainly two reasons:

- A large number of parameters affect the crystallization outcome, including purity of proteins, super-saturation, temperature, pH, time, ionic strength and purity of chemicals, volume and geometry of samples;

- We do not fully understand correlations between the variation of a parameter and the propensity for a given macromolecule to crystallize.

Conceptually, protein crystal growth can be divided into two phases: *search* and *optimization*. Search phase pursues a subset of all possible crystallization conditions, to determine conditions that yield promising crystallization outcome, i.e., it determines approximate conditions that may result in "high quality" crystal once optimized (i.e., crystal that ideally diffracts x-rays to high resolution). These conditions are varied during the optimization phase to produce diffraction-quality crystals. Neither of the two phases are trivial to execute. If we consider only 15 possible conditions, each having 15 possible values, the result would be $4.3789e + 017$ possible experiments; impossible to test exhaustively. Even a broad search phase may not produce any promising conditions, and many of the promising leads may elude optimization strategies.

To diminish the problem, it is useful to apply high-throughput robotic screens with knowledge management to improve crystallization search and optimization phases with a goal to accelerate protein structure resolution and function determination. Eventually, discovering the principles of crystal growth should diminish protein crystallization as a bottleneck in modern structural biology. Fortunately, other parts of the pipeline, including protein purification and production [3], are also being automated to match the screening throughput.

```
Tube: 1
     0.02M Calcium chloride dihydrate
     0.1M Sodium acetate trihydrate pH4.6
     30%v/v 2-methyl-2,4-pentanediol
Tube: 2
     0.4M Potassium sodium tartrate tetrahydrate
Tube: 3
     0.4M Ammonium dihydrogen phosphate
Tube: 4
     0.1M Tris hydrochloride pH8.5
     2M Ammonium sulfate
Tube: 5
     0.2M tri-Sodium citrate dihydrate
     0.1M Sodium HEPES pH7.5
     30%v/v 2-methyl-2,4-pentanediol
Tube: 6
     0.2M magnesium chloride hexahydrate
     0.1M Tris hydrochloride pH8.5
     30%w/v polyethylene glycol 4000
...
```

FIGURE 5.1: An example of first set of conditions in Hampton Research Crystal Screen I (HSCI).

5.1.1 Protein Crystallization Data

Protein crystallization remains primarily empirical. It has been considered an art rather than a science [143]; in general, the process is unpredictable and suffers high irreproducibility. Experience alone has produced experimental protocols for crystal growth that are effective in many settings. For example, an experimentally determined set of 48 agents proposed by Jancarik and Kim is, in general, highly successful, and, thus, often used during crystallization search phase [266]. An example of Hampton Research Crystal Screen I (HSCI)[1] is presented in Figure 5.1. This suggests that past experience will lead us to the identification of initial conditions favorable to crystallization. However, the challenge is the diversity of proteins, which in turn requires diversity of crystallization conditions.

Despite advances through practical experience, we need systematic and principled studies to improve our deep understanding of the crystallization process, and to provide a basis for the efficient automated planning of new experiments. One of the challenges in planning crystal growth experiments is the nonsystematic approach to knowledge acquisition: "the history of experi-

---

[1]http://www.hamptonresearch.com/hrproducts/screens.html

ments is not well known, because crystal growers do not monitor parameters" [143, p.14]. Even if the parameters are recorded, such information is often "read only" or highly inaccessible, e.g., laboratory log-books, or incomplete, such as crystallization conditions described in primary literature.

The trend is to assemble large data bases of crystallization experiments, and to apply statistical and data-mining techniques to unravel principles of protein crystal growth, and to optimize set of conditions to be used for an effective search phase [164, 242, 243, 285, 287, 308, 429, 467]. Following this direction, Edwards *et al.* attempted to optimize the industrial screen with 48 conditions from crystallization data on 755 different proteins [308]. Not surprisingly, the study showed that one can eliminate certain conditions and still not lose any crystal. More specifically only 9 conditions are required to crystallize all 755 proteins in this study, and only 6 conditions are needed to detect crystals for 338 proteins, i.e., the 6 conditions have a crystal detection rate of 60.6%.

Although these results are encouraging, we will likely never achieve "universal crystallizing conditions" since proteins have different properties. The structural genomics approach is to fill the structure space, and minimal screen approaches are suitable for this overall strategy. However, more disease and drug-discovery related protein structure determination approaches require that we determine the three-dimensional structure for any protein, and, thus, each crystallization screen result may need to be followed, regardless of how well or poorly the screen results are, or how many homologues the given protein has. In addition, even if a particular condition produces different results (i.e., not just crystals or just empty wells) across many proteins, it still may provide valuable information. Further, eventually we will be concerned with the quality of a crystal, and thus although crystallization condition CC_x may generate crystallization for most of the proteins screened, it may not be the effective crystallization–optimization starting point, or produce high-quality crystals.

Tracking detailed information about conditions that lead to successful structure determination is the first step in understanding the principles of crystallization. The Biological Macromolecular Crystallization Database (BMCD)[2] [201] is a repository for data from published crystallization experiments, i.e., positive examples of successful plans for growing crystals. It comprises information about the macromolecule itself, the crystallization methods used, and the crystal data. Because it is primarily literature-based, it stores only results for successful crystallization experiments, and does not record failed screens and optimizations.

As an important resource of positive crystallization experiences, this data base has been subjected to multiple statistical and machine-learning analysis. These efforts include approaches that use cluster analysis [164], inductive

---

[2]http://wwwbmcd.nist.gov:8080/bmcd/bmcd.html

learning [243], and correlation analysis combined with a Bayesian technique [242] to extract knowledge from the data base. Previous studies were limited because negative results are not reported in the data base (or the literature) and because many crystallization experiments are not reproducible due to an incomplete method description, missing experimental details, or erroneous data (which is the result of often skimpy and vague description of the experimental setup available from the primary literature). Consequently, although a useful resource, the BMCD is not currently being used in a strongly predictive fashion.

5.1.2 High-Throughput Protein Crystallization

Significant advancement in engineering and crystallization techniques has resulted in the use of high-throughput (HTP) robotic setups for the search phase of crystal growth. This massively increased the number of conditions that can be initially tested, and improved systematicity and reliability for approximating crystallization conditions in the search phase.

However, introducing robots and throughput also results in thousands of initial crystallization experiments carried out daily that require expert evaluation based on visual criteria. Clearly, solving the throughput bottleneck in the search phase created a bottleneck for the analysis of screen results.

HTP robotic protein crystallization systems are now capable of testing thousands of protein-cocktail combinations per day, for example using $1,536$-well plates [351]. Although these systems are highly useful, there are several problems and challenges that one has to be aware of [188]. These challenges also depend on a particular HTP platform is being used. There are several variants of HTP crystallization platforms, each having different focus and benefits:

1. **Nano-crystallography.** This is a novel technology that enables fast screening of crystallizing conditions to identify crystal-bearing cocktails by biorefrigence, i.e., analyzing video streams from a digital camera with rotating polarizing filter [68, 147, 324]. Crystal identification is then simplified, as it is sufficient to detect if the light in the well changes. As a result, crystal detection algorithms can achieve high both specificity and sensitivity.

   Although this is a highly accurate setup for crystal detection, there are two main drawbacks of the setup. First, a smaller number of conditions is usually used, which results in a less comprehensive search phase. Second, the imaging setup is not readily usable to distinguish different crystal forms, and it cannot identify various types of precipitates. An example of such setup has been presented in [68, 324]. Besides substantially reducing the amount of protein required to 1 nanoliter volumes, which enables crystallographic studies of rare proteins, this technology also extends super-saturation. Additionally, there seems to be some evidence that crystals grown in small volumes may be better ordered.

2. **HTP pipetting setup.** This is the most flexible system, mainly used for systematic crystallization optimization phase, although it can also be used during the search phase. The main advantage is the flexibility in terms of cocktails that can be used. It is also advantageous that the cocktails are prepared from basic stock just before the experiment, so there is no degradation of the cocktail. The capacity enables testing several hundred conditions. Different crystallization protocols can be used, and results can be analyzed and scored by human experts, or paired with an imaging setup for automated image classification. An example of such system is described in [394]. The emerging variant of this approach is *microfluidics* protein crystallography, where subnanoliter volumes are used in microcapillaries to achieve crystallization.

3. **Systematic robotic screen.** This setup uses similar pipetting robots as the HTP pipetting setup, except that the set of cocktails is larger and is relatively static over time. The reasons are to provide rapid but systematic screening of a broad range of conditions under which different crystallization events occur. The interest is not only to identify crystals, but to characterize different precipitates as well. The set of these systematic solubility experiments can then provide a quantitative measure of similarity among proteins. Using this protein "signature," one can compare different proteins in terms of their crystallization potential. The approach exploits past experience to identify initial conditions favorable to crystallization, such as the setup introduced in [287, 351].

We further focus on systematic robotic protein crystallization screens. There are two main variants of such HTP setups: (1) using micropipettes and (2) using high-density plates. The two methods differ in terms of robotics they use, throughput they can achieve, archival potential, and imaging challenges. While both methods achieve throughput of $\sim 400,000$ crystallization experiments per month [483], high-density plates are currently not an archival medium. However, high-density plates are easier to automate to higher throughput, if more then 200 proteins can be put through the screen per month; and they are also easier to integrate with user-directed and distributed interfaces for remote screen inspection. The micropipette robotic setup is less standard compared to HTP pipetting robots used in the high-density plate setup. Imaging, however, poses quite different challenges, as highlighted in Figure 5.2.

Next we describe features of a specific HTP protein crystallization system at the HTP screening lab of the Hauptman-Woodward Medical Research Institute (HWI)[3], Buffalo, NY [351]. Although, individual HTP setups differ in crystallization and imaging methods used, most HTP system need to address

---

[3]http://www.hwi.buffalo.edu/Research/Facilities/CrystalGrowt.html

FIGURE 5.2: (See color insert following page 72.) (a)–(c) Crystallization experiments in micropipettes (crystal, precipitate, clear). (d) Subset of a 1,536-well plate with highlights of recognized drop and area of interest for further image analysis.

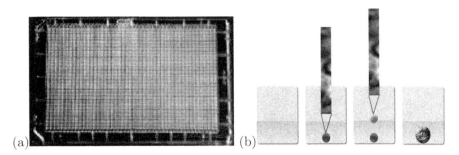

FIGURE 5.3: (a) 1,536-well plate. (b) Microbatch crystallization setup. Protein drop is placed under the oil, then mixed with crystallizing agent.

similar challenges. Although we focus on microbatch under oil, other crystallization methods are used with different setups, such as vapor diffusion, dialysis, thermal methods, and liquid diffusion.

5.1.2.1 Systematic HTP Crystallization Screen

Each protein crystallization experiment is conducted in a 1,536-well plate (see Figure 5.3(a)), which supports simultaneous evaluation of the crystallization potential of many distinct precipitant cocktails [351]. The experiments are microbatch setups under paraffin oil [100, 129] (see Figure 5.3(b)). Although other ratios are possible, the current setup uses 1:1 protein:cocktail volume ratio.

1,536 crystallization cocktails cover a broad range of conditions:

- Many salts and organics identified as protein precipitants in the BMCD [201];

- A broad range of pH, ionic strength, excluded volume, surface tension, viscosity, and chemical potential of water.

The concentrations of these agents range from very low to very high, e.g., from 0.1 M to 10.0 M for the salts (where solubility allows), and from 2%(w/v) to 40%(w/v) for the polyethylene glycols.

The robotic crystallization system at HWI has been evolving over the years, improving the robotics, imaging setup, reducing required volumes of protein, and optimizing cocktail selection [351, 352]. The drop volume per well has decreased from 1.0 μL to 400 nL. The plates are photographed from below, which modifies image analysis requirements (see Figure 5.4(a)) compared to images photographed from above (see Figure 5.4(b)). The camera position and reducing the drop volume affect automated image analysis the most:

1. Reversing the camera position affects mainly contrast of objects within the drop. Photographing experiments from the bottom of the well makes

crystal identification by crystallographers easier, while producing artifacts that complicate automated image classification. Images from the bottom eliminate many image distortions, caused by viewing the images through a layer of oil. However, the main drawback of images from the bottom is the high contrast center of the well, caused by a curved surface of the oil barrier in the well. This creates bright background and dark objects, and low contrast edges of the well with dark background and light objects. Photographing the drops from below also filters the light through the nonoptical-grade plastic floor of the well, introducing rippled bands of alternating light and dark into the image, as shown in Figure 5.5.

2. Reducing the drop volume while using the same well affects the drop shape. With larger volumes, drops fill the well, and, thus, have convex shape. Decreased volume changes surface tension, resulting in irregular, nonconvex contours, as depicted in Figure 5.5. Edge detection is then more challenging due to introduction of many unique artifacts and complicated drop edges. Nonuniform light and edges make drop segmentation more complex and increase both false positives and false negatives during crystal classification. To cope with the challenge, we have to introduce more complicated analyses, which are computationally intensive, such as a probabilistic model of the drop boundary described in Section 5.2.2.

In addition, evolving imaging setup resulted in increased image quality and reduced light variance across an image (see Figure 5.4(c)). Significant improvements were also realized by developing newer, optical-grade plates with round bottom (see Figure 5.4(d)).

Screening each protein using 1,536 different crystallization solutions (i.e., cocktails or reacting agents) [351] results in a set of precipitation reactions, called "the precipitation index." Images from individual precipitation reactions are captured and analyzed to determine the outcome of the crystallization process [127, 285]. An ontology used to describe the results of image classification comprises clear drop, amorphous precipitate, phase separation, microcrystal, crystal, and unknown. A set of ideal image examples is presented in Figure 5.9. The precipitation index of each protein contains six time points of all 1,536 image classifications. This information, combined with chemical and physical information about individual proteins is the main part of a crystallization knowledge base (see Figure 5.6). For most applications, however, further details will be useful or required; for example, a protein may be described by additional attributes, as in Figure 5.7. In addition, one needs to store detailed information about crystallization method, such as presented in Figure 5.8.

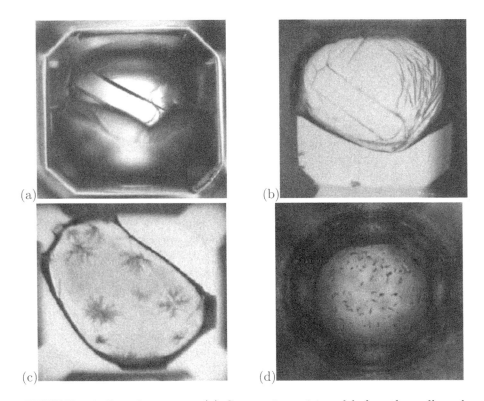

FIGURE 5.4: Imaging setup. (a) Camera is positioned below the well, and, thus, the drop is photographed through the bottom of the well. (b) Camera is positioned above the well, and, thus, the image is taken through the paraffin oil. (c) Improved imaging setup and light dispersion makes the drop illumination more even and, thus, reduces shadows and other artifacts that contribute to many false-negative and false-positive results. (d) Optical-grade, round bottom well plates improve image analysis by making the drops more round and centered.

FIGURE 5.5: A sample image illustrating complicating factors in image analysis. Back-lighting causes nonuniform lighting (a bright center surrounded by darkness) and reveals ripples on the surface of the well bottom (introducing spurious edges in the image). Small drop volumes cause drops to form irregular, nonconvex contours.

5.1.2.2 Systematic Knowledge Management for Protein Crystallization

Reasoning in the protein crystallization domain is often based on experience and analogy, rather than on general knowledge and rules. Experts remember positive experiences and apply analogy to form new solutions; negative experiences are used to avoid potentially unsuccessful outcomes. The challenge is to create computational systems to empower the experts. The main idea is to use an automated system that interfaces with the HTP robotic screen, to provide image classification, data management, and assist crystallization optimization planning, such as highlighted in Figure 5.10.

Systematic knowledge management using techniques from artificial intelligence proved useful in many engineering and scientific domains, and thus also apply to problem solving in protein crystallography. Case-based reasoning (CBR) is particularly applicable to this problem domain [281, 285, 286, 287] for the following reasons:

1. CBR supports rich and evolvable representation of experiences, stored and organized as triplets: problems, solutions, and feedback;

2. CBR provides efficient and flexible ways to retrieve the experiences, using similarity-based retrieval;

3. CBR applies analogical reasoning to solve novel problems.

```
Protein
    Organism
    Swiss-Prot/TrEMBL ID
    Amino acid sequence
    EC number
    Interpro domain
    Molecular weight
    Estimated pI
    Estimated charge at pH 7.0

Cocktail
    Cocktail Number
    Chemical Additive
    Chemical Formula
    Concentration
    Buffer Type
    Buffer
    Concentration
    pH

Precipitation reaction
    ProteinPlate
    Well
    Datetime
    Image Attributes
    Image Classification
```

FIGURE 5.6: The basic components of a crystallization knowledge base include information about proteins, cocktails, and precipitation reactions.

```
Molecular weight, calculated
Molecular weight, measured
Glycosylation
Phosphorylation
Lipidation
Monomer/multimer
Homomer/heteromer
Association complex for multimer
Isoelectric point, calculated
Isoelectric point, measured
Retention time on chromatography columns
Hydrodynamic radius
Polydispersity
Internal ions
Bound ions
Cofactors
Inhibitors
Oxidation state
Sensitivity to air
Stability
Membrane bound
Cytosolic
Species origin
Purification history
SDS-PAGE results
PAGE results
Storage state of the protein
Activity assay
```

FIGURE 5.7: The basic set of attributes describing protein properties.

```
Temperature
Chemical components (type & concentration)
Crystallization agent(s) (type & concentration)
Additive(s) (type and concentration)
Buffer (type & concentration)
Macromolecule concentration
Pressure
Gravity
Time to observation of crystals
Level of supersaturation
Seed (type, concentration, method)
Accuracy of pipetting

Volume of protein
Volume of cocktail
Container type
Container material
Geometry of drop
Temperature
```

FIGURE 5.8: The basic set of attributes describing crystallization method, and extra attributes relevant to batch method.

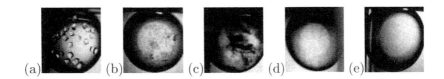

FIGURE 5.9: (See color insert following page 72.) The set of images ranging from crystal, through precipitates to clear drop show crystallization results from a large drop volume experiment. Although useful for manual inspection, possibly x-ray diffraction experiment, and definitely computational image analysis, these examples will likely not be used in a screen due to prohibitive large volume of protein required.

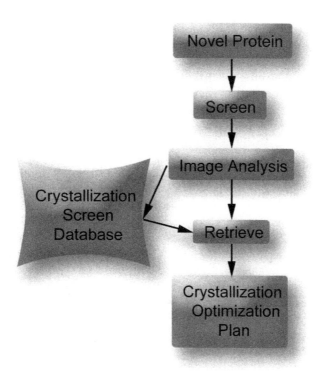

FIGURE 5.10: Automated image analysis can achieve higher throughput, consistency and objectivity in the HTP robotic crystallization. The system can also then automate data management and assist in crystallization optimization planning.

CBR is founded on the premise that similar problems have similar solutions. CBR systems comprise two essential components: (1) a repository of solved problems, called a case-base, and (2) a reasoning engine that uses the case-base to provide a solution for a new problem. The reader may find detailed descriptions of the CBR process and systems in [319]. More recent research directions are presented in [331] and practically oriented descriptions of CBR can be found in [57, 539]. We discuss one specific system in more detail in Section 5.3.

A case generally comprises an input problem, an output solution, and feedback in terms of an evaluation of the solution. The reasoning engine has several components, including a similarity-based retrieval algorithm and analogy-based reasoning system. The main goal of a CBR system is to retrieve the most relevant cases from a casebase, given a description of a problem. For example, case-based planning of protein crystallization experiments uses similarity-based retrieval to locate information about both successful and failed crystallization experiments. The precipitation index (PI) of a problem protein is compared to PIs of all proteins in the case-base. The proteins with the most similar PIs are considered for planning crystallization of a novel protein: successful crystallizations are used as positive planning experiences, while failed crystallization plans are considered to potentially avoid negative results. This basic process is described in Figure 5.11.

The effectiveness of CBR reasoning depends on the quality and quantity of cases in a case-base. Although even small number of cases provides accurate solutions in some domains, complex domains usually require an increased number of unique and relevant cases to improve their problem-solving capabilities by increasing the coverage. It is often useful to think of *typical* and *prototypical* cases, the former being representative of an experience that often occurs, while the other can represent a rare experience. Usefulness of the CBR system depends on how the cases cover problems in the domain. In addition, larger case-bases usually decrease system performance and thus efficient algorithms and implementation is required in complex domains.

It needs to be emphasized that CBR alone is not sufficient for solving problems in HTP protein crystallography, as we need to analyze images from large numbers of crystallography experiments, and also apply data mining for the analysis of the resulting data sets (see Figure 5.12). Thus, we extend basic CBR functionality by providing:

- Image-based processing to extend the expressibility of case representation, to provide for a protein similarity measure, and to assure a scalable and objective classification of crystallization experiment outcomes with appropriate image-feature extraction (see Section 5.2 for more details);

- Data base techniques for scalable case retrieval;

- Knowledge–discovery techniques to support domain-knowledge evolution and system optimization (see Section 5.4 for more details).

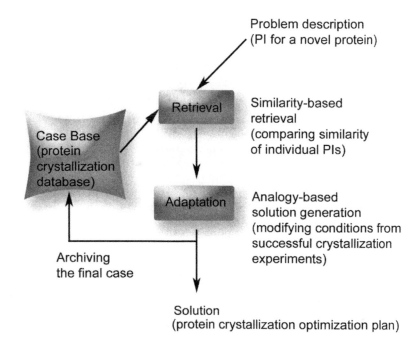

FIGURE 5.11: Case-base stores successful and failed crystallization trials. Each protein is first screened using the same set of crystallization conditions, which produces a precipitation index (PI). The PI of a problem protein is compared to PIs of proteins in the case-base. Proteins with the most similar PIs are retrieved. Successfully crystallized proteins are used as positive planning experiences, while failed crystallization plans are used to potentially avoid negative results in the future.

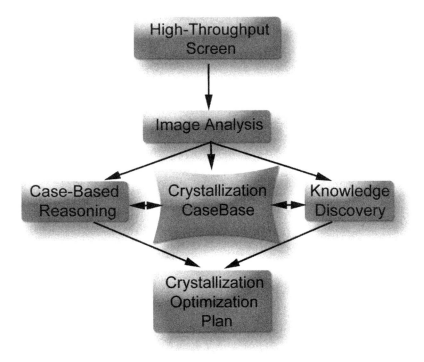

FIGURE 5.12: Image analysis, case-based reasoning, and knowledge discovery constitute the main components of an integrated protein crystallization system.

Next we introduce individual components in detail.

5.2   Image Analysis in HTP Protein Crystallization

Many problem domains require or at least benefit from automated image analysis, including machine vision, medical image interpretation, geographic information systems, and molecular biology. The goal of protein crystallization is understanding the three-dimensional protein structure, which is often essential to the understanding of its function and essential during drug design. Automated image analysis in HTP protein crystallization is particularly important. There are two main reasons why human analysis and intervention is either not desirable or possible:

1. HTP image acquisition requires processing thousands of images daily,

TABLE 5.1: Convolution filters affect matrix convolution; blur smoothes out details, while Laplacian approximates the image gradient.

Smoothing	Gaussian blur
Horizontal gradient	Vertical gradient
Laplacian filter: $\begin{smallmatrix} 0 & -1 & 0 \\ -1 & 4 & -1 \\ 0 & -1 & 0 \end{smallmatrix}$	Correlation filter

and the manual image classification will not scale up. Screening over 200 proteins a month requires processing and classifying $1,843,200$ images. Our current data base contains over $4,500$ proteins, each being screened with $1,536$ different conditions and photographed 6 times over a period of 2 weeks. Data storage requires approximately 1 DVD disk per protein experiment.

2. Image complexity makes the manual morphological analysis subjective and imprecise, which in turn, complicates the systematic processing and subsequent analysis of image information.

There is a lack of general approach to quantitatively evaluating crystallization reaction outcomes under the microscope. As a result, the major weakness of existing scoring methods is the tendency to confuse diverse forms of precipitates [143]. Several publications discuss automatic classification of images from 96-well crystallization screens [498, 551], and show high image classification accuracy. Standard image analysis techniques have been successfully applied to crystallization image classification, including convolution filters and edge detection methods.

Convolution filters are "a workhorse" of image analysis. Matrix convolution of an image U with a filter V produces a third matrix W:

$$W(i,j) = \sum_{i'} \sum_{j'} U(i',j')V(i-i',j-j').$$

The choice of filter determines the effect of the convolution; see Table 5.1.

- **Blur** smoothes out details in an image by convolution with a bivariate Gaussian function. Parameters include standard deviation σ, window size $w \times w$. This is equivalent to serially convolving the image with a $1 \times w$ and $w \times 1$ univariate Gaussian functions. Zuk *et al.* use a simpler, uniform 3×3 smoothing filter [577].

- **Laplacian** operator approximates the directionless second-order gradient of an image by convolution with a filter:

$$
\begin{array}{rrr}
0 & -1 & 0 \\
-1 & 4 & -1 \\
0 & -1 & 0
\end{array}
$$

Edge detection is usually achieved using the Canny algorithm [94] (used by [58, 287, 498]), or Sobel (used by [551]). Zuk *et al.* used Sobel for crystal-edge detection, but did not address segmentation [577].

- **Canny** edge detection algorithm first applies Gaussian filter to smooth the image, then creates the gradient of the image. There are three main advantages of Canny edge detection: it is sensitive, specific, and detects continuous edges as one object (see Figure 5.13(b)).

- **Sobel** edge detection algorithm uses a pair of 3×3 convolution masks, one estimating the gradient in the X-direction, and the other estimating the gradient in the Y-direction (see Figure 5.13(c)).

An example edge detection process is shown in Figure 5.14.

Although standard image analysis tools can be applied to protein crystallization image analysis and classification, reducing drop volumes to the sub-microliter and increasing number of cocktails creates challenges (see Figure 5.5) that require further specialization and integration of image analysis methods [126, 127, 285].

Modular system. In order to implement a flexible image classification system that will support changing image analysis setups over time, it is useful to implement a modular system for automated image classification that comprises the following stages:

1. *Well registration* locates and eliminates the boundaries of the well;

2. *Drop segmentation* eliminates the edges of the drop;

3. *Feature extraction* extracts descriptive image features used during image classification;

4. *Image classification* uses extracted features to label the image with one of the outcome classes — unknown, clear, precipitate, crystal. The goal is to further subdivide precipitate and crystal classes, and allow for multiple classes to be used simultaneously.

Different algorithms can be used to implement individual stages. To improve performance, it may also be useful to apply *a boosting* approach by combining results from multiple techniques. Next, we focus on describing individual stages and different approaches.

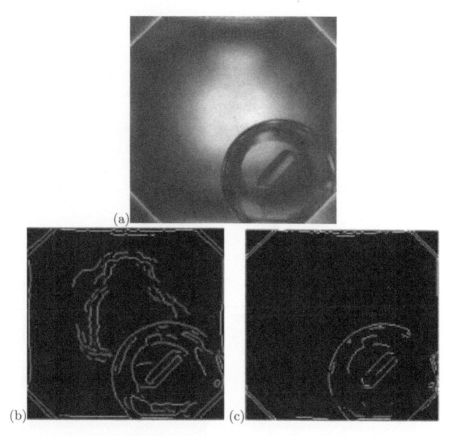

FIGURE 5.13: (a) Edges in the original image can be detected using (b) a Canny or (c) Sobel algorithm.

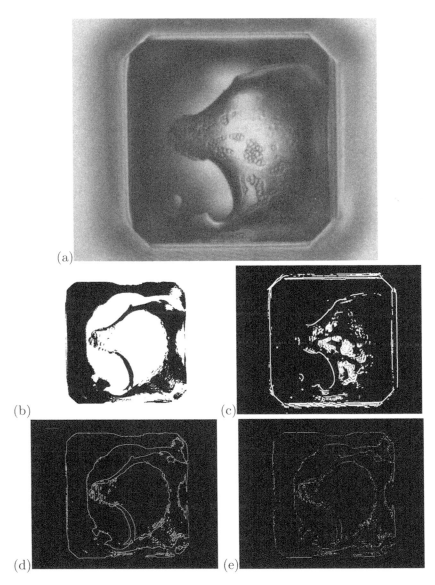

FIGURE 5.14: Simple crystal detection strategy (a) shows an original image of a crystal form, (b) depicts a binary image produced by converting a grayscale image to binary by thresholding, (c) shows edges in the image located by Sobel method, which finds edges at the points with the maximum gradient, (d) locates perimeter pixels in a binary image, and (e) depicts edges located using Sobel method, but only for the perimeter pixels.

5.2.1   Well Registration

The image analysis process starts by locating the boundaries of the well within the image, so that subsequent analysis may be done on its contents. This not only increases speed of subsequent analysis, but frequently improves classification accuracy by eliminating irrelevant artifacts.

Each well in the $1,536$-well plate has the same dimensions, and is photographed individually. The result is a grey-scale 632×504 image, with the well occupying a 425×425-pixel square roughly in the center [127]. The edges of the well are straight and approximately aligned within a degree or two of the image axes.

The algorithm can locate vertical well boundaries by finding the pair of pixel columns separated by the expected well width with the closest average pixel intensity. The horizontal well boundaries are obtained analogously, searching though pixel rows, separated by the expected well height [285].

Although robotic imaging setups improve precision of capturing individual images, a small registration error is still introduced from two main sources:

1. Robotic camera motion. As the robotic camera moves on the XY table, slight miscalibrations in positioning system introduce a consistent drift in (x, y) coordinates of subsequent wells.

2. Table/plate relative height variation. As the robotic camera moves on the XY table, slight changes in z axis due to adjusted camera height results in perceived plate-to-plate variation in well diameter.

To diminish the registration error, we can introduce the following correction to the image registration algorithm:

- Multiple well-diameter sizes are used in parallel, to compute a first approximation of the coordinates for all $1,536$ wells;

- The well diameter minimizing the variance in measured $(\Delta x, \Delta y)$ is selected;

- Coordinates for each well are corrected using a uniform $(\Delta x, \Delta y)$ between adjacent wells.

5.2.2   Drop Segmentation

The second step in image analysis usually involves identifying edges of the drop within the well, i.e., the drop segmentation step. Elaborate image-analysis may or may not be required to segment the drop from its surroundings, depending on the variation in appearance of the droplet in images from the imaging system. Ideally, the HTP system will produce images of droplets with consistent circular shape and radius, perfectly centered in the image, with sharp edges distinguishing the drop from a perfectly uniform, evenly lit

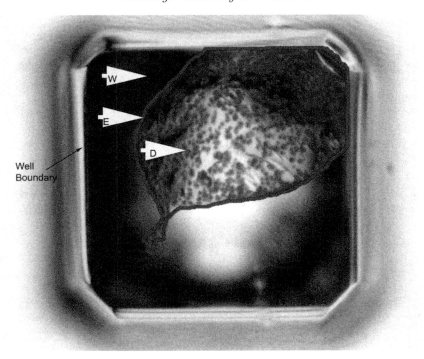

FIGURE 5.15: (See color insert following page 72.) Drop segmentation eliminates the edges of the drop before feature extraction and image classification steps. The algorithm divides drop into the empty well — W, the inside of a drop — D, and the edge of a drop — E.

background, with no distortion of any glass, plastic, or other medium between the drop and the photograph.

Images from the Veeco Optimag-1700 [58, 498], and other systems [287, 551] stray somewhat from this ideal, producing images with approximately circular or elliptical drops against a near-uniform background. Droplet segmentation divides the well interior into three pieces (see also Figure 5.15):

1. The empty well (W);

2. The inside of a drop (D);

3. The edge of a drop (E).

This segmentation enables us to exclude irrelevant artifacts from the image (i.e., the empty regions of the well and the edges of the drop) before further analysis.

In most images, drop edges have the most extreme intensity values and the highest intensity gradients. Thus, it is important that we exclude the drop edges from image preprocessing and analysis, by applying the image smoothness and straight-edge analyses described in Section 5.2.3. The droplet boundary can be pieced together from fragments uncovered by an edge-detection algorithm, such as Canny and Sobel discussed above.

The detected edges (the edge set) may originate from inside the droplet, along its boundary, or occasionally outside the droplet. The droplet edge must be reconstructed from a subset of the edge set, based on the central assumption of the elliptical or circular form of the boundary. Wilson [551] assumes a circular drop and uses gradients measured along the droplet edge to estimate (and iteratively re-estimate) the center and radius. Spraggon *et al.* also assume a circular drop, and assume the largest curved member of the edge set is part of its boundary [498]. Bern *et al.* use an iterative curve-tracking technique to fit a 60-sided, star-shaped polygon to the droplet boundary, beginning with a seed edge from the edge set [58]. Jurisica *et al.* select edges so as to reconstruct a curve that minimizes mean-squared deviation from a conic section [287].

Edge detection and circularity assumptions do not suffice when there is high variation in droplet shape and a nonuniform background, such as depicted in Figure 5.16. The central region of the well can be segmented using a probabilistic graphical model, as illustrated in Figure 5.17. The well is first divided into a grid of 17×17 regions, (i, j). Each region in the grid is represented by a pair of variables in the graphical model, forming a 60-component mixture of multivariate Gaussian distributions:

- One *latent, discrete variable* $x_{i,j} \in (1, \ldots, 60)$ controlling the active mixture component;

- One *observed, vector-valued variable* $\mathbf{y_{i,j}}$ representing local image intensity and gradient values, computed directly from the image.

Each mixture models the local state of the well, representing empty well W, drop interior D, or drop edge E. The state of region (i, j) is represented by the value of the latent variable $x_{i,j}$, with 20 mixture components assigned to each possibility. Each possible state has multiple mixture components in order to cover diverse textures in an image region with a given state. Each latent variable $x_{i,j}$ is linked to its neighbors by conditional probability relationships:

$$P(x_{i,j} \mid x_{i-1,j}), P(x_{i+1,j} \mid x_{i,j}), P(x_{i,j} \mid x_{i,j-1}), P(x_{i,j+1} \mid x_{i,j}).$$

These probabilities, together with the conditional probability $P(\mathbf{y_{i,j}} \mid x_{i,j})$ of the mixture, ensure that the inferred value of $x_{i,j}$ depends both on the local region of the image and the values of its neighbors. Latent and observed variables represent two layers of the model:

1. *The latent layer of the model* is represented by the complete set of latent variables: $\mathbf{X} = \{x_{i,j}\}$;

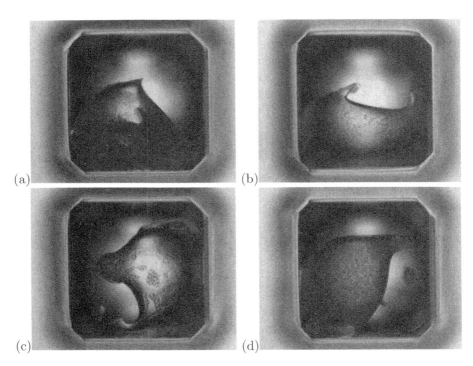

FIGURE 5.16: Irregular shapes, variable size, and nonuniform background make automated image analysis and crystal detection challenging. Despite progress to diminish these problems both by evolving imaging setup and improving image analysis algorithms, irregularity of drops still significantly contributes to false positives and false negatives.

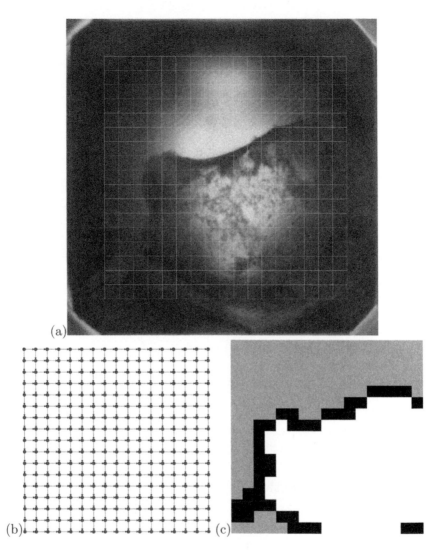

FIGURE 5.17: The Bayes net segmentation model. (a) We divide the well into a 17×17 grid of tiles and compute a vector of image intensity gradients for each tile. These vectors are the observed values of variables $y_{i,j}$. We model each tile's vector as a sample drawn from a mixture of multivariate Gaussians. (b) A mesh of conditionally interdependent latent variables, one per tile, controls which component is active in each mixture. (c) The active component in each tile determines the segment (one of W, D, E) assigned to that patch of the image.

2. *The observed layer of the model* is represented by the complete set of observed variables:

$$\mathbf{Y} = \{\mathbf{y_{i,j}}\}.$$

A *configuration* $\mathbf{X} = \mathbf{x}$ of the latent-variable layer represents a possible segmentation of the well. A given image is segmented by inferring the most likely configuration from the probability distribution:

$$P(\mathbf{X} \mid \mathbf{Y} = \mathbf{y}),$$

where $\mathbf{y}$ is computed from the image.

Inference in graphical models is exact and efficient for graphs with a tree topology. Unfortunately, our model necessarily has a highly connected grid topology. Thus, the most likely segmentation must be inferred using an inexact Loopy Belief Propagation algorithm [396]. We use the forward-backward algorithm to iteratively update each row of $\mathbf{X}$, and then each column.

To train the segmentation model, we used 45 images segmented by an expert. We have minimized required parameters in the model by using a single distribution for all horizontal conditional probabilities $P(x_{i,j} \mid x_{i-1,j})$. Analogously, we used a single distribution for all vertical conditional probabilities $P(x_{i,j} \mid x_{i,j-1})$. To reduce training, we tied all mixture model parameters together, i.e., we used a single mixture prior $P(x)$ and 60 component distributions of the observed variable $P(\mathbf{y} \mid x)$. Thus, we needed only train a single 60-component mixture model.

We have evaluated performance of the image segmentation model using a set of 50 hand-segmented images containing $4,319$ empty-well regions, $1,348$ drop-border regions, and $8,783$ intra-droplet regions. The segmenter correctly identified 96% of all well regions, 69% of all border regions, and 95% of all intra-drop regions, for a weighted mean of 93% overall accuracy [127].

5.2.3   Feature Extraction

Protein crystallization experiments have many possible outcomes, and many events can occur simultaneously in a single experiment: diverse types of precipitation, crystal formation, microcrystal formation, skin effect, and phase separation.

Fine-grained classification systems such as Wilson's [551] attempt to classify each discrete object visible in the droplet. Other systems classify the entire image as a whole [127, 498]. Depending on the goals and its sophistication, a whole-image classifier will report a binary outcome (crystal/no crystal), or multiple categories (clear drop, microcrystal, needle crystal, precipitate, phase separation, etc.).

The basic setup for whole-image classification has two phases: morphological characterization and image classification. *Morphological image characterization* applies image segmentation and feature extraction techniques to determine and quantify image texture, distinctive objects and their properties within the image. Feature extraction computes a set of numeric scores from the droplet image, with each score measuring an aspect of the texture of the droplet, such as the mean grayscale intensity, the presence or absence of straight lines, statistics on the distribution of light and dark points in the image, etc. An image can be characterized by a large number of descriptive features. However, many of the extracted features would be redundant or completely irrelevant to the classification. Thus, the goal is to extract a minimal number of features that discriminate among different crystallization experiment outcomes with high accuracy during image classification.

Image classification uses the extracted features to automatically classify an image into one of the possible categories, including clear drop, precipitate, and crystal. Further subclasses, as well as multiple classes can be distinguish for individual images. Common methods used during classification include decision tree, neural net, support vector machine, or simple linear discriminant. Classified images can then be readily used by crystallographers (e.g., identified crystals in the screen are direct candidates for optimization phase), or they can be stored in the crystallization data base and later used for case-based planning of crystallization optimization (see Section 5.3) and for automated knowledge discovery (see Section 5.4), as outlined in Figure 5.12.

Many approaches have been proposed for image feature extraction, including:

- Polynomials of fitting flexible curves, called snakes [375] or planes [332];

- Similarity invariant coordinate systems (SICS) that represent images as points and vectors [339];

- Attributed relational graphs (ARG) [442];

- Set of transformations [419], including Fourier [287], Hough [498, 551] and Radon [127];

- Mathematical models for detecting morphological structure and topology, such as quad-tree decomposition [280] and Euler number [2].

Most complex domains may suffer from a potential disconnect between what experts perceive as important features and what is objectively a computationally viable image feature with high discriminatory potential. For that reason, it is useful to approach the problem in iterations, applying multiple algorithms to compute diverse features, collecting the data, and then thoroughly evaluating which combination of features provides best accuracy. Importantly, specific features may discriminate only a certain subclass of all images, and, thus, a combination of features is necessary [127].

While certain feature–extraction methods are suitable for overall image characterization (e.g., Fast Fourier transform, quad-tree decomposition, Euler number), other methods are more suitable for individual object characterization (e.g., Radon transform). These features can also be characterized as: (1) *spatial-domain features* extracted from the two-dimensional intensity map of the image, and (2) *frequency-domain features* extracted from the two-dimensional frequency spectrum of the image.

5.2.3.1   Spatial-Domain Features

Spatial-domain features correspond to the quad-tree decomposition, Euler number, length of edges, the number of bends found in edges, the ratio of the length of edges to their minimum length, and the intensity of edges.

Spraggon *et al.* [498] compute edge scores per contiguous group of pixels in the edge image:

- **Area** is a generalization of the formula for computing the area of a circle:

$$length^2 \cdot (1 + closedness^2)^2 \cdot (2 - compactness)/4\pi,$$

  where:

    - *length* is the count of pixels in the edge;
    - *closedness* is the ratio of Euclidean distance between start and end pixels to the edge length; and
    - *compactness* is the ratio of maximum to mean Euclidean distances between edge pixels and the center of the edge.

- **Straight-edge score** measures the length of the longest straight line in the edge, taken to be the square of the maximum value of a discrete Hough transform computed on the edge pixels.

Spraggon *et al.* summarize the edge content of each image with five global edge features: number of edges, total edge area, total straight-edge score, maximal edge area, maximal straight-edge score, and maximal *area × straight − − edge score* value. Wilson [551] uses six image features:

- **Straight lines of constant intensity:**   length, angles (4 scores);

- **Shape — area:**   ratio of object area to that of the minimum bounding rectangle;

- **Shape — perimeter:**   ratio of the object's perimeter to its area;

- **Curvature:**   measures $\sum_i |\Delta\theta_i|$, the sum of absolute change in angle between consecutive pixels in the edge, measured in multiples of 45 degrees, expressed in the interval $\{-3,\ldots,4\}$;

- **Straightness:**   based on curvature codes above, measures the consecutive zeroes;

- **Boundary-shape descriptors:** graph of distances from a starting point to every other; difference between adjacent maxima and minima, area under graph (integration), normalized by size of object.

Spraggon *et al.*'s texture features are all computed from Gray-Level Co-occurrence Matrices (GLCM). A GLCM P for an image I and offset o summarizes the co-occurrence of pixel intensities; element p_{ij} counts the number of pixels p in I with value i whose neighbor pixel $p + o$ has value j. Spraggon *et al.* propose several values of o with which to compute GLCMs, as well as an isotropic GLCM computation. Their texture features are drawn from the original GLCM literature [233], and include contrast, correlation, variance, sum average/variance, entropy, energy, etc.

Related to the standard deviation score, Wilson also measures variation in raw pixel intensity and pixel-intensity gradient magnitudes [551].

5.2.3.2 Frequency-Domain Features

Frequency-domain features are calculated using a Fourier transformation to find the two-dimensional frequency spectrum of the image, which is then normalized to yield a two-dimensional map of intensity values ranging from 0 to 1. The map is converted to a boolean mask by comparing each value to a threshold constant selected empirically so that the shape of the spectrum is well characterized by the boolean mask. For example, frequency domain for crystal, precipitate, and clear images from Figure 5.2(a–c) are presented in Figure 5.18.

As seen in Figure 5.18, features of the frequency spectrum can be used to classify crystallization results. To compute the features of the "cross-like" shape, the following parameters are computed:

- The height of the horizontal bar and the width of the vertical bar of the cross are measured at multiple locations to capture their variability;

- The ratio of the height and the width of the corresponding bars, and the ratio of the length of the horizontal bar to the length of the vertical bar;

- Radial measurements made from the center of the mask to the edge of the cross at varying degrees, along with their variance, are computed and stored as parameters;

- The number of pixels in the mask;

- The circular average, i.e., the two-dimensional frequency domain is reduced to a one-dimensional vector by taking the average intensity of all values at different radial distances from the center of the image, and the resulting vector is sampled at three different locations. A fifth-degree polynomial is then fit to the curve and its third derivative and roots are calculated.

FIGURE 5.18: Frequency-domain features can be used to distinguish (a) crystal, (b) precipitates, and (c) clear crystallization results.

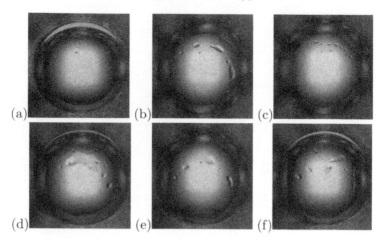

FIGURE 5.19: Clear drops range from almost completely uniform (a) to various images with diverse morphological features that are neither indicative of crystals nor precipitates (b–f).

The majority of images in the screen are noncrystal, but even the variability of clear drops is large, which makes image analysis challenging. In addition, $1,536$-well plates with 15 wells containing crystals may be classified with an accuracy exceeding 99% by a method claiming that no wells contain crystals. One approach to deal with this is to remove noncrystal images prior to classification. However, removing noncrystal droplets from the screen without losing any crystals is challenging due to the image variability, as shows in Figures 5.19 and 5.20, and also due to presence of local vs. global features as discussed in Figure 5.21.

5.2.3.3 Crystal Detection Methods

In protein crystallization, the most important images represent crystals. Protein crystals come in diverse forms: microcrystals, microneedles, needle crystals, and larger faceted crystals, as shown in Figures 5.22 and 5.23.

Considering all the complexities of structural proteomics and HTP protein crystallography, ideally image classification would achieve 0% false negative rate for crystal identification. However, in a regular imaging setup, this can be only achieved by increasing the false positive rate. Next, we focus on approaches to increase both specificity and sensitivity for crystal identification (see Section 5.2.3.5).

It should be noted that depending on the original image, certain crystal forms may or may not be detectable. For example, Figure 5.24 shows how use or absence of a polarizing filter affects observable features of a crystal within the image.

Even when the imaging setup supports crystal detection, other factors make

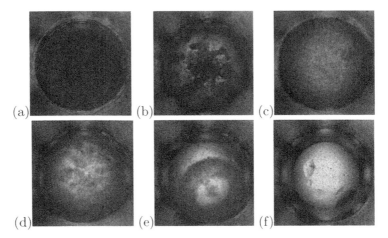

FIGURE 5.20: Diversity of precipitates ranges from dark (a) to light (f), with diverse number and granularity of identifiable objects with the drop.

accurate detection of diverse crystal forms challenging. Although most crystals exhibit straight edges in a droplet, this important differentiating feature in general may not be detectable in microcrystals in a HTP setup, due to insufficient size. Microcrystals may not have any straight edges of significant length, considering the size of images at least tens of pixels. To diminish this problem, we need to extract a set of image features specifically for detecting microcrystals (see Section 5.2.3.6).

Similarly, diverse subgroups of precipitates can be characterized by the level of pixel intensity, and homogeneity of the structure, as further discussed in Section 5.2.3.4.

5.2.3.4   Measuring Local Image Smoothness

Image smoothness can be detected by three main methods: the Laplacian operator, quad-tree decomposition, and Euler number.

The Laplacian operator computes the second derivative of the intensity of an image, which is the rate of change of the intensity gradient in an image. It is approximated by the convolution that computes the summed differences between each pixel's intensity and its neighbors' intensities:

$$L(p_{i,j}) = 4p_{i,j} - p_{i-1,j} - p_{i+1,j} - p_{i,j-1} - p_{i,j+1}.$$

The Laplacian operator can be used to identify discontinuity in an image, i.e., edges and local image smoothness [127, 19]. The Laplacian of a point (x, y) in an image with a locally uniform gradient is zero, even if the gradient magnitude at that point is high. Thus, we can use the squared Laplacian values in an image region to measure the local smoothness of the image. First the droplet is subdivided into 16 pixel blocks, and then the Laplacian values

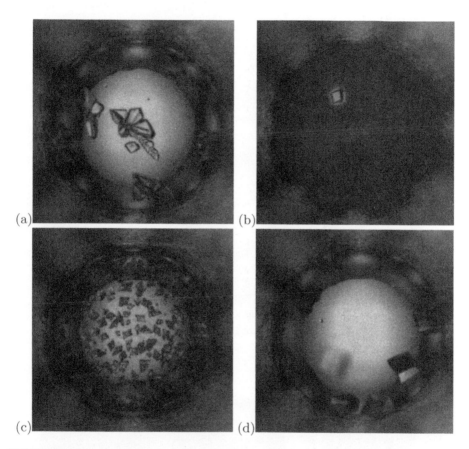

FIGURE 5.21: Sometimes, images belong clearly to one (a), while at other times the classification requires two or more classes (b). Sometimes the most important class (crystal) is the "majority" object (c), while at other times, it occupies an insignificant portion of the image (d).

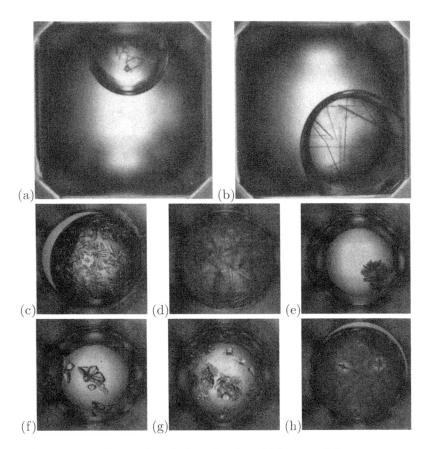

FIGURE 5.22: Examples of diversity by which crystal forms appear.

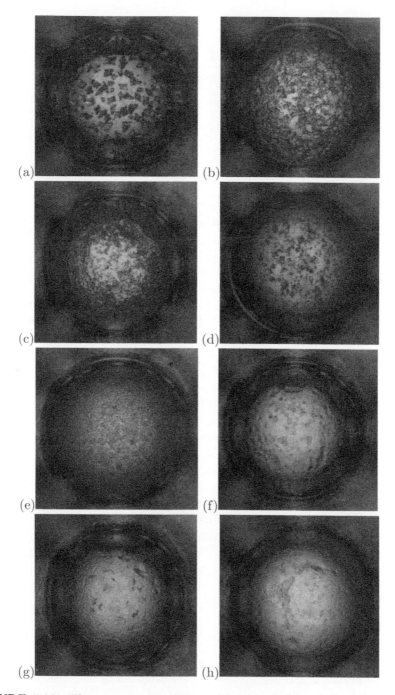

FIGURE 5.23: There are many diverse forms and shapes of microcrystals, which makes their identification challenging.

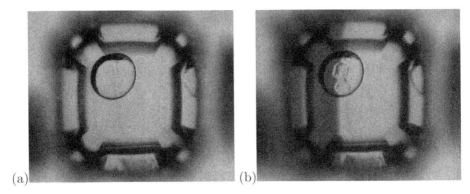

FIGURE 5.24: Polarizing filters affect crystal detection, since (a) without it the crystal form is not visible (b) while adding a polarizing filter clearly enables crystal detection.

within each block are squared and summed. This yields a smoothness measure for each block. We select the highest score to use as the measure of image smoothness (see Figure 5.25).

Quad-tree decomposition comprises splitting the image into four parts and examining the difference between the minimum and maximum pixel values in each quadrant. If the difference exceeds a predefined threshold then the quadrant is further subdivided into four parts, and the analysis repeats for each square. This process repeats until the minimum and maximum value in each square differs by less than the threshold. The image smoothness is then characterized by the number of squares examined (see Figure 5.26).

Euler number is based on measuring image topology by counting the total number of objects in the image minus the number of holes in those objects. Thus, methods such as quad-tree decomposition and Euler number can create a partial order of precipitates from more clear ones to more cloudy ones, or more smooth ones to speckled (see Figure 5.27).

However, precipitates could also include crystals (see Figure 5.21(b)), and thus one has to also search for straight edges. Similarly, a clear drop may include a single small crystal (see Figure 5.21(d)). Therefore, global filtering has only a limited utility, and it is useful to implement a multiscale analysis of image texture.

5.2.3.5 Detecting Intra-Drop Straight Edges

After eliminating the drop border and the surrounding empty well from the image (see Section 5.2.2), the area inside the drop can be searched for straight edges by applying a Radon transform to the Laplacian of the Gaussian-blurred image. The Laplacian-of-Gaussian (LoG) operation suppresses high-frequency image noise and reveals edge pixels (see Section 5.2.3.4 and Figure 5.25(b));

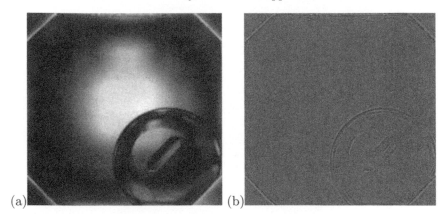

FIGURE 5.25: The Laplacian operator can be used to detect local image smoothness and identify discontinuity in an image by revealing edge pixels. (a) shows the original image, and (b) depicts the edges after applying the Laplacian operator.

the Radon transform, applied to edge data, detects straight lines (see Figure 5.28(b)).

The Radon transform [542] is related to the Hough transform, and detects straight lines in an image. Thus, it is useful to detect crystals and needle crystals. The input image must be pretransformed to highlight discontinuities. Thus, the input transformation can be done by an edge-detection algorithm, for example, a Laplacian-of-Gaussian convolution (see Figure 5.25(b)).

The Radon transform of an image $I(i, j)$ is an integral transform defined as follows:

$$R(\theta, k) = \int_{-\infty}^{\infty} I(k \cos \theta - l \sin \theta, k \sin \theta + l \cos \theta) dl.$$

$R(\theta, k)$ is the integral of $I(i, j)$ along all points (i, j) on a line of orientation θ and of distance k from a parallel line passing through the origin $(0, 0)$. Any large-magnitude $R(\theta, k)$ value can be interpreted as evidence of a straight edge in I, oriented at angle θ from the vertical, with distance k from a parallel line passing through the origin.

Notably, the straight lines detected by the Radon transform need not comprise contiguous pixels, since only the total number of pixels and their intensities matter. This is advantageous when analyzing an image with a sharp, straight crystal edge periodically obscured along its length by image noise, precipitate, skin, or external artifacts (e.g., dust). However, this characteristic may also complicate analysis when many unrelated discontinuities lie co-linearly in the same image.

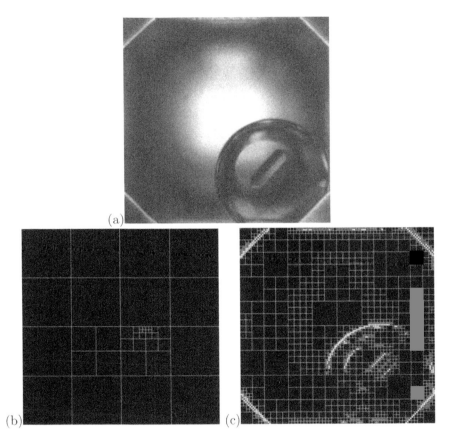

FIGURE 5.26: A given image is iteratively subdivided into parts, until each block satisfies the specified threshold. (b) shows the quad-tree decomposition of an image (a) using a threshold 0.7, while (c) shows a quad-tree decomposition with a threshold 0.2.

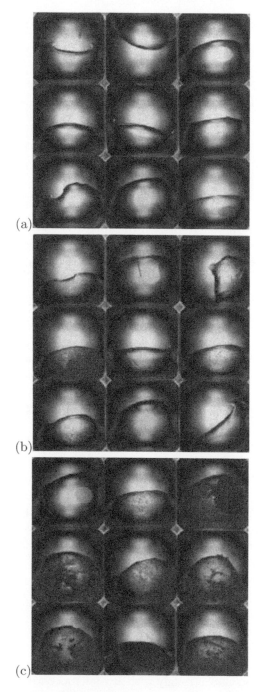

FIGURE 5.27: Sorting of precipitates can be done by using pixel intensity from light (a) through medium (b) to dark (c), or by using image granularity from smooth to speckled.

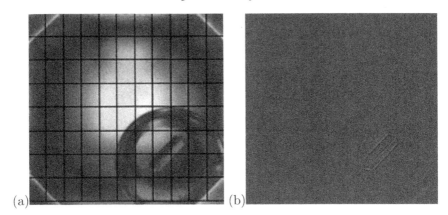

FIGURE 5.28: The Radon transform can used to detect straight lines in an image, such as crystal edges and needle crystals. (a) shows the original image, and (b) depicts the edges after aplying the Radon transform.

5.2.3.6 Detecting Microcrystal Features

One approach to detect crystals that are too small is to create a set of proto-typical images (see Figure 5.29. This library of 10 microcrystal exemplars can be selected by an expert crystallographer, and used for scoring their presence in the target image. To further increase the potential of this approach, 10 additional scores can measure the presence of target image features matching the inverse-image of the exemplar (see Figure 5.30).

The 10 individual microcrystals were cropped from a set of example micro-crystal-bearing images to produce a library of 10 images between 12 and 18 pixels on a side (see Figure 5.30). Prior to cropping, the example images were transformed using the Laplacian operator.

To score the target image, the 20 microcrystals from the library were then applied as a linear filter against the Laplacian of the target image, effectively computing the correlation between every possible superposition of the Lapla-cians of the two images.

A single microcrystal may appear as its inverse image under different light-ing conditions. An image feature correlating negatively with a microcrystal exemplar is equivalent to the same feature correlating positively with the inverse-image of the same exemplar. Thus, each exemplar is labeled with two correlation values, the maximum and the minimum.

Since the set of exemplars is limited, no individual microcrystal in a new image is likely to correlate strongly with any microcrystals in the library. This may not decrease the classification accuracy much, because usually several microcrystals appear at a time, and thus at least one or two should match an exemplar from the library.

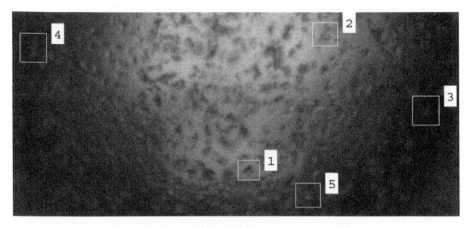

FIGURE 5.29: To cope with variety of crystal form, one can collect several representative exemplars to be used during image recognition.

FIGURE 5.30: An extracted set of microcrystal exemplars can be used as correlation filters during image recognition.

5.2.3.7 Image Feature Vector

Following the overall image analysis and multiple feature extraction approaches, we can now reduce each image to a vector of n features, which is then used to classify the image content; for example, 59 features have been used in [126]:

- 20 features characterize microcrystals;

- 21 features measure the presence of straight edges detected within a drop;

- 3 feature measures the smoothness of the drop content using quad-tree decomposition;

- 2 Euler-number features characterize the image topology;

- 10 features measure local extrema;

- 3 features characterize image energy.

Considering that all $1,536$ wells are photographed several times over a two-week period, we can also consider the time-course analysis by classifying the state of a single well at multiple time-points. Knowledge of the state of a

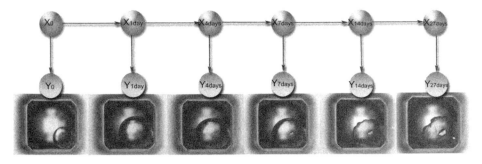

FIGURE 5.31: Image classification can be optimized by incorporating time-series data. A hidden Markov process models the changing state of the drop contents in latent variables x. The observed variables y are the feature vectors computed from their respective images. The value of each x is inferred from the value of its corresponding y and its latent-variable neighbors. As seen in the figure, drop morphology changes, but importantly, crystal cannot be detected until day 4, but is easily recognizable on day 27.

drop in past or future time points can improve the accuracy of classifying the present time point, using a probabilistic approach (see Figure 5.31).

5.2.4 Image Classification

The goal of image classification is to use the image feature vector to predict the class label of a target image. Once the feature vectors of a set of images have been computed, a classification system can be trained, using a set of preclassified images as a "ground-truth."

A variety of standard classification methods have been used for classifying crystallization images: Fisher linear discriminants [127], self-organizing maps [498], naive Bayes classifiers [551], decision trees [58, 576], support vector machines [576], and k-nearest neighbors [126]. The optimal method can be selected by considering several factors:

- **Number of training examples.** Simpler models with fewer parameters require fewer training examples. A good rule of thumb suggests that each model parameter requires at minimum of 20 values from the training data. Thus, a training set of 500, 10-dimensional feature vectors can be used to train a model with a maximum of 250 parameters. Linear discriminant analysis (LDA) requires $F + 1$ parameters per outcome; a full-covariance Gaussian model requires $F^2 + 1$ parameters per outcome.

- **Categorical features.** Categorical features are not directly representable by most models, with the exception of decision trees or case-based reasoning systems. For most models, a feature space with categorical variables $c_1, \ldots, c_C$ requires that a different model be trained for each possible configuration of those variables; using the remaining $F - C$ continuously valued feature dimensions.

- **Distribution of each feature.** LDA models each class as a multivariate-Gaussian distribution.

- **Conditional independence of features.** Naive Bayes models assume independence of the feature dimensions f_i given the class of the image or object.

- **Linear separability of the clusters.** LDA applies a linear decision boundary to the feature space.

- **Number of possible outcomes.** SVMs and LDA are naturally binary classifiers, though both are readily adapted to multiclass problems.

One approach to implement image classification is to apply *Linear Discriminant Analysis (LDA)* using n-dimensional image feature vectors, constructed during feature extraction (see Figure 5.2.3). Before any images may be classified, the linear discriminant must first be trained using a set of labeled images.

Given two sets of n-dimensional data points (crystal X and noncrystal N), LDA computes the vector $\mathbf{v}$ that maximally separates the points in X and N when projected onto $\mathbf{v}$. Further, LDA fits Gaussian distributions to X and N and calculates the distance along direction $\mathbf{v}$ that serves as the equiprobability boundary between the means of X and N. Effectively, LDA produces a n-element vector $\mathbf{v}$ and a numeric threshold t such that, given an image with feature vector $\mathbf{f}$,

$$\mathbf{f} \cdot \mathbf{v} > t \Leftrightarrow \text{the image is in class } X.$$

To determine how useful such image classification is in the real setting, one has to perform multiple tests, measuring classification consistency, sensitivity and specificity. However, it should be emphasized that since most images are likely noncrystal the image classification process is highly imbalanced.

5.2.4.1 Image Classification Consistency

Machine consistency of an image classification method can be evaluated by analyzing two plates containing independently run, identical crystallization trials, photographed after the same time lapse. A new linear discriminant was computed for each plate, using LDA and training data drawn from the remaining hand-labeled plates. A confusion matrix for the two plates displays the number of images assigned to each class (see Table 5.2).

Table 5.2 shows that the machine classification for both plates agreed that 399 wells contain crystals, and that 923 wells contained no crystals. The two classifications disagreed on the contents of the remaining 214 wells in each plate. This achieves machine-assessed consistency rate of 86%.

TABLE 5.2: A confusion matrix displays machine classification consistency, by comparing classification results on the two plates containing independently run, identical crystallization trials, photographed after the same time lapse.

Machine classification (glucose isomerase)		Plate 2 N	X
Plate 1	N	923	123
	X	91	399

TABLE 5.3: A confusion matrix displays human classification consistency, by comparing classification results on the two plates containing independently run, identical crystallization trials, photographed after the same time lapse.

Expert classification (glucose isomerase)		Plate 2 N	X
Plate 1	N	970	110
	X	84	372

Similarly, we can evaluate *human classification consistency* by having the same plates reviewed by a human expert. Corresponding confusion matrix is presented in Table 5.3.

Table 5.3 shows the expert labeling of both plates agreed that 372 wells contained crystals, and that 970 wells contained no crystals, while the remaining 194 wells showed conflicting results. These numbers yield a consistency rate, and thus an experimental repeatability rate, of 87%. Thus, machine consistency reaches limits of experimental repeatability.

To measure *classification accuracy* we compare the human expert's assessment of each plate with the corresponding machine classification, summarized by the confusion matrices in Table 5.4.

By assuming perfect accuracy of the human expert, we obtain accuracy scores for the automated image classification: for plate 1, precision 0.77, recall 0.69, and accuracy 85%; for plate 2, precision 0.83, recall 0.76, and accuracy 88%.

To measure the overall accuracy of the system, we split the set of 18 expert-

TABLE 5.4: Comparing human and machine classification on the two plates shows the accuracy of image classification.

Glucose Isomerase Plate 1		Plate 1 Machine classification	
		N	X
Human expert	N	983	97
	X	139	317

Glucose Isomerase Plate 2		Plate 2 Machine classification	
		N	X
Human expert	N	981	73
	X	115	367

TABLE 5.5: Assessing overlap of human and machine classification on $1,112$ crystal and $12,712$ noncrystal images measure the overall accuracy of image classification.

Test set images		Machine classification	
		N	X
Human expert	N	11547	1165
	X	333	779

reviewed plates into training and test sets of equal size, each with $1,112$ crystal and $12,712$ noncrystal images. Using the set we perform feature extraction on each as described in Section 5.2.3. After computing the optimal linear discriminant using LDA on the training set, we can measure the accuracy of the resulting classifier on the test set. The confusion matrix for our test set is shown in Table 5.5.

From Table 5.5 we obtain a test-set accuracy score of 89%. However, accuracy scores in this context may be deceiving, due to an imbalanced classes. Thus, the crystal-detection task is better viewed as an information retrieval task. In retrieval terms, our method has a precision of 0.40 and a recall of 0.70. Importantly, significantly extending diversity of crystal forms on a larger set of test images achieved test-set accuracy of 85%, a precision of 0.24 and a recall of 0.66. This slight drop in classification accuracy, especially in the precision of the method can be explained in part by a skewed distribution of crystal forms in the smaller data sets. The population of crystal images in this set is dominated by crystals from the glucose isomerase plates. The updated data set contained a wider variety of crystal forms and a smaller proportion of crystals.

This highlights the dependency of image classification results on the images

TABLE 5.6:   The performance of Fisher LDA on the
59-feature dataset against the original 23-feature dataset, and
the k-nearest neighbor classification (for $k = 50$).

Algorithm	Precision	Recall	Accuracy	ROC score
23 features LDA	0.10	0.62	0.83	0.80
59 features LDA	0.12	0.68	0.85	0.85
59 features KNN	0.13	0.76	0.85	0.87

being analyzed. Due to diversity of crystals, extended training is beneficial, and better proportion of crystal vs noncrystal images also improves the accuracy of the algorithm.

Linear discriminant analysis of the training set established a linear combination of the image feature vector to produce a single numeric score that maximally (with respect to the training sets) discriminates between crystal-bearing and noncrystal-bearing images. LDA also produces the numeric threshold used to classify new images, e.g., the test sets. Choosing a threshold makes a trade-off between sensitivity and specificity (or precision and recall), i.e., including as many crystal-bearing images, while excluding as many noncrystal-bearing images, as possible. Thus, the confusion matrices and the derived corresponding precision and recall scores change as the threshold changes: greater precision is achieved at the expense of lower recall, or *vice versa*.

The Receiver Operating Characteristic (ROC) curves of the retrieval exercises illustrate this trade-off in Figure 5.32. For the original retrieval evaluation, the area under its ROC curve, called the ROC score, is 0.875. The ROC score of the retrieval on the updated data set is 0.84 (see Table 5.6).

5.2.4.2   Characterization of Misclassified Images

To aid in system optimization, we have manually reviewed 101 images falsely classified as crystal-bearing images (i.e., false positives), and 204 images falsely classified as noncrystal-bearing (i.e., false negatives).

Characterizing features of the *false positive* examples generated by the image classification system revealed distinct trends in the data. Out of several themes present in these images, one is dominating: misclassification of microcrystals, as seen in Table 5.7.

The classifier could not differentiate microcrystals from some grains of precipitate with a microcrystal-like appearance, and images with wrinkles in the skin that resemble crystal edges (see Figure 5.33). The four believed crystals identified, shown in Figure 5.34, were retrospectively confirmed as genuine crystals by HWI crystallographers.

Microcrystals typically have a sparkle, i.e., adjacent points of intensity significantly above and below the local mean intensity, while grainy precipitates do not. An improved method would use this intensity difference as a discriminant between microcrystals and speckled precipitates, as shown in Figure 5.35.

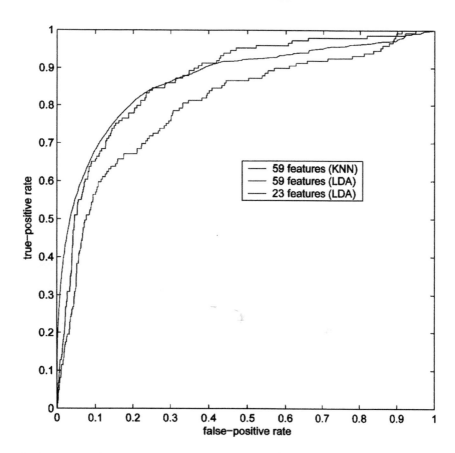

FIGURE 5.32: (See color insert following page 72.) The ROC curves of image retrieval exercises for the three classifiers — different set of image features, and comparing LDA with k-nearest neighbor algorithm (with $k = 50$). The evaluation is conducted on 127 human-classified protein screens containing 5,600 crystal images and 189,472 noncrystal images.

TABLE 5.7: Microcrystals are frequently misclassified and confused with speckled precipitates, skin effect, and phase separation.

Theme	# Images	Comments
speckled precipitate	92	> 25% visually similar to micro-crystals
skin effects	20	
contaminated wells	6	hair or fiber
mottled phase separation	1	
genuine crystals	4	

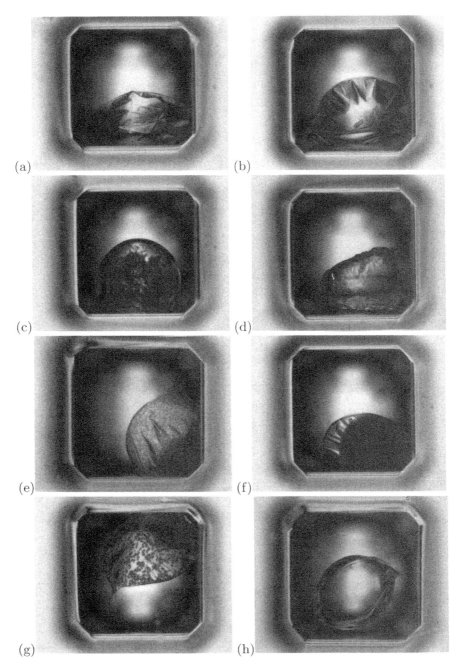

FIGURE 5.33: Misclassified noncrystals: skin effects, speckled precipitate.

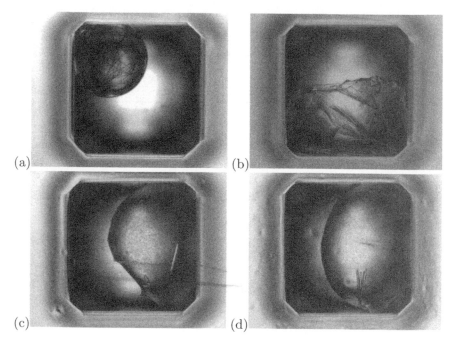

FIGURE 5.34: Genuine crystals detected by our method, erroneously labeled as noncrystals.

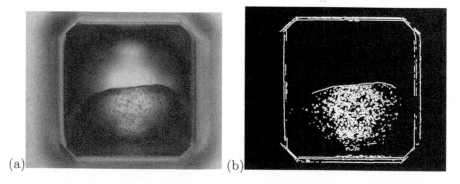

FIGURE 5.35: One of the approaches to detect microcrystals is to first identify edges and objects within an image, using Euler number (i.e., counting the total number of objects in the image and subtracting the number of holes in those objects). (a) shows the original image, and (b) depicts the identified edges and objects.

TABLE 5.8:   The image classifier fails to correctly label fine crystals or crystal forms without straight edges as crystals, resulting in false positives.

Theme	# Images	Comments
blurry crystals	17	
feathery crystals	55	crinkled, even sweeping, curvy, feathery edges
finely textured precipitate	22	
piles of small crystals	28	similar to those described in [551]
very fine, small needle crystals	56	
no obvious crystal	3	unusual drop textures
crystals near drop edge	9	suspected segmentation error

Characterizing 204 images falsely classified as noncrystal-bearing suggests several themes: (1) crystals too fine for detection and (2) crystals without straight edges. Examples of these false negatives are shown in Figure 5.36, and themes are summarized in Table 5.8.

5.3   Case-Based Planning of Crystallization Experiments

Once all images are scored and stored in the crystallization data base with information about individual cocktails and protein information, we can start using analogy to plan and optimize crystallization plans for novel proteins, as described in Section 5.3.1.2 and depicted in Figures 5.11 and 5.12.

Crystallization search and optimization phases are planning tasks. A single experiment corresponds to a simple plan and a series of experiments for a given protein corresponds to a more complex plan. Having information from crystallization screens enables a more refined and optimized plan for crystallization.

Planning in artificial intelligence involves techniques for selecting a sequence of actions, and reasoning about their positive and negative influences on the overall goal [469]. In case-based reasoning (CBR), details about experiments (the problem), and the crystallization plans (the solution) are stored. The problem comprises *"a precipitation index"* (PI), i.e., classified images for all 1,536 conditions across all time-points. Since PI measures how each protein reacts across a wide range of conditions, it can be used to objectively measure similarity among proteins, and thus guide case retrieval during CBR. Once the system identifies k similar proteins in the crystallization data base, the system can form a new plan using analogy.

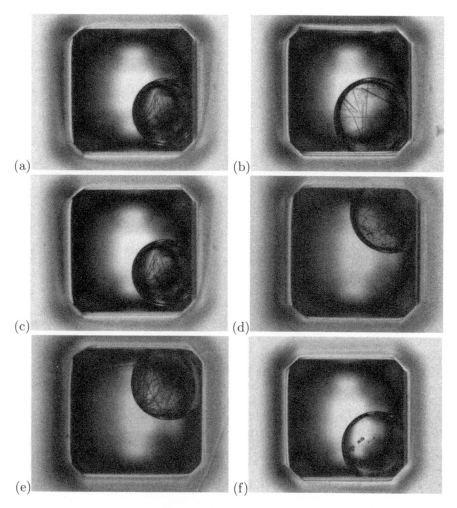

FIGURE 5.36: Several genuine crystals are missed either because they are fine grained or because fail to have straight edges.

CBR is a paradigm where experiences are represented as cases in a case base. The decision-making process finds similarities between the new problem and cases in the case-base, and applies analogical reasoning to reuse stored solutions to solve new problems.[4] The system comprises three main parts:

1. *Crystallography data base* stores information about proteins, cocktails, crystallization method, precipitation index and optimization plan (as shown in Figures 5.6–5.8). This repository of crystal growth experiments addresses the two main shortcomings of the BMCD: it comprises a comprehensive data base of crystallographic experiments and contains both positive and negative experiment outcomes.

2. *CBR system* uses the crystallization data base to plan new experiments. Case-based planning of crystallization experiments requires three main components:

   (a) A repository of past experiments — case-base (see Section 5.3.1);

   (b) A retrieval algorithm that can locate the most similar past experiments to a new protein (see Section 5.3.1.1);

   (c) An adaptation algorithm that can use crystallization plans of the most similar past experiments and domains knowledge to suggest optimization plans for the new protein (see Section 5.3.1.2).

3. *Knowledge discovery* is used to optimize the HTP screen, to remove conditions that provide little or no information, and to optimize crystallization plans by detecting and then applying trends observed across many conditions and protein properties (see Section 5.4).

Case-based planning of protein crystallization experiments uses the precipitation index (PI) of a novel protein as a problem description, compares it to PIs of all proteins stored in the case-base using a k-nearest neighbor similarity algorithm. Crystallization plans of proteins with the most similar PIs are considered for planning crystallization of a novel protein — successful crystallizations are used as a positive planning experience, while failed crystallization plans are used to potentially avoid negative results in the future Figure 5.11.

5.3.1   Crystallization Case Representation

Knowledge in the crystallography planning system is represented as cases and rules. *Cases* store information about individual experiments with diverse crystallization outcomes. We record both successfully crystallized proteins and those that resist crystallization. *Rules* are general principles acquired

---

[4]The CBR process and systems are described in detail in [319]. More recent research directions are presented in [331] and practically oriented descriptions of CBR can be found in [57, 539].

from crystallographers, or principles derived using knowledge–mining tools. Rules are used during analogy-based reasoning, to devise an optimization plan by reusing past successful and failed crystallization experiments.

Cases corresponding to individual experiences are stored in $\mathcal{T}\mathcal{A}3$[5] as a collection of attribute-value pairs. Although any analogy-based reasoning system (or other reasoning system) can be used to support crystallization experiment planning, $\mathcal{T}\mathcal{A}3$ provides specific advantages that makes it suitable for the task; namely, flexibility of k-nearest neighbor retrieval, and scalability of its data management approach. Our approach builds on a flexible computational framework for CBR called $\mathcal{T}\mathcal{A}3$. This system uses a variable-context, similarity-based retrieval algorithm, and a flexible case representation language [279, 282].

A *case* is a representation of a real world experience in a particular representation scheme. A case will be represented as a finite set of attribute/value pairs (descriptors):

$$case = \{\langle attribute_0 : value_0 \rangle, \langle attribute_1 : value_1 \rangle, \ldots, \langle attribute_n : value_n \rangle\}.$$

A collection of cases is called a case-base. Attributes in a case are grouped into categories to bring additional structure to a case representation. Information about the usefulness of individual attributes and information about their properties is used to perform this attribute clustering. Attribute membership in a particular category is defined either using domain knowledge or using a knowledge–discovery algorithm [180, 283]. This grouping allows for ascribing different constraints to different groups of attributes, and, thus, the retrieval process can be described as a constraint satisfaction process [185, 317, 518]. Selectively using individual categories during retrieval reduces the impact of irrelevant attributes on system performance. An example case representation is presented below:

```
class ProteinProperties {
    String      LatinSpeciesName;
    String      Genus;
    int         PNumber;
    String      AminoAcidSequence = {0 - 200000};
                   [Alphabet = 'ARNDBCQEZGHILKMFPSTWYV'];
    choice      Glycosylation = {'none','low','medium','high'};
    choice      Phosphorylation = {'none','low','medium','high'};
    choice      Lipidation = {'none','low','medium','high'};
    real        Retention time on chromatography columns =
                   {1 minute - 120 minutes};
    real        Hydrodynamic radius = {0 - 10000};
```

[5]The name $\mathcal{T}\mathcal{A}3$ (pronounced tah-tree) has been created as a phonetic equivalent to the name of the highest mountain range in Slovakia, Tatry.

```
choice      InternalIons = {'none','Fe2+','...'};
boolean     Oxidation state;
choice      SensitivityToOxygen = {'none','low','med','high'};
choice      SensitivityToPh = {'none','low','med','high'};
choice      SensitivityToProtyalysis = {'none','low',
                'med','high'};
choice      SensitivityToTemperature = {'none','low',
                'med','high'};
choice      ProteinCategory = {'Integral Membrane','Membrane',
                'Associated','Cytostolic'};
String      AdditivesRequiredForStability;
image       SDS-PAGE_Results;
image       PAGE_Results;
real        Activity assay = {?};
choice      StorageState = ('lyophilized','glycerol',
                'salt precipitation'};
real        StorageTemperature = {-100 - 40};
choice      MonomerMultimer = {'A1','A2','A1B1','A2B2',
                'A1B1C',...};
real        MolecularWeightCalculated =
                {1,000 dalton - 1,000,000 dalton};
real        MolecularWeightMeasured =
                {1,000 dalton - 1,000,000 dalton};
real        IsoelectricPointCalculated = {1.0 - 14.0};
real        IsoelectricPointMeasured = {1.00 - 14.00};
...
}
```

5.3.1.1 Retrieval Algorithm

Crystallization experiment retrieval involves partial pattern matching of a precipitation index for a new protein to precipitation indices in the crystallization data base. A similarity function is used to determine ranking of retrieved experiments; experiments with the PIs of minimum distance are considered most relevant, regardless of the screen outcome.

TA3's retrieval algorithm is based on a notion of relevance assessment [279, 471] and on a modified nearest neighbor matching [545]. In general, k-nearest neighbor matching algorithms suffer from a challenge to determine effective value for k, reduced performance in high-dimensional problem domains, and scalability problems for iterative queries. To cope with these challenges, we have modified the algorithm as follows:

- Grouping attributes into categories of different priorities so that different preferences and constraints can be used for individual categories during query relaxation;

- Using an explicit context, a form of bias represented as a set of constraints, during similarity assessment;

- Using an efficient query relaxation algorithm based on incremental context modifications.

$\mathcal{T}\!A\mathbf{3}$ differs from other case-based reasoning systems mainly by combining semantic measures of relevance (or similarity) into a k-nearest neighbor retrieval algorithm, and integrating data base management techniques for improving scalability of retrieval. $\mathcal{T}\!A\mathbf{3}$ achieves retrieval flexibility by using explicitly defined context (i.e., query constraints) in similarity-based retrieval. Although simple `<attribute, value>` representation is frequently sufficient for CBR systems, using a richer representation language, such as *Telos* [399], provides additional flexibility for the retrieval algorithm. For example, treating objects and attributes uniformly simplifies queries and enables easier control of k in k-nearest neighbor algorithm. Flexibility is also beneficial to support the following aspects of $\mathcal{T}\!A\mathbf{3}$:

- Depending on the resources available the system supports variable classification accuracy. This supports the principle of *anytime* algorithms [131, 183], i.e., algorithms that produce correct answer at any time, but using further computational resources may improve the solution. Such algorithms are essential for domains where the system must optimize the use of resources.

- Based on a given task, the system may change the retrieval objective as follows:

  1. *Recall-oriented retrieval* guarantees that all relevant cases are retrieved, although some irrelevant ones may also be included. This corresponds to high sensitivity.

  2. *Precision-oriented retrieval* assures that only relevant cases are retrieved, although some may be missed. This corresponds to high specificity.

- The system supports imprecise query specification. This enables the retrieval of cases even if the query is not specified completely. Moreover, similar cases are retrieved in addition to exact matches.

- Considering the result from the retrieval, the system can either manually or automatically perform query modifications through context restriction or relaxation. Overall, this framework supports query-by-example and query-by-reformulation.

These features are desirable for supporting complex domains and tasks, since the task requirements may change over time and individual users may have different preferences. In addition, the classification process must be efficient, i.e., the system must scale up as more cases are acquired.

Intuitively, we cannot measure similarity if we do not define the context of the comparison. To define k-nearest neighbor retrieval explicitly, we define query constraints, i.e., we define a context for retrieval. The goal is to provide unbiased retrieval but support measure of similarity (relevance) of retrieved cases to the query case. Using constraints, we define context as a parameter of a relevance relation, which maps a case-base onto a set of relevant cases.

A *context* is defined as a finite set of attributes with associated constraints on the attribute values:

$$context = \{\langle attribute_0 : constraint_0 \rangle, \ldots, \langle attribute_k : constraint_k \rangle\},$$

where $attribute_i$ is an attribute name and constraint $constraint_i$ specifies the set of values for $attribute_i$ that are considered a match. A *context* specifies the important attributes and how "close" an attribute value must be in order to satisfy the *context*. A constraint set $constraint_k$ must be a subset of the domain $domain_k$ for an $attribute_k$:

$$constraint_k \subseteq domain_k.$$

Similarity is defined as a relation between cases with respect to a specified *context*. It is defined to supplement equivalence by supporting partial matches. Two cases are considered similar if they satisfy a given *context*. A *case satisfies* a *context*, denoted $sat(case, context)$, if and only if for all pairs $\langle attribute_i : constraint_i \rangle \in context$, there exists a pair $\langle attribute_i : value_i \rangle \in case$ such that $value_i$ is in $constraint_i$:

$$sat(case, context) \; iff$$

$$\forall attribute_i \langle attribute_i : constraint_i \rangle \in context \;\rightarrow$$

$$\exists value_i \langle attribute_i : value_i \rangle \in case \;\wedge\; value_i \in contraint_i.$$

We say that the *context* is *satisfiable* if and only if there exists at least one case in a case base that satisfies it:

$$context \; is \; satisfiable \; iff \; \exists case \in CaseBase \; : \; sat(case, context).$$

Using the context, the input problem and cases in a case-base are interpreted and their similarity is assessed. A $case_1$ is *similar* to a $case_2$ with respect to a given *context*, denoted: $case_1 \sim_{context} case_2$, if and only if both $case_1$ and $case_2$ satisfy the *context*:

$$case_1 \sim_{context} case_2 \;\leftrightarrow\; sat(case_1, context) \wedge sat(case_2, context).$$

For the purpose of case retrieval, the similarity relation maps a *context* and a *CaseBase* onto the set of similar cases *SimilarCases* in the *CaseBase* that satisfy the *context*. Given a *CaseBase* and a *context*, the function returns a nonempty set of similar cases: $SimilarCases \subset CaseBase$, if and only if all cases in *SimilarCases* are similar to each other with respect to *context*, i.e., if they satisfy the given *context*. The retrieval function is complete in the sense that it returns all relevant cases and only relevant cases:

Input:	*context, CaseBase*
Output:	*SimilarCases*
Description:	$retrieve\,(context, CaseBase) = SimilarCases;$
	$\forall case \in CaseBase, case \in CaseBase\ \ iff\ sat(case, context)$

Explicitly defined context is then used to control the closeness of retrieved cases. If too many or too few relevant cases are retrieved using the initial context, then the system automatically transforms the context. There are two possible transformations: *relaxation* — to enable retrieving more cases and *restriction* — to allow for retrieving fewer cases. These context transformations are a foundation for supporting iterative retrieval and browsing.

A $context_1$ is a *relaxation* of a $context_2$, denoted $context_1 \succ context_2$, if and only if the set of attributes for $context_1$ is a subset of the set of attributes for $context_2$ and for all attributes in $context_1$, the set of constraints in $context_2$ is a subset of the constraints in $context_1$. As well, $context_1$ and $context_2$ are not equal:

$$context_1 \succ context_2\ iff$$

$$\forall \langle attribute_i : constraint_i \rangle \in context_1,$$

$$\exists \langle attribute_i : constraint_j \rangle \in context_2 \ :$$

$$constraint_i \supseteq constraint_j\ \wedge\ context_1 \neq context_2.$$

A $context_1$ is a *restriction* of a $context_2$, denoted $context_1 \prec context_2$, if and only if $context_2$ is a relaxation of $context_1$:

$$context_1 \prec context_2\ iff\ context_2 \succ context_1.$$

There are two possible implementations of context relaxation: reduction and generalization:

1. *Reduction* removes constraints by reducing the number of attributes required to match. Given *m_of_n* matching, the required number of attributes is reduced from m to p, where $0 < p < m \leq n$.

2. *Generalization* is a context transformation that relaxes the context by enlarging the set of allowable values for an attribute (see Figure 5.37).

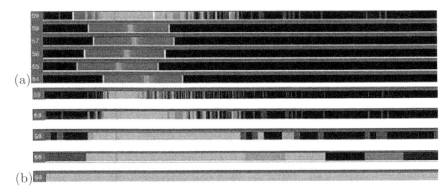

FIGURE 5.37: (See color insert following page 72.) One of the methods for context relaxation is generalization. Iterative generalization results in extending interval of values that are considered a match (a). (b) depicts a hierarchy of one relaxed attribute that was obtained by iteratively generalizing the attribute values.

Figure 5.38 presents partial representation of an input case that is used to specify context, and value relaxation that is used during case retrieval.

Analogically, there are two possible implementations of context restriction — *expansion* and *specialization*. *Expansion* is a context transformation that strengthens constraints by enlarging the number of attributes required to match: given m_of_n matching, the required number of attributes is increased from m to p, where $0 \leq m < p \leq n$. Thus, expansion is the inverse operation to reduction. *Specialization* strengthens constraints by removing values from a constraint set for an attribute. This may lead to a decreased number of cases that satisfy the resulting context.

As pointed out by Gaasterland in [185], the relaxation technique can advantageously be used for returning answers to a specific query as well as returning related answers. Without such automatic relaxation, users would need to submit alternative queries. The restriction technique works analogously, but is used mainly for controlling the amount of returned information, keeping it in manageable levels for the user. Because the search for relaxed or restricted contexts can be infinite, there must be a mechanism for controlling it, either by user intervention via user preferences or by other means.

Context relaxation and restriction are used during retrieval to control the quantity and quality, or closeness, of cases considered relevant. Thus, when modifying the context, the system may return an approximate answer quickly or may spend more resources getting an accurate answer. An approximate answer can be iteratively improved, so that the change between an approximate and an accurate answer is continuous.

If $context_1$ is a relaxation of $context_2$ then all cases that satisfy $context_2$ also satisfy $context_1$:

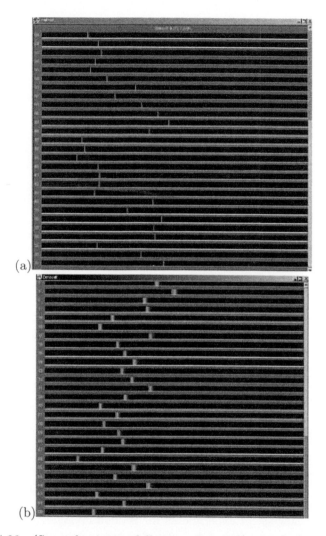

(a)

(b)

FIGURE 5.38: (See color insert following page 72.) An input case is used to specify the initial context (a). If the initial retrieval does not return any similar cases, the algorithm relaxes value constraints for individual attributes, resulting in context relaxation (b).

$$\forall case \in CaseBase:$$

$$(sat(case, context_2) \wedge context_1 \succ context_2) \rightarrow sat(case, context_1).$$

If $context_1$ is a restriction of $context_2$ then the set of cases that satisfy $context_1$ is a subset of the set of cases that satisfy $context_2$.

$$\forall case \in CaseBase:$$

$$(sat(case, context_1) \wedge context_1 \prec context_2 \rightarrow sat(case, context_2)).$$

Monotonicity is an important property of a flexible retrieval system and can advantageously be used in case-based classification — the more constraints are specified, the fewer cases satisfy it. Thus, the retrieval function based on context-based similarity assessment is monotonic. Given a *CaseBase*, $context_1$ and $context_2$:

$$(context_1 \succ context_2)$$

$$\rightarrow retrieve(context_1, CaseBase) \supset retrieve(context_2, CaseBase);$$

$$(context_1 \prec context_2)$$

$$\rightarrow retrieve(context_1, CaseBase) \subset retrieve(context_2, CaseBase).$$

Monotonicity feature is useful during the retrieval process. On the one hand, if not enough cases have been retrieved from the case-base, the initial context may be relaxed to retrieve additional "less" similar cases. On the other hand, if too many cases are retrieved, the initial context may be restricted to lower the number of retrieved cases. Restricting the context increases the similarity between cases since more attributes are required to match.

5.3.1.2 Crystallization Plan Optimization and Visualization

Once similar cases have been retrieved, the next step in case-based planning of crystallization experiment is adaptation. This is the process of modifying previous plans to fit the new protein. The most relevant retrieved cases, along with domain knowledge, are incorporated to determine appropriate parameters for a new set of experiments for protein crystallization. The most similar proteins are considered for planning crystallization conditions for a new protein. The system constructs a solution for the current crystallization problem by using appropriate descriptors from relevant experiments, i.e., both successful and failed crystallization experiments of proteins that have similar precipitation indices. If no protein is sufficiently similar to a novel protein, then standard crystallization optimization is used, using information from the screening outcome. The solution is a recipe for crystallization: crystal growth method, temperature and pH ranges, concentration of protein, and crystallization agent. Once a novel set of experiments for a protein has been planned, executed and the results recorded, a new case, which reflects this new experience, is added to the case-base. Cases with both positive and negative outcomes are equally valuable for future decision-making processes and for the application of machine-learning techniques to the case-base.

To support experiment planning, we define a case-based classification algorithm as follows [279]. The input to a classification problem comprises a *CaseBase*, a *context* and a problem $case_p$ (also called an input case, i.e., a case without a solution or appropriate class). An output is a class $class_k$ for a problem $case_p$, defined as follows:

$$(classify(case_p) \rightarrow \langle case_p, class_k \rangle) \leftrightarrow$$

$$(\exists case_r \in CaseBase \; : case_p \sim_{context} case_r \; \wedge \; \langle case_r, class_k \rangle),$$

where $\langle case_r, class_k \rangle$ denotes $class_k$ is the class for $case_r$. The system uses results from the precipitation index to suggest 80% of the solution, while 20% of the solution is determined from the knowledge obtained by data mining the repository for general trends.

To aid in planning, we also visualize the screen results, and the process of planning the optimization screen. To visualize time dependency of the crystallization process and to begin the process of crystallization plan optimization, we show precipitation index by projecting individual results to a color scheme: white squares denote crystals, green squares represent precipitates, black squares show clear drops, and red squares are unknowns (see Figure 5.39). Considering that each protein crystallization results in $9,216$ images, the main goal of the visualization system is to provide easier navigation through experimental results.

Selecting a specific well displays cocktail information and view of screen results for individual time points (see Figure 5.40). Crystallization results are automatically (or interactively) selected for optimization, using selected crystallizing conditions and optimization criteria (see Figure 5.41). At this stage the cocktail setup is optimized into 24-well plate optimization screen (see Figure 5.42). We combine information from the successful/failed crystallizations in a given screen with historical patterns obtained from data mining to plan novel experiments. This effort can be described as a combination of traditional crystallization optimization techniques with machine learning approaches.

5.4   Protein Crystallization Knowledge Discovery

To optimize crystallization screen selection and to aid crystallization optimization planning, we can apply knowledge–discovery algorithms. Knowledge discovery is the process of extracting novel, useful, understandable and usable

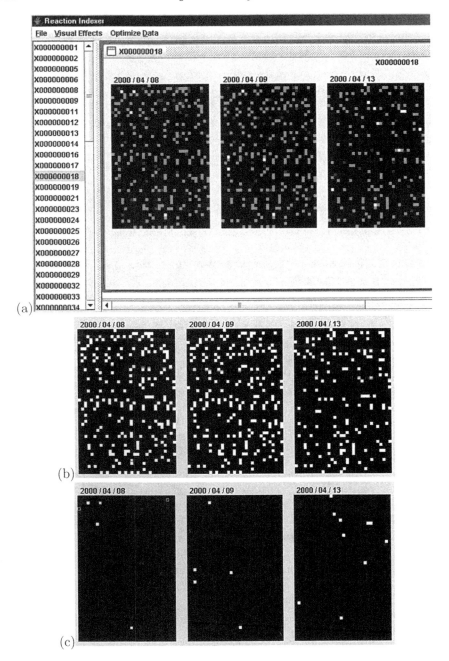

FIGURE 5.39: (See color insert following page 72.) Visualization and optimization of the initial crystallization plan. White squares denote crystals, green squares show precipitates, black squares represent clear drops, and red squares are unknowns. (a) depicts full view, (b) shows precipitates and crystal hits, and (c) highlights only crystals.

X000000018 □ ☒

File

	Reactions	Well #	Cocktail #	Chemical Additive	Chemical Formula	Concentration	Buffer Type	Buffer Concentration	pH
▣ ▣ ☐ ▣ ▣		13	2_C0013	Ammonium chloride	NH4Cl	2.5M	TAPS	0.1 M	9
☐ ☐ ▣ ▣ ▣		72	2_C0072	Lithium bromide	LiBr	1.87M	TAPS	0.1 M	9
▣ ▣ ☐ ▣ ▣		272	2_C0272	Ammonium thiocyanate	NH4SCN	2.7M	CAPS	0.1 M	10

FIGURE 5.40: (See color insert following page 72.) The panel displays time course data and crystallization conditions for selected wells.

Optimization □ ☒

File   Optimization Options   Probabilistic Prediction

Protein Name	Time (days)	Reactions	Well #	Cocktail #	Chemical Addi..	Chemical For...	Concentration	Buffer Type	Buffer Concen..	pH
X000000018	Sat Apr 08 00:... 1		68	2_C0068	Lithium bromi...	LiBr	3.73M	MOPS	0.1 M	7
X000000018	Sat Apr 08 00:... 1		72	2_C0072	Lithium bromi...	LiBr	1.87M	TAPS	0.1 M	9
X000000018	Sun Apr 09 12... 2		72	2_C0072	Lithium bromi...	LiBr	1.87M	TAPS	0.1 M	9
X000000018	Thu Apr 13 02:... 3		13	2_C0013	Ammonium c...	NH4Cl	2.5M	TAPS	0.1 M	9
X000000018	Thu Apr 13 02:... 3		206	2_C0206	Sodium nitrate	NaNO3	2.6M	TAPS	0.1 M	9
X000000018	Thu Apr 13 02:... 3		272	2_C0272	Ammonium th...	NH4SCN	2.7M	CAPS	0.1 M	10

Protein Name	Time (days)	Reactions	Well #	Cocktail #	Chemical A...	Chemical F...	Concentration	Buffer Type	Buffer Conc..	pH	PEG	PEG % (w/v)
X000000018	Sat Apr 08 ... 1		295	2_C0295	Ammonium...	NH4NO3	0.1 M	MES	0.1 M	6	20000	20
X000000018	Sun Apr 09 ... 2		771	2_C0771	Potassium ...	KSCN	0.1 M	Tris	0.1 M	8	4000	40
X000000018	Sun Apr 09 ... 2		815	2_C0815	Ammonium...	NH4Cl	0.1 M	Acetate	0.1 M	5	1000	20
X000000018	Sun Apr 09 ... 2		899	2_C0899	Magnesium...	Mg(NO3)2*...	0.1 M	MOPS	0.1 M	7	1000	20
X000000018	Thu Apr 13 ... 3		313	2_C0313	Lithium chl...	LiCl	0.1 M	CAPS	0.1 M	10	20000	20
X000000018	Thu Apr 13 ... 3		314	2_C0314	Lithium chl...	LiCl	0.1 M	MES	0.1 M	6	20000	20
X000000018	Thu Apr 13 ... 3		433	2_C0433	Sodium nitr...	NaNO3	0.1 M	HEPES	0.1 M	7.5	20000	40
X000000018	Thu Apr 13 ... 3		511	2_C0511	Sodium chl...	NaCl	0.1 M	HEPES	0.1 M	7.5	8000	20
X000000018	Thu Apr 13 ... 3		728	2_C0728	Ammonium...	NH4Br	0.1 M	Citrate	0.1 M	4	4000	40

Protein Name	Time (days)	Reactions	Well #	Cocktail #	Commercial C...	Chemical Addi...	Concentration	Chemical Addi...	Concentration	pH
X000000018	Sat Apr 08 00:... 1		1394	2_C1394	HR Quik Scre...	Sodium dihydr...	0.032 M	di-Potassium...	0.766 M	8.2
X000000018	Sun Apr 09 12... 2		1394	2_C1394	HR Quik Scre...	Sodium dihydr...	0.032 M	di-Potassium...	0.766 M	8.2

Protein Name	Time (days)	Reactions	Well #	Cocktail #	Chemical A...	Chemical F...	Concentration	Buffer Type	Buffer Conc..	pH	PEG	PEG % (w/v)
X000000018	Thu Apr 13 ... 3		1122	2_C1122	Potassium ...	KCl	0.1 M	TAPS	0.1 M	9	400	80

Protein Name	Time (days)	Reactions	Well #	Cocktail #	Chemical A...	Chemical F...	Concentration	Buffer Type	Buffer Conc.	pH	MPD	MPD % (w/v)
X000000018	Thu Apr 13 ... 3		1296	2_C1296	Potassium ...	KNO3	0.1 M	Acetate	0.1 M	5	MPD	60

FIGURE 5.41: Selected experiments for the optimization. One can identify both the time dependency of crystallization, as well as view the details about crystallizing conditions.

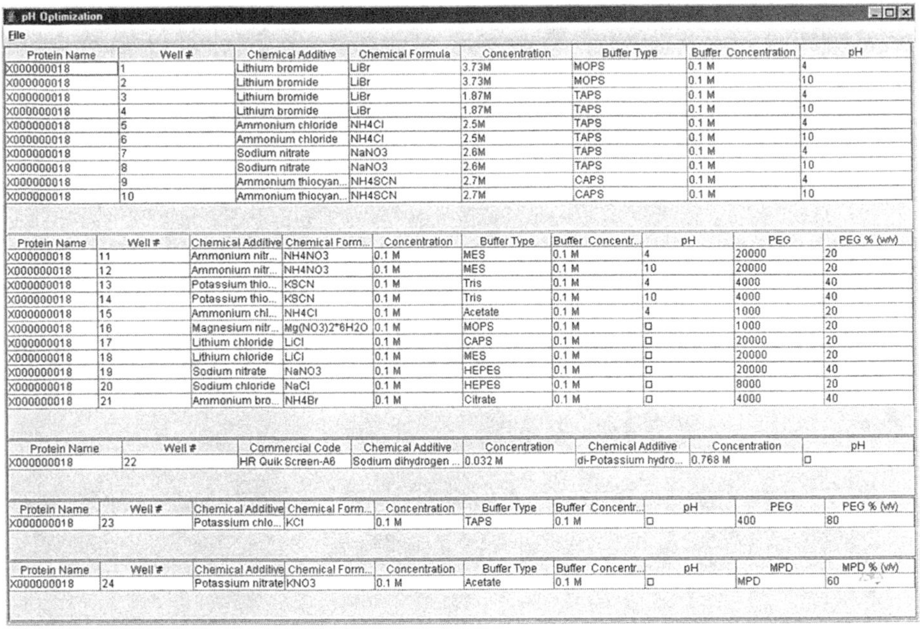

pH Optimization

File

Protein Name	Well #	Chemical Additive	Chemical Formula	Concentration	Buffer Type	Buffer Concentration	pH
X000000018	1	Lithium bromide	LiBr	3.73M	MOPS	0.1 M	4
X000000018	2	Lithium bromide	LiBr	3.73M	MOPS	0.1 M	10
X000000018	3	Lithium bromide	LiBr	1.87M	TAPS	0.1 M	4
X000000018	4	Lithium bromide	LiBr	1.87M	TAPS	0.1 M	10
X000000018	5	Ammonium chloride	NH4Cl	2.5M	TAPS	0.1 M	4
X000000018	6	Ammonium chloride	NH4Cl	2.5M	TAPS	0.1 M	10
X000000018	7	Sodium nitrate	NaNO3	2.6M	TAPS	0.1 M	4
X000000018	8	Sodium nitrate	NaNO3	2.6M	TAPS	0.1 M	10
X000000018	9	Ammonium thiocyan...	NH4SCN	2.7M	CAPS	0.1 M	4
X000000018	10	Ammonium thiocyan...	NH4SCN	2.7M	CAPS	0.1 M	10

Protein Name	Well #	Chemical Additive	Chemical Form...	Concentration	Buffer Type	Buffer Concentr.	pH	PEG	PEG % (w/v)
X000000018	11	Ammonium nitr...	NH4NO3	0.1 M	MES	0.1 M	4	20000	20
X000000018	12	Ammonium nitr...	NH4NO3	0.1 M	MES	0.1 M	10	20000	20
X000000018	13	Potassium thio...	KSCN	0.1 M	Tris	0.1 M	4	4000	40
X000000018	14	Potassium thio...	KSCN	0.1 M	Tris	0.1 M	10	4000	40
X000000018	15	Ammonium chl...	NH4Cl	0.1 M	Acetate	0.1 M	4	1000	20
X000000018	16	Magnesium nitr...	Mg(NO3)2*6H2O	0.1 M	MOPS	0.1 M	□	1000	20
X000000018	17	Lithium chloride	LiCl	0.1 M	CAPS	0.1 M	□	20000	20
X000000018	18	Lithium chloride	LiCl	0.1 M	MES	0.1 M	□	20000	20
X000000018	19	Sodium nitrate	NaNO3	0.1 M	HEPES	0.1 M	□	20000	40
X000000018	20	Sodium chloride	NaCl	0.1 M	HEPES	0.1 M	□	8000	20
X000000018	21	Ammonium bro...	NH4Br	0.1 M	Citrate	0.1 M	□	4000	40

Protein Name	Well #	Commercial Code	Chemical Additive	Concentration	Chemical Additive	Concentration	pH
X000000018	22	HR Quik Screen-A6	Sodium dihydrogen ...	0.032 M	di-Potassium hydro...	0.768 M	□

Protein Name	Well #	Chemical Additive	Chemical Form...	Concentration	Buffer Type	Buffer Concentr.	pH	PEG	PEG % (w/v)
X000000018	23	Potassium chlo...	KCl	0.1 M	TAPS	0.1 M	□	400	80

Protein Name	Well #	Chemical Additive	Chemical Form...	Concentration	Buffer Type	Buffer Concentr.	pH	MPD	MPD % (w/v)
X000000018	24	Potassium nitrate	KNO3	0.1 M	Acetate	0.1 M	□	MPD	60

FIGURE 5.42: 24-well plate optimization screen for the selected conditions using specified optimization criteria.

information from large data sets. A data base of protein crystallization outcomes has three parts (see Figure 5.43):

1. *Protein dimension*, which stores physicochemical properties of proteins. Finding patterns in this dimension enables us to group proteins based on diverse measures of similarity, e.g., precipitation index, amino acid sequence homology, function, structure, etc.

2. *Crystallization condition dimension*, which provides ingredients and properties of individual crystallization conditions used in the screen. Analyzing the properties of cocktails enables us to identify cocktail components that are supportive of the crystallization process, or those conditions that never (or extremely rarely) result in useful outcome.

3. *Precipitation index*, which represents cocktail outcomes for individual proteins.

When a time axis is incorporated, the protein-cocktail matrix becomes a three-dimensional tensor: protein × cocktail × time. The elements of this tensor may be binary (crystal/no crystal), multinomial (crystal/precipitate/clear/...), or more complex representations of experimental

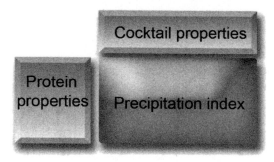

FIGURE 5.43: Protein crystallization outcome data base comprises three main parts: physicochemical protein properties, ingredients and properties of crystallization conditions, and precipitation index.

outcome (real- or binary-valued vector of crystal, precipitate, clear, etc., outcomes).

The precipitation index by itself can answer important questions, such as:

- Which proteins crystallize under similar conditions? Which cocktails are redundant? Correlation analysis, association mining, or clustering may be used effectively to obtain such information. Figure 5.44 shows two clusters of proteins, represented by self-organizing maps [316, 514]. Intuitively, similar color patterns correspond to similar crystallization results, i.e., crystallization occurs under similar conditions, and similarity is also achieved for other outcomes. A specific area within the component plane represents the same cluster of cocktails across all proteins in the study. The color renders individual outcomes, ranging from red for crystal to blue for clear.

- How can we minimize a crystallization screen, while maximizing its effectiveness? What is the minimal screen that produces a crystal in at least $n\%$ of proteins crystallizable in the larger screen? Kimber *et al.* computed a set of minimal screens using a combination of exhaustive search of all possible six-cocktail sub-screens, and a greedy algorithm combining theses sub-screens into size-12 and -24 sub-screens [308].

- Can the crystallization matrix be explained by a small number of underlying hidden variables? Does each protein respond uniquely and independently to each cocktail? This question may be answered by applying singular value decomposition to factor the $m \times n$ precipitation matrix into an $m \times k$ protein matrix, a $k \times k$ singular value matrix, and a $k \times n$ cocktail matrix. A singular value decomposition concisely describes each protein and each cocktail in terms of the strength of their interactions with a set of k hidden variables.

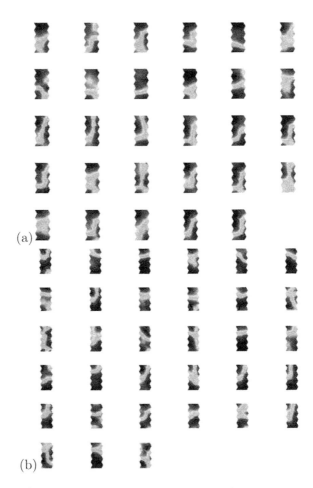

FIGURE 5.44: (See color insert following page 72.) Two clusters of proteins ((a) and (b)) show different profiles across clusters, but homogeneity within clusters. Self-organizing maps are used to render the similarity among the proteins. A specific area within the component plane represents the same cluster of cocktails across all proteins in the study. The color renders individual outcomes, ranging from blue that represents clear, through yellow that corresponds to precipitate, to red that corresponds to crystals.

When the precipitation index is supplemented with protein and cocktail data, more complex questions may be answered:

- What proteins crystallize under a common set of chemical conditions?

- What cocktails consistently cause crystallization reactions for proteins with certain physicochemical characteristics?

- What patterns can we find that link protein properties with cocktail properties in successful crystallization trials?

Quantitative frequent-itemset and association-rule mining answer these questions, e.g., Gopalakrishnan *et al.* [210] and Cumbaa and Jurisica [126]. Kantarjieff *et al.* [291] used correlation and basic statistics to study the relationship between pI and pH of successful protein-cocktail combinations. Transactions in the crystallization data base correspond to facts about proteins, cocktails, and particular experiments, e.g.:

```
Cocktail #222, pH < 6.0,
contains: PEG 20000, CAPS,
crystallizes proteins: glucose isomerase,
  Q8U2K6 (N-type ATP pyrophosphatase).
```

An *itemset* is a subset of a transaction. Frequent-itemset discovery [207] finds all itemsets in a data base whose frequency or support exceeds some defined minimum. An *association rule* describes the co-occurrence of two disjoint itemsets (i.e., antecedent $\Rightarrow$ consequent) in the transactions in the data base [7], such as:

```
{protein concentration >10 mg/mL,
   medium molecular weight, low pI,
   organism: A. aeolicus,
   cocktail contains CaCl2*2H2O}
=> {crystal}.
```

Clustering can be used to identify similarly behaving cocktails (see Figure 5.45(a)) and proteins that crystallize under similar conditions (see Figure 5.45(b)).

5.5 Conclusions

X-ray crystal structure determination is a linear process. The protein is must first be purified, screened, and crystallized. Second, x-ray diffraction

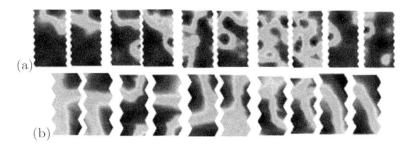

FIGURE 5.45: (See color insert following page 72.) Clustering cocktails identifies both similarities and differences among individual conditions (a). Analogously, one can cluster proteins based on their propensity to crystallize under the same conditions (b).

data can then be collected, followed by structure determination and refinement. Each step must be completed before the next can begin. Solving a bottleneck in this pipeline usually creates a bottleneck in the next stage.

Robotic systems now routinely handle most phases of protein crystallization experiments: cocktail preparation, cocktail and protein solution mixing in sub-microliter volumes, and imaging of each experiment. Typically, this pipeline ends with a human expert manually assessing an image from each experiment photograph for a positive crystallization outcome. Thus, the screening bottleneck is now replaced by an image classification bottleneck.

One of the current challenges in the field is to replace the expert with a high-accuracy automatic image-classification system (e.g., [19, 58, 126, 127, 285, 498, 551, 576, 577]). Another challenge is to discover the factors that effect crystallization of individual proteins under diverse chemical conditions. Albeit higher throughput means more proteins can be screened and eventually crystallized, related challenges are to prioritize proteins that are of high importance and keep track of all required information. This also relates to data management and quality assurance.

HTP protein crystallization screens combined with automated image classification, reasoning and data mining can diminish screening as a major bottleneck, and decrease the time required to determine crystallization conditions, and thus potentially advance the structural studies for a growing number of research projects. Introducing automated image analysis and classification achieves two important goals: (1) it improves throughput, and (2) generates consistent and objective results. Achieving these goals enables us to use the screen to identify crystals in the HTP screen so optimization strategies can be followed. Further, we can feed these results to data mining and reasoning algorithms. Case-based planning can improve crystallization optimization planning. Data mining can integrate results from HTP screens with information about crystallizing conditions, intrinsic protein properties, and results from crystallization optimization.

Chapter 6

Integration of Diverse Data, Algorithms, and Domains

Max Kotlyar and Kevin Brown

High-throughput (HTP) methods are providing a comprehensive description of cells at several levels (see Figure 6.1). HTP sequencing has deciphered the genomes of multiple species and DNA microarrays have provided extensive information about several transcriptomes. More recently, methods have been developed to analyze the proteome and interactome. In yeast, several HTP methods have been used to determine the expression levels [196, 500] and subcellular localization of proteins [256, 325]. The interactome of yeast is being determined by HTP methods such as yeast 2-hybrid [261, 526], tandem affinity precipitation (TAP) [190], and HTP mass spectrometric protein complex identification (HMS-PCI) [247].

Although HTP methods have provided a large amount of data at the genomic and proteomic levels, our understanding of these levels remains limited. At the genome level an important unresolved issue is the identification of individual genes. At the transcriptome level one of the main questions is how gene expression is regulated. The same question can be asked about the proteome, but currently there is insufficient data to address it. At the level of the interactome the main question is how to reliably assign interaction partners to proteins and how to measure the direction, affinity and context of those interactions. Although there are several HTP methods for determining protein–protein interactions (PPIs), all of them have limited accuracy. Also, a number of questions occur across all levels — primarily, how information about one particular level is related to phenotype and to other levels.

One of the obstacles to answering any of these questions is the low overlap of HTP results. Regardless of the domain, different HTP platforms provide different biases and different coverage of genomes. Similarly, different computational analysis generates different results. Thus, results from many existing studies have little overlap [383, 547]. This does not necessarily mean that all nonoverlapping results are false positives, as under-representation due to sampling is a major issue.

This chapter focuses on addressing these issues at the interactome level, but the same issues are faced by all of the systems biology approaches discussed in Chapter 7. It is clear that no single database or algorithm will be successful

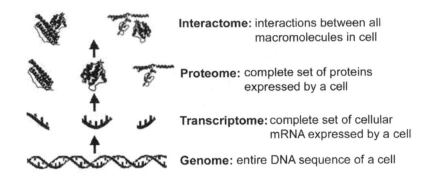

FIGURE 6.1:   Cellular contents are commonly divided into distinct subpopulations, such as the genome, transcriptome, proteome, and interactome [172]. These divisions correspond to levels of information transfer in a cell.

at solving the complex analytical problems at the interactome level. Thus, we need to integrate different tools and approaches, multiple sources of data, and diverse data types. Along with this integration, we need to further develop tools for the management and analysis of the resulting underlying datasets.

A large variety of data types and integration methods have been used to predict PPIs. We first review relevant public data repositories, databases and basic tools for data integration and integrated data analysis. We then review machine-learning approaches to data integration, and discuss their strengths and limitations.

6.1   Integration of Data and Tools

Bioinformatics has provided a wealth of opportunities to traditional biological laboratories, bringing with it novel challenges and benefits through the integration of disparate data sources. Whereas the traditional biologist could manually compare their gene- or protein-of-interest across databases, visually inspect results, and select best matching records accordingly, HTP analysis requires batch processing of large datasets, and optimization of queries across distributed data sources with high accuracy and precision. Unfortunately, databases have not traditionally been designed with interoperability in mind, nor were the original biological databases set up to handle cross-database queries.

While these challenges arise from the technical aspects of databases, further challenges arise from the biological design of HTP platforms. For instance, though DNA microarray technology allows us to probe entire genomes for expression levels across many experimental conditions, annotation errors, noise, and proprietary chip designs make integrating individual experiments a nontrivial task.

Data integration is helpful for addressing questions about all levels of information in a cell. In some cases, a single type of data cannot fully answer a biological question. For example, the problem of finding coregulated genes can be partially addressed by searching for genes with correlated expression profiles. However, certain genes can have similar profiles while being controlled by different regulatory networks [39]. To account for such cases we may locate not only genes that have a similar expression profile but also a common set of transcriptional regulators, identified by DNA binding sites [39].

Data integration can also improve the reliability of HTP results. For example, HTP methods that determine PPIs currently identify many false positives. However, results found by any two methods have a much lower false-positive rate since the chances of random or systematic errors are lower [531].

A wide variety of approaches have been used for integrating HTP biological data, including simple union, Bayesian frameworks, and meta-analysis.

The simplest approach consists of taking the union or intersection of results from different HTP methods [362, 531]. The obvious caveats are the potential exclusion of true positives, and unfair equal voting from methods with different error rates and systematic biases.

One of the common approaches to integration is a Bayesian framework. This framework determines the probability of a hypothesis based on a set of evidence. For example, one can determine the probability of two proteins interacting based on a set of evidence that includes gene expression data, gene functional annotations, and results from HTP interaction detection methods [268].

An approach called meta-analysis has been used by a number of studies to predict disease state from multiple gene expression datasets [393, 459].

Many studies have used machine-learning and data-mining approaches such as support vector machines [67], neural networks [573] and association mining [427, 123]. In addition, some approaches are unique to individual studies and were developed for integrating specific types of data [39, 116]. Table 6.1 summarizes some of the common approaches for integrating HTP biological data, and we discuss them further in Section 6.2.

6.1.1 Public Data Repositories

The modern bioinformatics era has witnessed a paradigm shift in the collection and dissemination of data. The need for public data repositories has begun to pervade every recess of biology, and, following the lead of the computer science industry, biologists have begun to adopt industry-wide standards. We will present many of the public biological databases, as well as standards that will aid in the sharing of new experimental data in the years to come.

TABLE 6.1: Common approaches for integrating HTP data.

Problem Addressed	Data Types	Method	Studies
Identification of genes	output from gene prediction systems	Bayesian	Pavlovic 2003 [435]
Identification of transcription factor binding motifs	sequence and mRNA data	specific to study	Conlon 2003 [116]
Identification of coregulated genes and their regulators	mRNA data	Bayesian	Segal 2003 [480]
	DNA binding sites and mRNA data	specific to study	Bar-Joseph 2003 [39]
	mRNA data	machine learning	Creighton 2003 [123]
Identification of gene function	mRNA data, transcription factor binding sites, functional annotation of genes	Bayesian	Segal 2001 [481]
	mRNA data and phylogenetic profiles	machine learning	Pavlidis 2003 [434]
	mRNA data, transcription factor binding sites, physical and genetic interaction data	Bayesian	Troyanskaya 2003 [522]
	predictions of correlated evolution, gene expression profiles, gene fusion events	union and	Marcotte 1999 [362]
Identification of gene regulatory networks	mRNA data	Bayesian	Friedman 2000 [182]
Identification of PPIs	physical interaction data, mRNA data, gene function, process annotation, gene essentiality	Bayesian	Jansen 2003 [268]
	gene function, localization, process annotation, sequence motifs	machine learning	Oyama 2002 [427]
	topology of PPIs, physical interaction data	Bayesian	Goldberg 2003 [208]
	physical and genetic interaction data, correlation of gene expression profiles	union and intersection	von Mering 2002 [531]
	sequence data, properties of amino acids	machine learning	Bock 2001 [67]
Identification of disease state	mRNA data	machine learning	Crimins 2003 [124], Berrar 2003 [59]
	mRNA data	meta-analysis	Rhodes 2002 [459], Moreau 2003 [393]

6.1.1.1 Nucleic Acid Databases

The creation of the Human Genome Project (HGP) brought about an over-whelming amount of sequence data that had to be stored, managed, and analyzed. However, even before this time, researchers had identified and collected vast amounts of sequence data, which was deposited in the NCBI's Genbank database[1], EBI's EMBL-Bank[2], and the DNA Databank of Japan (DDBJ)[3] [154, 515, 556]. The vast number of individual efforts resulted in a large amount of redundancy in the sequence data collected. Additionally, noncontiguous sequences that were deposited did not represent complete gene sequences.

Computational genomics efforts have been of the utmost importance to analyze not only this historical data, but also the genome-wide data provided by the HGP. The identification of individual human genes and their corresponding sequences has been an ongoing challenge. The Unigene database[4] [546] continually aligns the sequences from the Genbank database into contiguous, gene-centric clusters, and adds information about chromosomal location, protein similarities, and expression information, along with the aligned gene sequence.

While Unigene provides a minimal amount of annotation about each clustered sequence, it is NCBI's Gene database that provides a fully annotated gene-centric record, including aliases, annotated functional information, literature surveys, Gene Ontology (GO) annotation [26, 234], and cross-references to other resources [356]. In a similar effort, the European Molecular Biology Laboratory (EMBL) provides the Ensembl database[5] as a comprehensive, integrated source of annotation of large genome sequences, which is based upon the EMBL-Bank database [254].

Although each of these web-based resources has provided tremendous information to the biologist about specific genes-of-interest, they have been less effective for HTP biology as they are not suitable for systematic computational analyses. The traditional biologist is interested in the most detailed description of an individual gene or small set of genes, complete with previous research findings, in a visually appealing and easy-to-read and searchable format. However, computational analyses require much more data about the entire collection of genes in a machine-readable format. To this end, each of NCBI's databases (Genbank, Unigene, Gene, etc.) are available through flat-file downloads. These can be parsed easily, and loaded into relational database systems. The eUtils[6] provided through NCBI also provide web-based appli-

---

[1]http://www.ncbi.nlm.nih.gov/Genbank/
[2]http://www.ebi.ac.uk/embl/
[3]http://www.ddbj.nig.ac.jp/
[4]http://www.ncbi.nlm.nih.gov/entrez/query.fcgi?db=unigene
[5]http://www.ebi.ac.uk/ensembl/
[6]http://eutils.ncbi.nlm.nih.gov/entrez/query/static/eutils_help.htm

cation program interface (API) to the databases. EMBL has also provided an API to access the wealth of information stored in Ensembl, including individual sequences and their translations.

6.1.1.2 Protein Databases

The main repository for complete protein information is UniProt[7], which is provided by the EBI. UniProt is the unification of the SwissProt protein database, TrEMBL, and the PIR-PSD databases [23]. SwissProt is a manually curated database of proteins, while TrEMBL is the translation of the DNA sequences in EMBL-Bank. PIR-PSD was an annotated resource formerly provided by the Protein Information Resource group at Georgetown University Medical Center. Together, UniProt currently provides the largest, minimally redundant collection of annotated protein information in the world [23].

The UniProt database, while web-accessible for the biologist, is also provided as a downloadable flat-file and in a more structured XML format. Complete protein sequences can be downloaded in FASTA format as well. These repositories are incrementally updated daily, with major builds provided quarterly.

UniProt has provided a rich resource to bioinformatics analysis through its cross-referencing to other datasources. For instance, a UniProt record provides links to PubMed records describing the identification and characterization of the protein, automatic and manual annotation using GO terms, and domain information from InterPro. It also provides cross-references to Genbank/EMBL protein and nucleotide sequences, which are useful to link to other NCBI and EBI resources.

6.1.1.3 Protein–Protein Interaction Databases

6.1.1.3.1 Known PPI databases. As noted in Chapter 4, there are many publicly available sources for protein–protein interaction data, both experimentally determined, and computationally predicted. Some are focused on a single organism, while others are multiorganism. Some of the larger databases consisting of known protein–protein interactions (PPIs) are listed in Table 6.2; however, for a more complete review of the current biological databases, see the annual database issue of *Nucleic Acids Research*[8], published as the first issue of each calendar year.

The databases listed in Table 6.2 are largely manually curated, using either paid curation teams or volunteer molecular experts. The goal of the projects listed is to provide centralized access to the historical protein interaction data that has been published and is covered in PubMed. Without such curation,

---

[7]http://www.uniprot.org/

[8]http://nar.oupjournals.org/

TABLE 6.2: Curated protein–protein interaction databases: BIND (http://www.bind.ca; [29]), DIP (http://dip.doe-mbi.ucla.edu; [561]), HPRD (http://www.hprd.org; [440]), IntAct (http://www.ebi.ac.uk/intact; [245]), MINT (http://mint.bio.uniroma2.it/mint/; [568]), MIPS (http://mips.gsf.de; [381]).

Database	Organism	Access	Primary ID
BIND	Multiple	Web, download, API	GI
DIP	Multiple	Web, download	SwissProt/PIR, GI
HPRD	Human	Web, download	RefSeq
IntAct	Multiple	Web, download, API	UniProt
MINT	Multiple	Web, download	SwissProt
MIPS	Mammalian	Web, download	SwissProt

systematic computational access to this historical data is limited. Further, now that the resources to store and maintain the PPI data are in place, any new PPI data that is generated can be collected in parallel, providing an ever-growing resource of PPI information.

At the time of this writing, the most comprehensive view of the human protein–protein interaction data is provided by the Human Protein Reference Database (HPRD). The interactions can be downloaded in the PSI-XML standard format [244], which currently contains nearly 13,000 PPI. However, an even more comprehensive view of the literature-based human PPI can be obtained by integrating all of the above databases into a data warehouse. Despite the fact that they are all are based on the published literature, the overlap between them is rather low (see Figure 6.2), with only 31 interactions in common between BIND, DIP, HPRD and MINT. This integration, as will be discussed later, is nontrivial; however, it can produce a known human PPI dataset that contains more than 18,300 PPIs. This provides much greater coverage of the known human interactions, which may be important in the analysis of disease-related datasets, or understanding the global physiology of the human cell.

6.1.1.3.2   Predicted PPI databases.   In addition to the databases of published human interactions, several databases have been developed to predict novel PPIs in a variety of organisms. The methods that have been employed to make these predictions include homology, orthology, domain fusion, chromosomal location, phylogenetic profiling, and gene coexpression. While the known PPI databases provide a central repository for the published interactions, the predicted interaction databases allow us to expand the protein interaction networks into areas not yet covered in a particular organism. For instance, since 2002, over 100,000 protein interactions have been published in HTP model organism studies, while fewer than 5,000 have been published in

TABLE 6.3:   Mapping genes to proteins — practical aspects.

An integral first step in combining gene expression data with protein–protein interaction datasets is making the connection between the protein products, and the genes that encode them. Microarray datasets contain one or more identifiers (e.g., Affymetrix Probeset ID, Genbank ID, etc.) for each transcript on the array, however these are gene identifiers. Similarly, the protein interaction datasets will often have a protein-based identifier, such as a Ref-Seq or SwissProt ID. Therefore, the path from the gene ID to the protein ID must first be established in order to perform any subsequent analysis. Although this has not been a problem in yeast, where systematic ORF names have been used in both cases, in higher organisms, such as mouse and humans, this can be a significant challenge.

As an example, a transcript from a typical human cDNA array may use Genbank ID AA167728. This particular ID maps to Unigene cluster Hs.112058. However, this cluster contains 318 Genbank ID's. Therefore, if any other transcript on that particular array, or a different array used one of the other Genbank ID's, the correspondence between transcripts encoding the same gene would not be made.

Although mapping between Genbank ID's is difficult, it is an even greater challenge mapping to proteins. However, if one were to construct a data warehouse, with local copies of Unigene and SwissProt, one could construct an intermediate table to map between SwissProt records and Unigene clusters, using the multitude of Genbank ID's stored in both SwissProt and Unigene. It is not, however, sufficient to try and map between the Genbank ID's on the microarray and the cross-references stored in the SwissProt records. In one study published by Garber *et al.* using a 24,000 cDNA array [186], none of the 22,681 Genbank ID's are found in SwissProt. However, if Unigene is used as an intermediate, 7,422 of the microarray transcripts can then be mapped to SwissProt records.

Another effective solution is to make use of BLAST, the Basic Local Alignment Search Tool, and the NCBI eUtils to compute the mapping between the protein and gene identifiers. In this scenario, the eUtils can be used to retrieve the nucleotide sequence corresponding to the Genbank ID. The sequence can then be used to query the SwissProt database through BLAST, using blastx, to retrieve the matching SwissProt ID.

In either case, there is one inherent difficulty. Databases, particularly Unigene, are updated frequently, so any precomputed mapping will require updating when the data warehouse is updated. While not technically difficult to do, it does require time and computing resources on a relatively frequent basis.

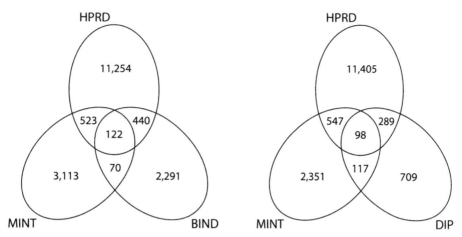

FIGURE 6.2: Venn diagrams showing overlap between known human databases. All PPIs were downloaded from BIND, DIP, HPRD and MINT, and compiled into a data warehouse. For the databases that contain multiple organisms (BIND, DIP, MINT), only the human components were extracted for the purposes of this overlap. Proteins were mapped to SwissProt to provide a common reference identifier before analysis (see Table 6.3). (Download dates: BIND — 25/02/05; DIP — 26/01/05; HPRD — 28/02/05; MINT — 01/02/05).

humans. Leveraging the model organism data allows us to predictively extend the human interaction network into new areas that have yet to be studied, providing new hypotheses and furthering experimental studies. Databases containing the predicted PPIs are summarized in Table 6.4, although the list is far from comprehensive. See the *Nucleic Acid Research* database issue for additional database references.

6.1.2 Interoperability

One of the major challenges in bioinformatics is in integrating these various datasources. Redundancies in the biological data, such as multiple Genbank identifiers referring to the same gene or protein sequence are challenging to deal with. However, versioning of the data is even more of a challenge. As more information is added to these public repositories, and the sequences are processed to remove errors or redundancies, records may be added or removed from the database. While this appears to be a problem for all public repositories, nowhere has this been more of an issue than with the Unigene database. Unigene provides complete builds monthly, in which many clusters may be split, joined, or deleted outright, as sequence alignments change.

While a database management system may theoretically be able to cope

TABLE 6.4: Predicted protein–protein interaction databases: HPID (http://wilab.inha.ac.kr/hpid/; [227]), OPHID (http://ophid.utoronto.ca; [84]), POINT (http://point.nchc.org.tw:3333/; [252]), Predictome (http://predictome.bu.edu; [379]), STRING (http://string.embl.de; [530]). Multiple methods* comprise chromosomal proximity, phylogenetic profiling, and domain fusion. Multiple methods** comprise known and genomic context, HTP experiments, gene coexpression, and known PPIs.

Database	Organism	Access	Primary ID	Prediction method
HPID	Human	Web	n/a	Domains, homologs
OPHID	Human	Web, download	UniProt	Known, orthologs
POINT	Human	Web	n/a	Orthologs
Predictome	Multiple	Web, download	Multiple	Multiple methods*
STRING	Multiple	Web, download	UniProt	Multiple methods**

with the addition of new data, and can be rebuilt to recompute relationships, the challenge lies in the data that has already been analyzed and stored in nonrelational formats. For example, the analysis of a set of microarray experiments that may rely on Unigene identifiers is likely stored in an Excel spreadsheet. Alternatively, a protein profiling experiment may be based on SwissProt identifiers, some of which may be deleted or converted to secondary accession numbers in subsequent releases. In order to avoid cross-references to identifiers stored in analyzed data, some labs have reported freezing their internal databases until analyses are published to avoid such issues.

In order to facilitate data exchange, standards are essential. To this end, the MIAME standard for the reporting and exchanging of microarray data [80], the HUPO PSI-MI XML format for reporting and exchanging molecular interaction data [244], and the MIAPE and PSI-MS data exchange formats are currently being developed for the reporting of mass spectrometry data [424].

These standards are an excellent starting point to facilitate future sharing of information between researchers and databases. However, while these standards provide guidelines about how the data should be described, what features are essential to include, and simplify sharing that data, there still exists an underlying complication; the use of common identifiers to describe genes and proteins.

6.1.3 Bridging the Gap

It is easy to see that with a large number of competing databases that are in a constant state of flux, it will be difficult to establish and maintain cross-links between data sources. In addition, if a dataset is published that is based on an earlier version of SwissProt (in the case of proteomics or protein–protein interactions) or Unigene (in the case of microarrays), there is the question of what happens to that data six months later, or one year later, when the database versions have changed. In order for future meta-analysis or data integration to be carried out, one first needs to verify the integrity of the data, and correct any inconsistencies. All datasets must be moved to a common identifier, or to common versions even if they use the same identifier.

6.1.4 Benefits of Data Integration

Integration of large amounts of biological data is a key aspect of bioinformatics. This allows one to validate biological data, form new hypotheses, and predict new properties of individual entities. This is particularly important in HTP experiments, such as yeast 2-hybrid (Y2H), mass spectrometry complex identification, or microarray analysis. Since noise is a dominant feature in these HTP datasets, integration of additional information may help to reduce the noise in the biological data.

6.1.4.1 Interaction Confidence Measures

Some of the earliest examples of this examined the integration of microarray expression data and protein–protein interactions [191, 216, 267, 301]. These works revealed that interacting proteins are more frequently coregulated at the transcriptional level than noninteracting or random protein pairs. Thus, if two genes are coexpressed, there is a greater likelihood that their gene products will encode interacting proteins. This may be particularly true in larger protein complexes [134, 531], where each component in the complex needs to be expressed in the same time and place (see Figure 6.3).

It has also been suggested that in order for two proteins to interact, they must be colocalized within the cell [268, 499]. The Gene Ontology (GO) cellular component aspect provides information about the localization of individual proteins. Integrating GO data with PPIs, one may conclude that two proteins that are not colocalized have a reduced likelihood of interacting. However, in order for this analysis to be carried out accurately, the complete set of localizations must be known for each protein, which is far from reality at this point in time.

Another use of the GO ontology is based on the assumption that interacting proteins are more likely to have similar biological functions [204, 341, 499], and computational methods to assess this functional similarity often rely on the GO terms assigned to the interacting proteins. There are several methods to use GO terms to add confidence to PPI:

1. *Simple overlap between terms*: look for a single GO term in common [132, 333];

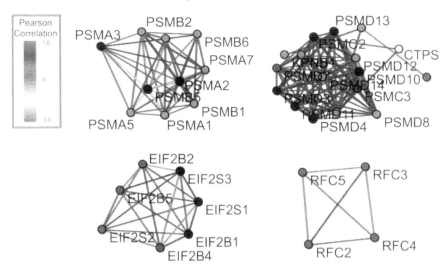

FIGURE 6.3:   (See color insert following page 72.) Protein–protein interactions identified through HTP screens (yeast high confidence [531]), mapped to human proteins, and integrated with the GeneAtlas mRNA expression data [513]. Nodes in graphs are proteins, while edges indicate interactions between proteins. The darker blue edges indicate a higher correlation in the gene expression data, suggestive of higher confidence in the individual protein interactions.

2. *Significant overlapping term*: count occurrence of terms in entire network; less common terms are more significant (more specific) [268];

3. *Jaccard coefficient*: the number of terms in common/total number of terms [33];

4. *Semantic similarity metric*: terms ranked according to frequency for describing all annotated proteins [84, 348].

The simple overlap method and the Jaccard coefficient fail to take into account the depth within the GO tree that terms occur, where increasing depth implies greater specificity of the descriptive term. However, significant overlap and the semantic similarity both determine the specificity of the overlapping terms by counting their frequency either within the entire PPI network (former) or all annotated proteins (latter), making these methods more specific. An example of the semantic similarity measure is shown in Figure 6.4, which illustrates the similarity between the proteins BIR4 and ICE3.

Finally, integration of domain information may provide clues as to whether or not two proteins are likely to interact. Domain information about each

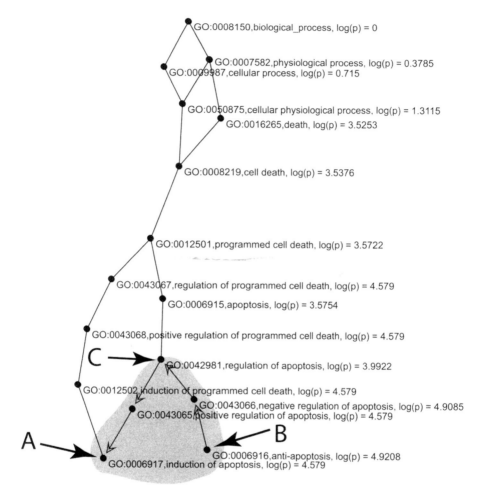

FIGURE 6.4: (See color insert following page 72.) The GO biological process tree is illustrated, where each node represents an individual GO term, and the labels include GO identifiers, description, and semantic similarity score (*log*(*p*-value)) for each node. The maximum semantic similarity for ICE3 (SP:P42574) and BIR4 (SP:P98170) are found between the GO terms "induction of apoptosis" (A) and "anti-apoptosis" (B), through the common parent term "regulation of apoptosis" (C). This provides a final semantic similarity of 3.992 for ICE3 and BIR4.

protein can be obtained from InterPro[9] [22], which is cross-referenced in the UniProt records. By characterizing the domain pairs that are more frequently found in known protein interactions, one can then assign a greater likelihood to proteins bearing those significant domain pairs. Deng *et al.* and Betel *et al.* used such an approach to characterize the domain interaction networks derived from PPI [62, 133], while Brown and Jurisica used the domain–domain networks derived from known human interactions to provide one source of validation for predicted PPI [84].

With these parameters established, it was then possible to start integrating such information to provide reliable predictors of protein–protein interaction. All HTP datasets are prone to noise, and thus integrating multiple verification methods will yield higher quality datasets. For instance, at the time of this writing, the single-largest dataset of Drosophila PPIs was by Giot *et al.* [204], which contained 20,405 PPIs, more than 75% of which were deemed to be of low confidence. To help deal with the inherent noise in this Y2H dataset, J. Bader *et al.* developed a linear regression model to rank each PPI [33]. This work used the GO biological process and cellular component data, gene coexpression, and gene expression level to validate their model, in the process reiterating that GO term similarity and higher expression correlation correlated with the higher confidence interactions.

In another study, Jansen *et al.* used Bayesian networks to integrate mRNA coexpression, coessentiality of proteins, and colocalization to predict the co-complexing of proteins *de novo* [268]. Of course, this approach could also be used to filter HTP PPI datasets, which was included as part of their analysis. In this work, the authors showed that the sensitivity (defined as the number of true positives over the number of total positive calls) of a predicted inter-action dataset using the aforementioned features and comparing to a "Gold Standard" dataset was 27%, while the experimentally determined HTP data had a sensitivity of 1%, at comparable error rates. This suggests that mRNA coexpression, coessentiality, and colocalization, while individually only weak-ly predictive of cocomplexing, together can reliably predict cocomplexing as accurately as the HTP experiments themselves.

6.1.4.2   Integration of Model Organism PPI with Validation

While the growth in the model organism predictomes has been exponen-tial, similar growth in the human interactomes has lagged far behind. We attempted to address this disparity by creating OPHID, the Online Predict-ed Human Interaction Database [84]. This work began by integrating all of the experimentally determined PPI data from model organisms into a data warehouse. A mapping between the model organism proteins and human pro-teins was established using BLAST, and computing protein orthologs using a reciprocal best-hit approach. Finally, the model organism PPI were mapped

---

[9]http://www.ebi.ac.uk/interpro/

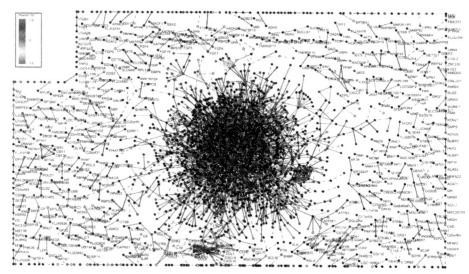

FIGURE 6.5: (See color insert following page 72.) The protein interaction data from OPHID was combined with the gene-expression data from GeneAtlas. The interactions were filtered to include only those with transcripts on the microarray, leaving 30, 495 PPIs. Shown here are 4, 674 high confidence interactions, including data from DIP, MINT, *C. elegans, D. melanogaster, S. cerevisiae,* and *M. musculus.* Each node represents a protein, edges indicate interactions, and the color indicates the degree of correlation between the gene transcripts.

to human proteins through the orthologs, and combined with known human interactions integrated from BIND, DIP, HPRD, and MINT, to create the largest human interactome currently available. Figure 6.5 shows 4, 674 PPIs with 3, 105 proteins obtained from OPHID, combined with gene expression data from GeneAtlas [513].

Since the transfer of the model organism data may introduce additional noise on top of that already present in the HTP datasets, we also needed to develop a scoring system for the predicted interactions. As has been discussed above, integrating domain information, gene coexpression, and GO terms help to validate protein interactions. The 16, 034 known human interactions provided a reliable training dataset to establish statistical cut-offs for each scoring metric, from which we were able to provide evidence of interaction for almost 5, 500 predicted interactions.

6.1.4.3 Interaction Network Dynamics

An interesting supposition was put forth by J. Han *et al.* with regard to coexpression of interacting proteins: In a high-confidence PPI dataset, the distribution of Pearson correlations between gene vectors should include both high and low correlation values [227], showing a bimodal correlation profile. It was suggested that those high-confidence interactions with a low average Pearson correlation coefficient (PCC) may indicate transient protein interactions, which occur in a single time and place. The authors referred to these as "Date Hubs." Those proteins with a high average PCC were termed "Party Hubs," since their respective protein interactions all occurred in the same time and space. An example is shown in Figure 6.6. This intriguing observation indicates that dynamics in PPI networks may be inferred through the integration of PPI and expression data.

Figure 6.6 shows an example, where the interactions surrounding Rpl25 have a high average PCC, suggesting a set of interactions that occur in the same time and place. Rpl25 is a ribosomal protein, and the ribosomes are ubiquitous protein complexes within the cell that carry out a distinct function: the translation of messenger RNA. These complexes have previously been shown to be highly coexpressed [267].

In contrast, the date hub Skp1, with a low average PCC, shows a highly connected protein that shares its associations between multiple distinct pathways or complexes. Skp1 associates with proteins involved in glucose transport and repression (Grr1), attachment of condensed chromosomes to spindle fibers (kinetochore subunits Sgt1, Ctf13), and ubiquitin conjugating enzymes (Sgt1, Ufo1, Rub1). It is easy to envision the transient and distinct interactions that take place during such disparate processes, such as proteolysis, glucose metabolism, and cell-cycle regulation and cell division.

6.1.5　Tools for Integration

Recent efforts have been devoted to reducing data-integration issues by developing comprehensive data warehouses, where the integration of disparate databases is precomputed, and updated frequently. These efforts include AliasServer, Atlas, and SeqHound [260, 382, 484]. IBM has presented an alternative approach with their middle-ware product, Information Integrator®, which facilitates data federation as opposed to data warehousing.

6.1.5.1　AliasServer

AliasServer is a an interactive web server that is designed to handle some of the multitude of aliases that refer to a single protein, such as RefSeq, SwissProt, and Ensembl ID's that each point to the same sequence [260]. The rationale behind AliasServer is to compute a 64-bit CRC key for each protein sequence, and map each alias to that key in a relational database. While the server is also offered with a SOAP interface, the concept may be easily implemented in-house to improve performance on large queries.

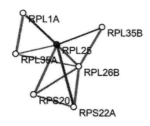

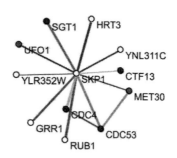

Party Hub: Rpl25
RPL25: ribosomal protein L23a.e
RPL1A: ribosomal protein L1
RPL35A: 60S large subunit ribosomal protein
RPL35B: 60S large subunit ribosomal protein
RPL26B: 60S large subunit ribosomal protein
RPS20: 40S ribosomal protein S20
RPS22A: 40S ribosomal protein S15a.e.c10

Date Hub: Skp1
SKP1: kinetochore protein complex CBF3, subunit D
SGT1: subunit of SCF Ub ligase complex/ essential
      component of kinetochore complex
CDC53: controls G1/S transition; non-catalytic scaffolding
      protein for Cdc34p to Skp/F-box proteins
CTF13: kinetochore protein complex, CBF3, 58kD subunit
      (chromosome segregation and movement of
      centrromeres along microtubules)
MET30: cell cycle progress; involved in degradation of the
      Cdk-inhibitory kinase
RUB1: ubiquitin-like protein
UFO1: involved in degradation of Ho protein
HRT3: putative nuclear ubiquitin ligase
YN51: putative member of Ub ligase complex Swe1p
YNL311C: putative member of Ub ligase complex
GRR1: req'd for glucose repression and glucose and cation
      transport

FIGURE 6.6: (See color insert following page 72.) A party hub, such as RPL25, is defined as a hub (a highly connected node) with a high average Pearson correlation (PCC). These proteins are assumed to interact with their partners in the same time and space. They are also more likely to represent stable protein interactions, such as those found in large protein complexes. Alternatively, the date hub is defined as a hub with a low average PCC. These proteins interact with one or few partners at a time, such as proteins that are involved in distinct cellular pathways. These are more likely to represent transient protein interactions.

6.1.5.2   Atlas

Atlas is a data warehouse service that is being provided by the University of British Columbia's Bioinformatics Centre [484]. The back-end of Atlas is a large data warehouse that incorporates many of the public databases, including gene, protein, protein interaction, gene ontology, and annotated resources such as OMIM, Entrez Gene, and HomoloGene, and provides API for easy access to the warehouse. The APIs are available for C++, Java, and Perl. The advantage to using a remote data warehouse such as Atlas is that it frees up local resources that might be devoted to maintaining current versions of in-house databases and maintaining relational integrity. However, like AliasServer, creation of an in-house version of Atlas may offset any trade-offs with using a remote data warehouse, such as transmission lag time and remote server loads.

6.1.5.3   SeqHound

Like Atlas, SeqHound integrates databases such as Entrez sequence databases, 3D structural data, and functional annotations in the form of GO terms and PubMed references [382]. Originally designed to support the BIND database, SeqHound provides public access via APIs in Perl, C or C++, and can also be installed locally. SeqHound is highly customizable and flexible, and provides an alternative approach to storing and accessing disparate datasources through a large data warehouse.

6.1.5.4   Information Integrator

Information Integrator is a middle-ware product from IBM that's been developed specifically with the Life Sciences in mind. The idea behind data federation, and hence Information Integrator, is to present multiple databases to the user as though they are a single local database. Thus, queries to these multiple databases can be written in simplified SQL statements, the queries can be optimized by the DB2 optimizer, and run on a combination of local and remote databases. The databases need not be strictly DB2 databases, or even relational databases at all. Information Integrator can access XML, flat files, Excel files, BLAST and Entrez, all through standard SQL queries. In addition, Information Integrator can be extended to include additional datasources through the use of relational wrappers.

An example of accessing a BLAST datasource is illustrated in Figure 6.7. Information Integrator greatly simplifies the development of applications that map homologous sequences by presenting BLAST as though it were a relational table, parsing the BLAST output, and filtering based upon predicates added to the SQL query.

```
Query:

SELECT spid, e_value, score
FROM SWISSPROT
WHERE
e_value < 1e-10 AND
     org = 'Human' AND
blastseq = 'CLVPNEVRAD ENEGYEHEET LGTTEFLNMT
            EHFSESQDMT NWKLTKLNEM NDSQVNEEKE'




Result:
SPID        E_VALUE                 SCORE
----------  ---------------------   --------------
095405      +3.2963900000000E-030 +1.286420E+002
```

FIGURE 6.7:   IBM Information Integrator and BLAST.

6.2   Characterizing and Predicting Protein–Protein Interactions

Most cellular processes are carried out through PPIs. These processes include intracellular communication, signal transduction and gene regulation [152]. The role of an individual protein in an organism is largely determined by its interactions with other proteins [349].

PPIs have traditionally been identified a few at a time, by a number of genetic and biochemical methods. These methods, known as small-scale screens, have reliably identified a large number of interactions but they cannot keep pace with the rate at which new proteins are being discovered. Consequently, in the last three years, HTP methods have been developed to identify PPIs on the scale of entire proteomes. These methods identify pair-wise interactions (yeast 2-hybrid [261, 526]), protein complexes (spectrometric protein complex identification (HMS-PCI) [247], and tandem-affinity purification (TAP) [190]) and functional relationships that may involve direct interactions (correlated mRNA profiles, HTP identification of synthetic lethal mutations [520]). In addition, a large number of computational approaches have been developed to predict such relationships between proteins.

Computational approaches can be divided into four main groups, based on the type of information they analyze: protein primary sequence, genomic data, protein structure, and topology of experimentally derived interaction networks.

		Actual Status	
		noninteracting	interacting
Predicted Status	noninteracting	True Negative (TN)	False Negative (FN)
	interacting	False Positive (FP)	True Positive (TP)

sensitivity: interactions predicted correctly
= TP/(TP+FN)

specificity: noninteractions predicted correctly
= TN/(TN+FP)

false negative rate: interactions predicted incorrectly
= FN/(TP+FN)
= 1 - sensitivity

false positive rate: noninteractions predicted incorrectly
= FP/(TN+FP)
= 1 - specificity

FIGURE 6.8:   Criteria for evaluating predictions of PPIs.

6.3   Approaches to Data Integration

One of the rationales for data integration arises because all of the current HTP methods for identifying PPIs have high error rates: they identify many false positives while failing to find many true interactions [531]. Results from any one method have little overlap with results from other HTP methods or from small-scale screens. HTP methods have identified over 80,000 interactions in yeast but only 2,400 of these were detected by more than one method. Also, interactions from HTP methods often show properties that suggest they are false positives. For example, many interactions involve proteins from different subcellular localizations.

Integrating results from unrelated methods can reduce the number of false positive and false negative cases (defined in Figure 6.8). False negatives can be reduced because each method is likely to find previously undetected interactions while false positives can be reduced if each method is used to confirm previously detected interactions. In addition to improving error rates, integration may highlight the relationships between factors that affect interaction [195]. For example, integrating information about protein domains and protein localization may show that certain pairs of domains facilitate interaction only in specific subcellular environments.

Union of Datasets

		Actual Status	
		noninteracting	interacting
Predicted	noninteracting	TN $= \cap_{i=1}^{n} TN_i$	FN $= \cap_{i=1}^{n} FN_i$
Status	interacting	FP $= \cup_{i=1}^{n} FP_i$	TP $= \cup_{i=1}^{n} TP_i$

low false negative rate (FNR), high false positive rate (FPR)

Intersection of Datasets

		Actual Status	
		noninteracting	interacting
Predicted	noninteracting	TN $= \cup_{i=1}^{n} TN_i$	FN $= \cup_{i=1}^{n} FN_i$
Status	interacting	FP $= \cap_{i=1}^{n} FP_i$	TP $= \cap_{i=1}^{n} TP_i$

high false negative rate (FNR), low false positive rate (FPR)

FIGURE 6.9: Effects of taking the union and intersection of datasets $D_1, ..., D_n$. Dataset, D_i, is assumed to have true negatives, TN_i, false negatives, FN_i, false positives, FP_i and true positives, TP_i.

6.3.1 Union and Intersection

The most straightforward approach to integration is to take the *union* or *intersection* of the HTP data. The union consists of all results, including interactions identified by only one method. Taking the union is effective if methods identify few false positives but many false negatives. The union usually gives a much lower false-negative rate than individual methods but a much higher false-positive rate. Taking the intersection produces opposite effects. The intersection consists of results that are common to all methods; it usually gives a much higher false negative rate than individual methods but a much lower false-positive rate. It can be effective if methods find many false positives and few false negatives (see Figure 6.9).

One of the most comprehensive evaluations of individual HTP methods and the intersections of their results is presented in [531]. Since individual methods had high false-positive rates the union of the methods was not considered. Results were evaluated by comparison with a reference set consisting of interactions identified by small-scale screens.

The evaluation criteria were *accuracy* and *coverage. Accuracy* represented

		Actual Status: Based on detection by small-scale screens (S)	
		Pairs not detected by S	Pairs detected by S
Predicted Status: Based on detection by high-throughput methods (H)	Pairs not detected by H	True Negatives: Not considered	False Negatives: Pairs detected by S but not by H
	Pairs detected by H	False Positives: Pairs detected by H but not by S	True Positives: Pairs detected by H and S

Accuracy: fraction of HTP data confirmed by small-scale screens
=TP/(TP+FP)

Coverage: fraction of small-scale screen data found by HTP methods
=TP/(TP+FN)

FIGURE 6.10: Calculation of accuracy and coverage for PPI data.

the reliability of the results (Figure 6.10) and was calculated as the percentage of results confirmed by the reference set. *Coverage* represented the percentage of true interactions that were identified. It was calculated as the percentage of interactions in the reference set, identified by HTP methods.

Among individual methods, the highest accuracy and coverage were approximately 16% and 21%, respectively. For most methods there was a tradeoff between accuracy and coverage. Only one method (TAP) had relatively high values for both accuracy and coverage — about 12.5% and 21%, respectively[10] Interactions found by any two methods had accuracy and coverage of 26.38% and 5.58% respectively, while interactions found by any three methods had accuracy and coverage of 52.45% and 0.69%, respectively.

---

[10]These values were determined by assuming binary interactions between all protein pairs in a complex. If binary interactions are assumed only between the bait protein and other complex members, then the accuracy and coverage are 27.8% and 8.5%, respectively. If interactions are considered only among the set of proteins present in the reference set, then the accuracy increases to 40.5%.

6.3.2 Bayesian Approach

Using the union or intersection of PPI datasets has serious limitations [531]. The union produces low accuracy (high false positive rate), while the intersection gives low coverage (high false negative rate). A better integration approach would avoid these extremes and would take into account the error rates of the datasets being integrated.

A Bayesian approach provides a systematic way of meeting these criteria. This approach integrates both numerical and categorical data, weights each dataset according to its reliability, and accommodates missing data.

A Bayesian approach gives the probability of an interaction based on available evidence. The approach is summarized by the following formula:

$$P(I = 1|e) = \frac{P(e|I = 1)P(I = 1)}{P(e|I = 1)P(I = 1) + P(e|I = 0)P(I = 0),} \tag{6.1}$$

where I is a random variable representing presence or absence of interaction $(0/1)$ between a pair of proteins; e is evidence for an interaction, e.g., detection of interaction by yeast 2-hybrid (0 if detected, 1 otherwise); $P(I = 1|e)$ is a posterior probability, the probability of interaction given the evidence; $P(e|I = 1)$, is a likelihood, the probability of the evidence given an interaction; and $P(I = 1)$, is a prior probability, the probability of interaction without reference to any evidence, e.g., in a given organism, the total number of interacting protein pairs as a percentage of all possible protein pairs.

The main idea of the equation is that an interaction is likely if the evidence is a property often found among interacting protein pairs and rarely found among noninteracting protein pairs. The term $P(e|I = 1)$ expresses the probability of finding the evidence among interacting protein pairs, while the term $P(e|I = 0)$ expresses the probability of finding the evidence among noninteracting protein pairs.

The equation can also be viewed in terms of the false-positive and false-negative rates of the evidence. The numerator of the equation takes into account the false-negative rate of the evidence — the numerator will have a high value if the false-negative rate is low. For example, if the evidence consists of yeast 2-hybrid results that identify most interactions, then the number of false negatives will be low (low false-negative rate) and the probability of detection given an interaction, $P(e|I = 1)$, will be high. The denominator takes into account the false-positive rate of the evidence — its value will be low if the false-positive rate is low. In other words, if yeast 2-hybrid rarely detects false positives, then the probability of detection given a noninteraction, $P(e|I = 0)$, will be low.

Although all Bayesian methods follow the idea of Equation 6.1, there are also several differences among the methods. One difference is the way terms in Equation 6.1 are calculated. In many cases, all of the terms are calculated from data [208] but priors are sometimes based on background knowledge

[268] or expert opinion [522]. Another difference is the way multiple evidence is combined.

In general, the same equation that is used with one type of evidence (Eq. 6.1), can be used with multiple types of evidence.

$$P(I = 1|e_1, e_2, ..., e_n) = \frac{P(e_1, e_2, ..., e_n|I = 1)P(I = 1)}{\sum_{i=0,1} P(e_1, e_2, ..., e_n)|I = i)P(I = i).} \quad (6.2)$$

However, determining the likelihoods, $P(e_1, e_2, .., e_n|I)$, becomes more difficult. For example, if the evidence is a single binary variable such as detection by yeast 2-hybrid, then the likelihood takes on just two values, $P(e = 0|I)$ and $P(e = 1|I)$; if the evidence consists of n binary variables then the likelihood takes on 2^n values, $P(e_1 = 0, e_2 = 0, ..., e_n = 0|I)$, $P(e_1 = 0, e_2 = 0, ..., e_n = 1|I)$, etc. The calculation of likelihoods can be simplified if the evidence is assumed to be *conditionally independent*. A set of evidence, $\{e_1, e_2, .., e_n\}$, is conditionally independent if, given the class I, the probability of one type of evidence is unaffected by the presence of any other evidence: $P(e_i|I) = P(e_i|I, e_j)$, $i \neq j$. For example, if an interacting pair has been detected by yeast 2-hybrid, the probability of the pair being detected by other methods does not change. Under the assumption of conditional independence, Equation 6.2 can be rewritten as follows:

$$P(I = 1|e_1, e_2, ..., e_n) = \frac{P(e_1|I = 1)P(e_2|I = 1)...P(e_n|I = 1)P(I = 1)}{\sum_{i=0,1} P(e_1|I = i)P(e_2|I = i)...P(e_n|I = i)P(I = i)}. \quad (6.3)$$

A Bayesian method that assumes conditional independence is referred to as a *Naive Bayes classifier*. A Bayesian method that makes no assumptions about conditional independence is referred to as a *fully connected Bayesian network*.

6.3.3 Bayesian Approach Assuming Conditional Independence of Evidence

An example of a Bayesian approach to data integration that assumes conditional independence of evidence is presented in [208]. This study determined PPIs in yeast from two types of data: yeast 2-hybrid results and information about the topology of a protein interaction network. More specifically, the aim was to determine the probability that small-scale screens had detected an interaction between proteins v and w, given yeast 2-hybrid results, Y_{vw}, and topology information, C_{vw}. The probability was determined by the following equation:

$$P(I_{vw} = 1|C_{vw}, Y_{vw}) = \frac{P(C_{vw}|I_{vw} = 1)P(Y_{vw}|I_{vw} = 1)P(I_{vw} = 1)}{\sum_{i=0,1} P(C_{vw}|I_{vw} = i)P(Y_{vw}|I_{vw} = i)P(I_{vw} = i)}, \quad (6.4)$$

where I_{vw} indicates the absence or presence (0/1) of interaction, as detected by small-scale screens. Detection by small-scale screens was assumed to indicate a true interaction. Y_{vw} indicates whether the pair (v, w) has been detected (0/1) by yeast 2-hybrid. C_{vw} is a *mutual clustering coefficient*, a variable that describes the topology of the PPI network close to proteins v and w. In this PPI network, vertices represent proteins and edges represent interactions detected by yeast 2-hybrid. The authors had found that in this network, protein pairs detected by small-scale screens were likely to be adjacent to the same vertices [208]. The authors developed the variable C_{vw}, in order to use this property for predicting pairs detected by small-scale screens. C_{vw} expressed the probability that the number of mutual neighbors observed for pair (v, w) would be matched or exceeded by chance. A low probability meant that (v, w) was more likely to have been detected by small-scale screens.

Equation 6.4 was used to calculate the probability of interaction for each pair of yeast proteins. The terms in the equation were calculated from all pairs of yeast proteins, except the pair being tested. $P(Y_{vw}|I_{vw} = 1)$ was calculated as the fraction of interacting protein pairs that were detected by yeast 2-hybrid (if $Y_{vw} = 1$) or not detected by yeast 2-hybrid (if $Y_{vw} = 0$). $P(C_{vw}|I_{vw} = 1)$ was determined by calculating values of C for all protein pairs (except the test pair), binning the values, and finding the fraction of interacting pairs in the bin corresponding to C_{vw}.

The naive Bayesian classifier used by the authors had two important benefits common to other Bayesian methods: the ability to combine diverse types of evidence and the ability to take into account the error rate of each evidence type. By assuming conditional independence, the posterior probability could be easily calculated and additional evidence would not increase the complexity of the calculations.

However, the assumption of conditional independence can also be a drawback. As shown in Figure 6.11, the assumption can hide important predictive relationships. In this example interacting protein pairs have correlated values for C_{vw} and Y_{vw} — both values are either low or high. In other words, C_{vw} and Y_{vw} are not independent, given information about interaction. With the assumption of independence, I_{vw} cannot be predicted from values of C_{vw} and Y_{vw}. Intuitively, this is because neither variable on its own can differentiate between interacting and noninteracting pairs; each variable has the same values among interacting pairs as it does among noninteracting pairs. More formally, if Equation 6.4 is applied to the data from Figure 6.11, any pair of values (c_{vw}, y_{vw}) gives a probability of interaction of 0.5:

- $P(Y_{vw}|I_{vw} = 1) = P(Y_{vw}|I_{vw} = 0) = 0.5$, since half of the interactions are detected by yeast-2-hybrid;

- $P(C_{vw}|I_{vw} = 1) = P(C_{vw}|I_{vw} = 0)$, since all intervals along the y-axis have the same number of interactions and noninteractions;

- $P(I = 1) = P(I = 0) = 0.5$.

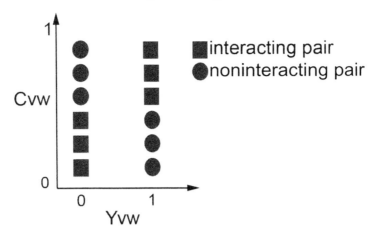

FIGURE 6.11: Example of evidence that is not conditionally independent.

If the assumption of independence is not made, then Equation 6.4 has to be written as follows:

$$P(I = 1|C_{vw}, Y_{vw}) - \frac{P(C_{vw}, Y_{vw}|I_{vw} = 1)P(I_{vw} = 1)}{\sum_{i=0,1} P(C_{vw}, Y_{vw}|I_{vw} = i)P(I_{vw} = i)} \qquad (6.5)$$

The term $P(C_{vw}, Y_{vw}|I_{vw} = 1)$ can describe the fact that among interacting pairs, values of C_{vw} and Y_{vw} are either both high or both low (Equation 6.6).

$$P(C_{vw}, Y_{vw}|I_{vw} = 1) = \begin{cases} 0.5 \ if \ C_{vw} > 0.5, \ Y_{vw} = 1 \\ 0.5 \ if \ C_{vw} \le 0.5, \ Y_{vw} = 0 \\ 0 \ if \ C_{vw} > 0.5, \ Y_{vw} = 0 \\ 0 \ if \ C_{vw} \le 0.5, \ Y_{vw} = 1 \end{cases} \qquad (6.6)$$

The term $P(C_{vw}, Y_{vw}|T_{vw} = 0)$ can be defined in a similar way. Using these equations T_{vw} can be predicted from C_{vw} and Y_{vw}.

Although the above example is artificial, many types of HTP evidence are not conditionally independent. This is often the case because essential processes in a cell, such as gene regulation and PPIs, only take place under specific sets of conditions. For example, PPI may only occur if two proteins have certain domains and are present in the same subcellular compartment at specific concentrations. Consequently, the probability of several conditions appearing together among interacting pairs is much higher than would be expected by chance. If evidence consists of conditions that are required for the outcome, then the evidence is not conditionally independent. Other types of evidence may not be conditionally independent either. For example, interaction detection methods may base their decisions on similar information and may have similar biases [531].

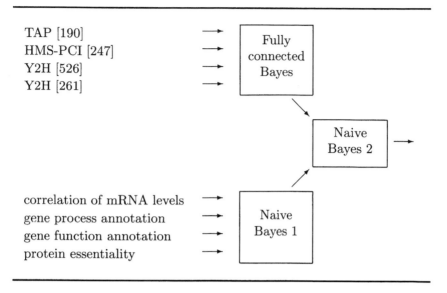

TAP [190]
HMS-PCI [247]
Y2H [526]
Y2H [261]

Fully
connected
Bayes

Naive
Bayes 2

correlation of mRNA levels
gene process annotation
gene function annotation
protein essentiality

Naive
Bayes 1

FIGURE 6.12:   Combination of classifiers used by Jansen *et al.* [268]

6.3.4   Bayesian Approach Not Assuming Conditional Independence of Evidence

A study by Jansen *et al.* is an example of a Bayesian approach that does not assume conditional independence of all variables [268]. The aim of the study was to predict whether a given pair of proteins were members of the same protein complex.

Eight sources of evidence were used for predictions: four datasets from HTP detection methods, one dataset describing the correlation of gene expression profiles (based on two studies), and datasets with gene functional annotation, process annotation, and information on whether given proteins are essential for survival. The prediction method involved three Bayesian classifiers (Figure 6.12). One classifier did not assume conditional independence among evidence (fully connected Bayes), and made predictions from the HTP datasets. A second classifier assumed conditional independence (Naive Bayes) and made predictions from the remaining datasets. A third classifier also assumed conditional independence and made predictions based on the output of the first two classifiers.

The classifiers calculated posterior probabilities as shown in Figure 6.13. In order to calculate likelihoods, any evidence with numerical (continuous) values was binned. The likelihood of a given value, $P(e_i = k|I)$, was calculated as the fraction of positive cases (positive if $C = 1$, negative if $C = 0$) in the bin containing the value k.

Fully connected Bayes	Naive Bayes 1

$$P(C = 1|e_1, e_2, e_3, e_4)$$

$$= \frac{P(e_1,e_2,e_3,e_4|C=1)P(C=1)}{\sum_{i=0,1} P(e_1,e_2,e_3,e_4|C=i)P(C=i)}$$

- $e_1, ..., e_4$ have binary values (i.e., interaction detected or not detected)

- likelihoods calculated as:

e_1	e_2	e_3	e_4	# pos pairs	$P(e_1, .., e_4\| C = 1)$
0	0	0	0	p_1	p_1/p_{total}
0	0	0	1	p_2	p_2/p_{total}
.	.	.	.	.	.
.	.	.	.	.	.
1	1	1	1	p_{16}	p_{16}/p_{total}

$$P(C = 1|e_5, e_6, e_7, e_8)$$

$$= \frac{P(e_5|C=1)P(e_6|C=1)...P(C=1)}{\sum_{i=0,1} P(e_5|C=i)P(e_6|C=i)...P(C=i)}$$

- $e_5, ..., e_8$ have numerical binned values (e.g., correlation coefficients) and categorical values

- likelihoods of categorical values calculated as:

e_6	# pos pairs	$P(e_6\|C = 1)$
$value_1$	p_1	p_1/p_{total}
$value_2$	p_2	p_2/p_{total}
.	.	.
.	.	.
$value_n$	p_n	p_n/p_{total}

Naive Bayes 2

$$P(C = 1|e_9, e_{10})$$

$$= \frac{P(e_9|C=1)P(e_{10}|C=1)P(C=1)}{\sum_{0,1} P(e_9|C=i)P(e_{10}|C=i)P(C=i)}$$

- e_9 and e_{10} have binned values

- likelihoods calculated as:

e_6	# pos pairs	$P(e_6\|C = 1)$
$value_1$	p_1	p_1/p_{total}
$value_2$	p_2	p_2/p_{total}
.	.	.
.	.	.
$value_n$	p_n	p_n/p_{total}

FIGURE 6.13: Computation carried out by classifiers in Figure 6.12.

The three classifiers were trained and tested on positive cases consisting of protein pairs from the same complexes, (based on results from small-scale screens) and negative cases consisting of protein pairs unlikely to be found in the same complexes due to different subcellular localizations. The testing was done by seven-fold cross-validation. In each iteration, the classifiers were tested on $1/7th$ of the data, and calculations were based on $6/7th$ of the data. The results had lower error rates than results from individual sources of evidence.

The combination of three classifiers had some of the same strengths as other Bayesian approaches: the ability to integrate and appropriately weight diverse types of evidence. In addition, the fully connected Bayesian classifier was able to capture relationships between specific combinations of evidence values and membership of proteins in the same complex. Finding such relationships can improve predictions and can indicate the most effective combinations of evidence. For example, if the classifier found that detection by both TAP and Y2H methods was the best combination of evidence, then future studies could confirm results of one method by using the other.

A fully connected Bayesian classifier also has a number of limitations. Its calculations may require a great deal of computing time, its results may not be very reliable, and it can combine only a relatively small number of evidence sources. These problems result from having to estimate a distribution with many dimensions, such as $P(e_1, e_2, ...e_n | C = 1)$. Even if the evidence consists of binary variables, estimating this distribution can be difficult because the set of variables can have many combinations of values. For example, with four binary variables there are 16 combinations (Figure 6.13). The number of combinations increases exponentially with the number of variables, so that thirty binary variables can have over a billion possible combinations of values. A fully connected Bayesian classifier has to calculate conditional probabilities for all of these combinations. Determining the number of positive and negative cases for each combination may require very long computing times.

An additional problem is reduced reliability due to overfitting. Overfitting can occur if a classifier estimates a large number of parameters from relatively few cases. A fully connected Bayesian classifier runs into this problem because the number of variable-value combinations can be very large and many of the combinations may be represented by just a few cases. Probabilities estimated from a small number of cases are unlikely to hold for new data.

Because of these limitations, studies often seek a compromise between a fully connected Bayesian classifier and a Naive Bayesian classifier. The classifier of Jansen *et al.* is an example of such a compromise [268]. A common approach is a network that contains both conditionally independent and dependent variables. Bayesian networks represent dependency relationships as a graph (Figure 6.14). Nodes in the graph represent variables and edges represent conditional dependency: an edge from node a to node b means that b is conditionally dependent on a. For each node there is a conditional probability function, often in the form of a contingency table (as in Figure 6.13), which

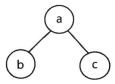

FIGURE 6.14: Example of a Bayesian network. Variables b and c are dependent on variable a, and will have conditional probability functions $P(b|a)$ and $P(c|a)$.

determines how values of the node depend on the values of input nodes. The graph structure and the conditional probability functions define the network. A number of algorithms are available that attempt to determine the structure and functions that are most probable given the data [270, 437]. However, the algorithms cannot guarantee that the optimum network will be found because the number of possible dependency relationships can be very large.

6.3.5 Other Machine Learning Approaches

According to [493] learning denotes changes in the system that are adaptive in the sense that they enable the system to do the same task or tasks drawn from the same population more efficiently and more effectively the next time. This might require not only new knowledge, but also increasing the effectiveness with which the system uses the knowledge it already possesses. Two distinct research areas have been developed:

1. Acceleration of problem solvers;

2. Inducing concepts from examples.

Machine learning refers to computational methods that determine rules underlying a set of data. Usually the aim is to solve classification or regression problems. Algorithms that determine Bayesian networks from data are considered machine learning methods. A large number of other machine learning methods have been developed and several have been used for integrating HTP biological data. Common machine learning methods include support vector machines, neural networks, decision trees and rule induction systems.

Support vector machines and neural networks can find nonlinear relationships between input and output variables. In the bioinformatics field these methods have mainly been used for classification. Support vector machines have been used to predict PPIs from protein primary sequence [67] and gene function from different types of data [434]. Support vector machines find the best decision boundary for separating positive and negative cases (Figure 6.15). If cases cannot be separated by a linear decision boundary they are mapped to a higher dimensional space where they can be separated by

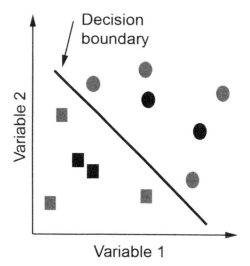

FIGURE 6.15: Support vector machines determine the best decision boundary for separating positive and negative cases (circles and squares, respectively). The boundary is placed so as to maximize its distance from the support vectors (outermost points) of each class — represented by shaded circles and shaded squares. This example could represent the problem of predicting PPIs from gene expression datasets, with variables 1 and 2 corresponding to correlation coefficients.

a hyperplane. The hyperplane corresponds to a nonlinear decision boundary in the input space. Among the strengths of support vector machines are fast learning, few user defined parameters, and confidence levels for predictions.

Neural networks have been used for problems such as prediction of protein secondary structure [249] and analysis of gene expression data [422]. Neural networks used for classification often consist of a graphical structure, similar to a Bayesian network. Inputs to a node are either variable values of training/test cases or outputs from other nodes. Each node takes a weighted sum of its inputs, and outputs a nonlinear function of this sum. The output from one or several "output" nodes represents the classification of a given case.

A drawback of both support vector machines and neural networks is the difficulty of interpreting their decisions. It can be hard to identify which variables (or combinations of variables) played important roles in decisions and how their importance was determined.

Decision trees and *rule induction systems* do not have this problem since their output includes easily interpretable rules for classification. A rule is simply a conjunction of variable values. For example, a rule for predicting PPIs might state:

```
if Y2H = 1 and TAP = 1 then interaction = 1.
```

Decision trees and rule-induction systems have been applied to problems such as prediction of protein secondary structure [482], analysis of gene expression data [346], prediction of genetic interactions [555], and prediction of cocomplexed protein pairs [571]. Rules are often determined by a "greedy" approach:

1. The best variable for differentiating positive and negative cases is identified;

2. Cases are separated into several subsets based on this variable;

3. Steps 1 and 2 are repeated on the subsets of cases.

The steps can be repeated until subsets have a specified minimum number of cases. A drawback of decision trees and rule-induction systems is that they are less effective at identifying nonlinear decision boundaries.

All machine learning methods also face the same limitations as Bayesian methods, in terms of finding dependencies among variables. If the methods assume that variables are conditionally independent then predictive relationships may be missed. If the assumption is not made, then prediction reliability and computational complexity can become problems. In practice, conditional independence is assumed for most variables.

6.3.6 Data Mining

Data mining is similar to machine learning in the sense that it also tries to find patterns underlying (sub)sets of data. However, data mining focuses only on identifying the patterns rather than solving classification or regression problems. For example, given clinical information about cancer patients, a data-mining task might be to identify sets of attributes occurring frequently in patients with a given remission time.

Association-mining is a data-mining approach that has been applied to a number of bioinformatics problems. The aim of association mining is to find associations between items in a database. Traditionally, it has been used for the problem of "market basket analysis:" determining which sets of items are frequently purchased together. The input for market basket analysis consists of a set of transactions, each transaction listing the items that are part of a single purchase. The natural example of a transaction is a supermarket basket holding the purchases of a single customer. The output from market basket analysis consists of association rules, which describe groups of items that are frequently purchased together.

An example of an association rule would be as follows:

```
pancakes & butter ⇒ maple   syrup,
```

meaning that a customer purchasing pancakes and butter is likely to purchase maple syrup as well. In general, an association rule has the form:

$$\text{LHS} \Rightarrow \text{RHS},$$

where LHS and RHS comprises disjoint sets of items. A rule states that whenever the LHS items are present in a transaction, the RHS items are likely to be present as well. The quality of a rule is described by the measures *support* and *confidence*.

Support describes how frequently the rule occurs among transactions; it is defined as the fraction of transactions that contain both the LHS and RHS items of the rule. Confidence describes the accuracy (precision) of the rule and is defined as the number of transactions with both the LHS and RHS items, divided by the number of transactions with LHS items.

Association-mining has been used to mine medical data and HTP biological data. Doddi *et al.* [138] applied association mining to medical data to find rules relating patient symptoms, diagnosis, and procedures performed on patients. Satou *et al.* [472] used association mining to find relationships between protein sequence, structure and function. Creighton and Hanash [123] mined gene expression data to find groups of positively and negatively correlated genes. Oyama *et al.* [427] aimed to find rules relating protein features to protein–protein interactions.

The study by Oyama *et al.* [427] is a particularly good example of association-mining used for integrating diverse HTP data. The study searched for rules relating a large variety of protein features to protein–protein interactions. The rules expressed relationships such as the following:

```
IF (protein A has KINASE_ATP motif)
   AND (protein B has KINASE_ST motif)
THEN (proteins A and B interact).
```

The actual format of the rules was slightly different; instead of the format:

```
IF (features of proteins A and B)
THEN interaction,
```

rules were formulated as follows:

```
(given interacting proteins A and B)

IF (features of protein A)
THEN (features of protein B).
```

However, these two formats express the same relationships between protein features and interactions.

The data for the study comprised 4,307 pairs of interacting proteins from yeast, detected either by small-scale screens or yeast 2-hybrid. Each protein pair was viewed as a transaction. The items in a transaction were features of the two interacting proteins, including functional annotation, structural

TABLE 6.5: Mapping between protein features and items. Each possible feature (feature type, feature value combination) is represented by a unique item. The possible features of a single protein are represented by 5,241 items, the features of a protein pair are represented by twice this number of items. For example, a protein pair, (A, B), with features $\{function(A) = translation,$ $structure(A) = SH2, function(B) = DNA\ repair\}$ would be represented as the item set $\{2, i, 5242\}$.

		Feature (type)	Feature (value)	Item #
1st Protein in Pair		Function	DNA repair	1
			translation	2
			.	.
		Structure	SH2 domain	i
			SH3 domain	i+1
			.	.
	.	.	.	5241
2nd Protein in Pair		Function	DNA repair	5241+1
			translation	5241+2
			.	.
		Structure	SH2 domain	5241+i
			SH3 domain	5241+i+1
			.	.
	.	.	.	10482

annotation, sequence motifs, and properties of the amino acids. Each possible annotation category was represented as a separate item (Table 6.5). A total of 10,482 items were used to represent the possible features of a protein pair. An association mining algorithm called *Apriori* [7] was applied to this dataset and found 6,367 rules with minimum a support of 9 and a minimum confidence of 75%. Some of the rules were confirmed by existing PPI knowledge, such as the importance of the SH3 domain in mediating interactions. However, the effectiveness of the rules for predicting interactions was not tested.

A key advantage of association mining over machine learning or statistical methods, is that it can identify more dependency relationships in data. Association-mining algorithms try to return all rules that meet minimum support and confidence requirements. These rules may reveal important synergistic relationships among variables (e.g., a set of protein features which individually do not predispose to protein–protein interaction, but in combination are an accurate indicator of interaction). Machine learning methods are likely to miss some of these relationships.

While association mining performs a more thorough search, it also has several drawbacks. Most association mining algorithms cannot be directly applied to numerical data; numerical variables have to be binned which entails some loss of information. At low support settings, association mining often returns an enormous number of rules, too numerous to be reviewed by a human user. Many of the rules have a large degree of overlap and many may not generalize to new data. Also, there are limits to the size of datasets that can be mined in a reasonable time. Run time can increase exponentially with the number of variables (items) in a dataset.

6.3.7 Meta-Analysis

Meta-analysis is an approach that uses classical statistical methods to combine data [239, 393]. It has been used for integrating gene expression datasets that were designed to answer the same question. For example, it was applied to datasets that measured gene expression levels in localized prostate cancer and benign prostate tissue [109, 459]. The aim of meta-analysis is to validate differentially expressed genes from individual datasets and to determine a set of differentially expressed genes that do not reflect the biases of individual datasets.

Approaches based on meta-analysis proceed in three main steps:

1. Testing each gene, in each study for differential expression. This is often done by t-tests but other methods are also available, such as ANOVA and Bayesian tests. These tests provide a p-value, which defines the probability that the differential expression of the gene could have occurred by chance.

2. Adjusting the p-values from these tests to account for multiple testing, since the tests in step 1 are repeated for all genes. Multiple testing can mean that some of the genes with significant p-values are likely to be false positives. For example, any group of 100 genes is likely to have 5 genes with p-values of 0.05. One method of adjusting p-values, called the Bonferroni correction, consists of multiplying each p-value by the number of tests.

3. Testing whether a gene is differentially expressed across datasets by using its p-values from different datasets. One approach for this step is to test the null hypothesis that statistically significant genes are different in each dataset [459]. This hypothesis is tested for each gene that is present in all datasets. A summary statistic, S, is calculated for each such gene, using the gene's p-values from different datasets:

$$S = -2log(p_1) + -2log(p_2) + ... + -2log(p_n), \qquad (6.7)$$

where p_i is the gene's p-value from dataset i.

A p-value for this summary statistic is then determined by a comparison to 100,000 summary statistics generated by randomly selecting p-values from individual datasets. The p-value is equal to the fraction of randomly generated statistics with lower values.

The meta-analysis approach can be very effective for filtering out false positives — genes that are differentially expressed due to biases in individual datasets. It can also identify genes that have consistently low p-values across datasets but are not among the most promising genes in individual datasets. Another strength of meta-analysis is that data from different microarray platforms can be readily integrated since all results are converted to p-values. One of the limitations of current studies using meta-analysis is that genes are only considered individually.

6.4   Conclusions

Integration of HTP data can be accomplished by using diverse methods, all of which face certain tradeoffs. The simplest integration approach, using the union or intersection of datasets, has obvious drawbacks but can provide quick estimates of the value of integration. Bayesian methods can also be fairly simple but are able to fulfill many objectives. The Bayesian approach provides a systemic, understandable way of weighting diverse types of evidence. Bayesian methods usually have to make assumptions about conditional independence, but other methods face this problem as well. Machine learning methods not based on Bayesian probability have been widely used for data integration but are not suitable for as many types of problems. Support vector machines and neural networks are often used when a complex, nonlinear relationship is expected between input and output variables. However, the reasons for their decisions cannot be easily interpreted. Decision trees and rule induction systems generate easily understandable rules but are less effective at identifying nonlinear relationships.

Data-mining and meta-analysis address somewhat different objectives compared to other data integration approaches. Data mining focuses on finding relationships in data, without trying to solve classification or regression problems. However, the dependency relationships that are identified can be helpful for solving these problems. Integrating data mining and machine learning approaches could reduce conditional independence assumptions. Meta-analysis is quite different from other integration approaches and may not be applicable to a wide variety of problems. Its aim is to integrate datasets designed to verify a common hypothesis. This means that the application of meta-analysis is currently limited to gene expression data.

Data integration is an essential component of bioinformatics, from the apparently simpler task of integrating datasets from similar types of experiments, to the more complex analysis facilitated by integrating different types of data. This chapter discussed how integrating microarray data with PPI data has been used to take a static protein interaction network, and transform it into a dynamic network regulated by different cellular conditions. We have also shown how integrating multiple datatypes has allowed the filtering of noise inherent in HTP datasets. The future will surely hold even greater promise in terms of identifying critical components or pathways involved in disease, identifying new drug targets for therapy, and elucidating how the cell responds to various cellular conditions, primarily from even further integration of additional data and datatypes. Of course, the past, present and future directions have only been made possible by concerted efforts to provide increasing amounts of biological data in publicly accessible databases.

Chapter 7

From High-Throughput to Systems Biology

What defines a system? Is it a union of its parts, its function(s), or some combination of both? A simplistic view of systems biology would define it as the study of individual cells, multicellular organs, or whole organisms as sums of their component parts. Such a definition includes aspects of system structure and function, design, dynamics, and regulatory control. It emphasizes an understanding of "wholeness" that supersedes understanding of the individual component parts comprising the system itself.

Why has systems biology received so much recent attention? In short, it is because the key first step of defining system structures has quickly advanced from fantasy to reality in the postgenomic era. The achievement of full genome sequencing projects in many organisms, including human, has defined the initial "parts list" encoded in the medium of hereditary information transfer, DNA. The technological development associated with these achievements has spawned the nascent fields of genomics, proteomics, and multiple "-omic" disciplines defined by their systematic, nonhypothesis-driven approaches to biological experimentation.

In this chapter, we will give an overview of systems biology and integrate this with a description of current efforts at understanding biological systems. We will introduce the concept of systematic biology and describe its critical role in defining the architectural framework of biological systems.

7.1 An Overview of Systems Biology

The life sciences are undergoing a profound transformation at the threshold of what is widely regarded as the century of biology [288]. From a collection of narrow, well-defined, almost parochial disciplines, they are rapidly morphing into domains that span the realm of molecular structure and function through to the application of this knowledge to clinical medicine. The results of teams of individual specialists dedicated to specific biological goals are providing insight into system structures and function not conceivable a decade ago.

System level understanding, the approach advocated in systems biology, requires a change in our notion of "what to look for" in biology [313]. While an understanding of individual genes and proteins continues to be important, the focus is superseded by the goal of understanding a systems structure, function and dynamics. Historically, the application of general systems theory to biology dates back as early as 1933 [60]. Bertalanffy suggested that old-fashioned science "tried to explain observable phenomena by reducing them to an interplay of elementary units studied independently of each other" [60]. In contrast, contemporary science recognized the importance of "wholeness", defined as "problems of organization, phenomena not resolvable into local events, dynamic interactions manifest in the difference of behavior of parts when isolated or in higher configuration, etc.; in short, systems of various orders not understandable by investigation of their respective parts in isolation." This remains an effective definition of systems biology as practiced today with the integration and application of mathematics, computer science, engineering, physics, chemistry and biology to understanding a range of complex biological regulatory systems [313].

Identifying all the genes and proteins in an organism is analogous to creating a list of all parts of an airplane. While such a list provides a catalog of individual components, it alone is not sufficient to understand the complexity underlying the engineered object [313]. One cannot readily build an airplane or understand its functional intricacies from a parts list alone. One needs to understand how these parts are assembled to form the structure of an airplane. This is analogous to drawing an exhaustive diagram of gene-regulatory interactions and their biochemical interactions, as these diagrams would provide a limited knowledge of how changes in one part of the system may affect other parts. To understand how a particular system functions, how the individual components interact during operation and under failure must be examined first. From an engineering perspective, answers to key questions become critical, such as, what is the voltage on each signal line? How are the signals encoded? How is the voltage stabilized against noise and external fluctuations? How do the circuits react when a malfunction occurs in the system? What are the design principles and possible circuit patterns, and how can they be modified to improve system performance [313]?

System level understanding of a biological system can theoretically be derived from insight into four key properties [313]:

1. *System structures.* These include the network of gene interactions and biochemical pathways, as well as the mechanisms by which such interactions modulate the physical properties of intracellular and multicellular structures.

2. *System dynamics.* How a system behaves over time under various conditions can be understood through dynamic analysis methods such as metabolic and sensitivity analysis, and by identifying essential mechanisms underlying specific behaviors.

3. *The control method.* Mechanisms that systematically control the state of the cell can be modulated to minimize malfunctions and provide potential therapeutic targets for treatment of disease.

4. *The design method.* Strategies to modify and construct biological systems having desired properties can be devised based on definite design principles and simulations instead of blind trial and error.

Despite the tremendous advances in our understanding of system structures, inspecting genome data bases alone will not get us very far in addressing system-based problems [416]. The reason is simple. Genes code for protein sequences. They do not explicitly code for the interactions between proteins and other cell molecules and organelles that generate function. Nor do they indicate which proteins are on critical pathways for supporting cell and organelle function in health and disease. Much of the logic of the interactions in living systems is implicit. Wherever possible, nature leaves that to the chemical properties of the molecules themselves and to the exceedingly complex way in which these properties have been exploited during evolution. Conceptually, it is as though the function of the genetic code, viewed as a program, is to build the components of a computer, which then self-assembles to run programs about which the genetic code knows nothing [416].

7.2 Systematic Biology

Attempts to derive systems-level understanding are still thwarted by the absence of required extensive and precise information of system structures. It is for this reason that "systematic" biology has rapidly become prominent as an experimental methodology. Different views on the emergence of this topic have been proposed [246]. Initially, the term was used to describe state-of-the-art experiments in yeast using such tools as microarrays of all open reading frames (ORFs), functional studies using bar-coded deletion mutant strains, and the construction of a mass-spec derived PPI network. Such work that is truly nonhypothesis-driven (in a traditional sense) produces the kind of data being rapidly assimilated from high-throughput experiments in all domains. Insight gained from such projects has demonstrated the power of discovery research versus traditional hypothesis driven approaches. We will review some of these efforts below, as they are the keystones in defining system structures for the study of biological systems.

7.2.1   Genome Sequencing

The full genome sequencing of what is now a relatively large number of vertebrate and invertebrate species has produced an explosion in available sequence data over the past 10 years. Comparisons between species have taught fundamental lessons about evolution and divergence. Although the term "blueprint of life" is frequently overused, these data have given us the template and building blocks from which life is constructed. This information includes the blueprints for all RNAs and proteins, the regulatory elements that ensure proper expression of all genes, the structural elements that govern chromosome function, and the records of our evolutionary history [538]. *Homo sapiens* [327, 529] and most major model organisms now have full genome sequence, including *S. cerevisiae* [380], *D. melanogaster* [5], *C. elegans* [118], *M. musculus* [538], and most recently, *R. norvegicus* [198]. Although some of the initial hyperinflated expectations regarding the accelerated pace of scientific discovery have not yet been fulfilled in the postgenomic era, a key early application has been the ability to find disease genes of unknown biochemical function by positional cloning [327]. This method involves mapping the chromosomal region containing the gene by linkage analysis in affected families and then examining the region to find the gene itself. Positional cloning is powerful, but it has also been extremely tedious prior to the knowledge of full genomic sequence. The sequence available in public databases now allows rapid identification in silico of candidate genes from an identified region. For a mendelian disorder, a gene search can now often be carried out in a matter of months as opposed to years. Knowledge of the complete set of human genes and proteins will also greatly expand the search for suitable drug targets in the fight against disease. A recent compendium lists 483 drug targets as accounting for virtually all drugs currently on the market from pharmaceutical companies [141, 142]. Although only a minority of human genes may be drug targets, it has been predicted that the number will exceed several thousand, and this prospect has led to a massive expansion of genomic research in pharmaceutical research and development.

7.2.2   Expression Profiling with DNA Microarrays

The use of DNA microarrays to measure expression levels of genes is the most prominent example of an experimental platform that has both enabled and forced biological research to shift from experiment-driven to data analysis-driven hypothesis generation. In *S.cerevisiae, C. elegans*, and *D.melanogaster*, microarray data have provided fundamental insight into basic biological processes [184, 255]. In the clinical domain, the derivation of molecular signatures of cancer subtypes has increased both our understanding of the biology of tumorigenesis, and provided new candidate markers for diagnostic and prognostic purposes [456].

A typical DNA microarray experiment involves fluorescent labeling of cDNA derived from one RNA sample, and hybridizing it competitively to cDNA derived from a second RNA sample labeled with a different fluor on the surface of a microarray slide. The microarray contains a series of cDNA sequences that can then be analyzed by laser scanning to quantify the relative amounts of fluorescent signal corresponding to the respective RNA samples. By using a "reference" pool of RNA as one of the labeled samples in each experiment, one can statistically make comparisons across samples in a data base of multiple experiments. In this platform, the data derived consists of ratios of expression levels of test compared to reference samples, and results in a matrix of experiments across one dimension, and genes along the other. The other major platform, popularized by Affymetrix, involves the deposition of oligonucleotide probes to a solid surface that fluorescently labeled cRNA from a sample of interest is hybridized to. The data generated is an absolute measurement of signal intensity rather than a ratio as in most cDNA microarray experiments.

Using these experimental platforms, the bottleneck then becomes to derive meaningful patterns from the data. In the simplest example, this can be a straight pairwise comparison of expression levels between two samples of interest, but more commonly the issue is one of deriving patterns from data where no *a priori* knowledge exists regarding what patterns or clusters are naturally present. This is the case in most clinical cancer microarray datasets, where little is known beforehand of what patterns will be present or even should be present. Some of the single largest datasets available currently include the Rosetta Inpharmatics yeast data, with over 300 experiments on all 6,800 *S. cerevisiae* ORF's [255], and the *C. elegans* data base from Stuart Kim, with 533 experiments using over 18,000 worm ORFs' [307]. Rosetta's approach involved construction of a reference data base or "compendium" of expression profiles corresponding to 300 diverse mutations and chemical treatments in S. cerevisiae, and correlation of the cellular pathways affected with patterns in expression profiles. The utility of the approach was validated by examining profiles caused by deletions of uncharacterized genes to identify and experimentally confirm eight uncharacterized open reading frames encoding proteins required for sterol metabolism, cell wall function, mitochondrial respiration, or protein synthesis. They also demonstrated that the compendium could be used to characterize pharmacological perturbations by identifying a novel target of the commonly used drug dyclonine.

In the cancer domain, datasets now exist for most hematologic malignancies and the majority of common solid tumors. The common theme emerging is the correlation of expression profiles with clinical outcomes not apparent in conventional staging systems. Although none of these discoveries has traversed the gap to routine clinical use, the potential for molecular substaging of tumors will undoubtedly revolutionize cancer diagnosis and treatment at some point in the future.

7.2.3   Functional Genomics

One of the major tasks remaining for systematic approaches to biological experimentation is the functional annotation of genomes. This has been tackled on a genome-wide scale in yeast, and in a limited manner in *C. elegans* and the mouse. Giaver *et al.* have described functional profiling of the yeast genome using a series of bar-coded deletion strains [197]. Interestingly, 18.7% of genes ($n = 1,105$) proved essential for growth on rich glucose medium, only about half of which were previously known to be essential. Fitness profiling in a number of different growth media revealed new genes involved in these processes. This type of data layered on top of protein interaction graphs can demonstrate the key hubs and nodes within a cellular protein network. In *C. elegans*, two studies have used the unique approach of RNAi [178, 209, 212, 290] to interrupt all genes along specific chromosomes. This strategy has also been employed to systematically catalog the function of mutations in the DNA damage response pathway [78, 213].

In the mouse, the most comprehensive study to date has been a systematic production of mutations along chromosome 11 using ENU-based mutagenesis against an inversion functioning as a balancer chromosome [305]. A total of 230 new recessive mutations were described, revealing novel gene function in hematopoiesis, fertility, craniofacial and cardiovascular development. Although much work remains to be done, the near future will conceivably bring an understanding of the mutational consequences of every gene in the genome for readily manipulable vertebrate model organisms such as the mouse.

7.2.4   Mass Spectrometry and the Evolution of Proteomics

Although we are only just beginning to appreciate the power and limitations of the genomic revolution, the nascent field of proteomics promises to deliver an even more radical transformation of biological and medical research. Encoded proteins carry out most biological functions, and to understand how cells work, one must study what proteins are present, how they interact with each other and what they do. The term proteome defines the entire protein complement in a given cell, tissue, or organism. In its wider sense, proteomics research also assesses protein activities, modifications and localization, and interactions of proteins in complexes.

Mass spectrometry (MS) is the study of gas phase ions as a means to characterize the structures, and hence identities, of molecules. Modern proteomics essentially began with the commercialization of soft ionization techniques in the 1990*s*, in particular electrospray ionization (ESI) and MALDI, which permitted analysis of proteins by MS for the first time. These instruments have largely replaced traditional Edman chemical sequencing for the analysis of proteins, and are capable of identification at picomolar to subpicomolar levels.

Tandem mass spectrometry (MS/MS) provides a means for fragmenting a mass-selected ion and measuring the mass-to-charge ratio (m/z) of the product ions that are produced during the fragmentation process, generating detailed structural information as a result. The mass selectivity of many commercial MS systems permits the isolation of single precursor peptide ions from mixtures, thereby removing the contribution of any other peptide or contaminant from the sequence analysis step. The product ion spectra can subsequently be interpreted to deduce the amino acid sequence of a protein.

A number of protein separation methods are available for purification prior to MS application, ranging from gel-based to liquid-chromatography systems. The computational analysis of the resulting data is not a trivial problem, as spectra obtained need to be analyzed against data bases of known peptides and proteins to obtain reliable identification. Most algorithms currently available are "home-grown," but the field is rapidly maturing as the benefit and utilization of MS technology expands.

The evolution of MS technology is beginning to have broad impact in biology and medicine. Some of the prominent examples to date include characterization of the mitochondrial proteome in mice [392], proteomic analyses of mouse mutant strains [430], and utilization of proteomic patterns in serum for the detection and diagnosis on tumors [443].

7.3   Protein–Protein Interactions

DNA is the ultimate repository of biological complexity. However, the spectrum of transcribed proteins or proteome is the most crucial medium for short-term information storage. In this view, `DNA = data`, and `protein = execution`. As a consequence, many researchers have viewed the next great milestone in biology to be the definition of all PPIs within an organism.

Two main experimental approaches have been used in this effort to date. One is mass spectrometry (as introduced above), while the other approach is called yeast 2-hybrid technology. This alternative provides a genetic approach to the identification and analysis of PPIs. It relies on the modular nature of many eukaryotic transcription factors, which contain both a site-specific DNA-binding domain and a transcriptional-activation domain that recruits the transcriptional machinery. In this assay, hybrid proteins are generated that fuse a protein X to the DNA-binding domain and protein Y to the activation domain of a transcription factor. Interaction between X and Y reconstitutes the activity of the transcription factor and leads to expression of reporter genes with recognition sites for the DNA-binding domain. In the typical practice of this method, a protein of interest fused to the DNA-binding domain (the so-called "bait") is screened against a library of activation-domain hybrids (referred to as "preys") to select interacting partners. The 2-hybrid system has evolved into a proteomic strategy by the construction of ordered arrays of strains expressing either DNA-binding domain or activation-domain fusion proteins, the implementation of improved selection methods and plasmids, the use of mating to introduce pairs of plasmids for testing, and the use of robotic-based automation.

Both MS and yeast 2-hybrid strategies are amenable to the high throughput required to make feasible the testing of all possible interactions for transcribed sequences. Both approaches have been used to catalog large volumes of PPIs in the yeast *S. cerevisiae* [190, 247]. The yeast two-hybrid strategy has been employed in a large-scale, high-throughput manner to define PPIs in *Drosophila melanogaster* [204] and the worm *C. elegans* [173, 337, 526]. In *C. elegans*, a yeast 2-hybrid strategy has also been used to define all interactions within the DNA damage response pathway and integrate this with functional data from mutant strains [78, 517].

7.4 Modular Biology from Protein Interaction Domains

The sequencing of complete genomes provides a list of genes and proteins responsible for cellular regulation. However, this does not reveal what these proteins do, nor how they are assembled into the molecular machines and functional networks that control cellular behavior. One of the more interesting concepts emerging from the study of protein sequence and structure is that of "modular domains" that are repeated throughout multiple different proteins and sometimes multiple times within the same protein. These protein interaction domains are involved in the regulation of cellular processes by directing the association of polypeptides either with one another or with phospholipids, small molecules, or nucleic acids. One can readily imagine how the modular nature of these domains, and the flexibility of their binding properties, have facilitated the evolution of cellular pathways [436].

Regulatory proteins are frequently constructed in a cassette-like fashion from domains that mediate molecular interactions or have enzymatic activity. Such interaction domains can target proteins to a specific subcellular location, nucleate the formation of multiprotein complexes, and control the conformation, activity, and substrate specificity of enzymes. These interaction domains can usually fold independently, and are readily incorporated into a larger polypeptide in a manner that leaves their ligand-binding surface available to recognize exposed sites on their protein partners. For example, 115 SH2 domains and 253 SH3 domains are encoded by the human genome. Some domains serve specific functions; for example, SH2 domains generally require phosphotyrosine sites in their primary ligands and are, therefore, dedicated to tyrosine kinase signaling. Other domains, such as SH3, can bind motifs found in a broader set of proteins and display a wider range of biological activities. A further means of building interaction surfaces is through the joining of repeated copies of a small peptide motif, yielding a much larger structure with multifaceted binding properties. Examples include HEAT (huntingtin, elongation factor 3, a subunit of protein phosphatase 2A, and TOR1), TPR, armadillo, and ankyrin sequences. Such examples demonstrate the function of interaction domains as scaffolds for complexes that control basic cellular processes. In this view, the cell uses a limited set of interaction domains, which are joined together in diverse combinations, to direct many of the actions of regulatory systems [436].

The above leads to a straightforward concept how the successive use of interaction domains can form linear signaling pathways. The prototypical example is that of the grb2 SH2- SH3 adaptor, which links pYXN motifs on RTKs to PXXP sites on Sos, a GEF for the Ras GTPase that stimulates MAP kinase pathway signaling. Such interactions can also potentially generate more complex networks that generate crosstalk between pathways and integrate signals from distinct receptors. For example, Grb2 also binds through its C-terminal SH3 domain to an RXXK motif in the docking protein Gab1, which is consequently tyrosine phosphorylated to create binding sites for the SH2 domains of cytoplasmic signaling proteins such as the p85 subunit of PI3K [347].

Another consequence of modular protein interaction domains is the potential use in evolution to produce new cellular functions by joining domains in new combinations to create novel connections and pathways within the cell. Interaction domains in this manner provide a way to increase the connectivity of existing proteins, and to create proteins with new functions. This is likely one of the fundamental reasons how the apparent complexity of an organism can increase so markedly with only a marginal corresponding increase in gene number [436].

Many examples exist where mutant cellular proteins that cause inherited disorders or malignancy exert their effects through either a loss of specific protein interactions or through the creations of aberrant protein complexes. The Bcr-Abl oncoprotein, for example, engages the Grb2 SH2 domain in leukemic cells. Another classic example is mutation in the *ras* guanine nucleotide exchange factor, occurring in a large number of solid tumors with varying frequency. Ki-*ras* is constitutively activated by mutations in $\sim$ 40–50% of colorectal cancers [365, 521] and $\sim$ 30% of lung adenocarcinomas. Ras mutation can lead to activation of several pathways, including the Raf/MEK/MAPK [226] and the PI3K/PKB [465] signaling cascades, both of which are capable of inducing transcriptional events.

7.5   Protein Interaction Networks

One of the more interesting and unpredicted consequences of the generation of large scale PPI data is the concept that networks of interacting proteins can be viewed and analyzed as large graphs, with nodes representing proteins and edges corresponding to interactions. Mathematical and computational methods have been derived for analyzing such networks in other subject areas, such as the Internet or communication and power grids. These techniques are being rapidly adapted and extended to study PPI networks. One of the startling revelations is the possibility that functional information (both in terms of individual proteins, and subgraphs) may be obtainable from graph–theory based approaches. We will briefly review below some of the emerging concepts from the study of PPI networks (further review is provided in Chapter 4).

7.5.1   Scale-Free Networks

One can find examples of networks everywhere. Power grids deliver electricity to our homes, land and wireless communication networks connect our telephone systems, and networks also describe our professional and social interactions. Systematic research into the structure of networks in general suggests that many have a "scale-free" structure, i.e., the connectivity of individual nodes in a network follows a power law, and thus some hubs (i.e., highly connected nodes in a network) have a seemingly unlimited number of links and no node is typical of the others (reviewed in [43]). In a scale-free network, new connections are added preferentially to nodes that already have a high degree of connectivity.

In contrast, random networks have nodes and connections characterized by a normal distribution, and thus do not contain hubs. Whether networks in biology truly follow such "scale-free" architecture is an open question, although there is considerable evidence that this may be the case. For example, analysis of yeast 2-hybrid networks demonstrate that the probability that a given yeast protein interacts with k other yeast proteins follows a power law with an exponential cut-off at k_c 20 [272], a topology that is also shared by the PPI network of the bacterium *Helicobacter pylori* [454]. This indicates that the network of protein interactions in two separate organisms forms a highly inhomogeneous scale-free network in which a few highly connected proteins play a central role in mediating interactions among numerous, less connected proteins. Alternatively, some of the common arguments for the reasons why the World Wide Web (WWW) is scale-free do not necessarily have biological correlates. For example, it is clear that new web pages have a predilection for more highly connected nodes within the WWW, but this is not typically the case for new proteins added to regulatory or signaling systems as we move up the chain of evolutionarily more complex organisms.

Alternative evidence suggests that the scale-free characteristics of current PPI networks may not be influenced by biology, but rather by human bias. Since false negatives are dominant in currently available PPI networks, it is feasible that hubs are hubs because many researchers study their neighborhood, and nodes with low connectivity are largely neglected (i.e., rarely selected as baits in MS experiments). It has been shown in social networks, that reinforcement learning can create scale-free structures even on a backbone of nonscale-free networks [20].

Since well studied areas have a lower likelihood of false negatives, some have argued that it makes more sense to focus on local features of PPI networks, such as graphlets [447] and motifs [387, 489, 566], rather than global properties [271, 274, 366, 450]. This research strongly suggests that current PPI networks can be more accurately modeled by geometric random graphs. The mismatch of current PPI networks with scale-free models is much bigger than PPI networks than with 30% perturbed geometric random graphs (i.e., removing and adding 30% of nodes, and rewiring 30% of edges) [447]. The future will tell if new experimental data will support or reject this new model.

7.5.2 Specificity and Stability in the Topology of Protein Networks

Robustness is a desired property of any engineered system, and it is an essential property of all biological systems [125]. The phenomenological properties exhibited by robust systems can be classified into three areas [313]:

1. *Adaptation*, which denotes the ability to cope with environmental changes;

2. *Parameter insensitivity*, which indicates a system's relative insensitivity to specific kinetic parameters;

3. *Graceful degradation*, which reflects the characteristic slow degradation of a system's functions after damage, rather than catastrophic failure.

When designing systems, engineers cover all three areas of robustness by applying several approaches [313]:

- Introducing a *system control* such as negative-feedback and feed-forward control to implement adaptation;

- Implementing *redundancy*, whereby multiple components with equivalent functions are introduced for backup;

- Promoting *structural stability*, where intrinsic mechanisms are built to promote stability;

- Fostering *modularity*, where subsystems are physically or functionally insulated so that failure in one module does not spread to other parts and lead to system-wide catastrophe.

Not surprisingly, these approaches used in engineering systems are also found in biological systems. Biologists and biophysicists new to studying complex networks often express surprise at a biological network's apparent robustness [125]. Some even conclude that these mechanisms and their resulting features seem absent in engineering [108, 311]. However, ironically, it is in the nature of their robustness and complexity that biology and advanced engineering are most alike. Good design in both cases (e.g., cells and bodies, engines and airplanes) means that users are largely unaware of hidden complexities, except through system failures. Furthermore, the robustness and fragility features of complex systems are both shared and necessary [125]. Although the need for universal principles of complexity and corresponding mathematical tools is widely recognized, sharp differences arise as to what is fundamental about complexity and what mathematics is needed [125].

Molecular networks operating in living systems are rapidly being defined, and direct physical interactions among proteins comprise one such network. Structural relationships between nodes define pathways for the propagation of signals. Not only are direct links between hubs suppressed, but hubs also tend to share fewer of their neighbors with other hubs, thereby extending their isolation to the level of next-nearest neighbor connections. Both the hub node itself and its immediate surroundings tend to separate from other hubs, reinforcing the picture of functional modules clustered around individual hubs. Because highly connected nodes serve as powerful amplifiers against deleterious perturbations, it is particularly important to suppress this propagation beyond their immediate neighbors. Scale-free networks in general are very vulnerable to attacks aimed at highly connected nodes. This may be why correlations between connectivity of a given protein and the lethality of the mutant cell lacking this protein are not particularly strong.

Protein interaction networks overall are characterized by a high degree of interconnectedness, where most nodes are reachable from each other, i.e., they are linked to each other by at least one path. The question arises how such a heavily intertwined and mutually dependent dynamical system can perform multiple functional tasks and still remain stable against deleterious perturbations. Comparisons with random network models suggest that for both interaction and regulatory networks, links between highly connected proteins are systematically suppressed, while those between highly connected and low-connected pairs of proteins are favored. This effect decreases the likelihood of cross-talk between different functional modules of the cell and increases the overall robustness of a network by localizing effects of deleterious perturbations Figure 7.1. The resulting conclusion is that molecular networks in a living cell appear have organized themselves in an interaction pattern that is both robust and specific [366].

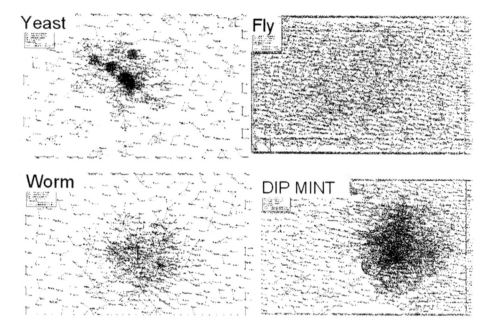

FIGURE 7.1: Rendering protein–protein interaction networks for different organisms immediately suggests that we have a highly incomplete system and that different experimental techniques used to elucidate interactions are biased. As a result, the least populated worm PPI data base is sparse, while yeast PPI network has many dominating hubs and complexes, similarly as human PPI data set from OPHID [84], shown in Figure 6.5.

7.6   Theoretical Models of Biological Systems

The accumulation of data from systematic experiments has surfaced the potential to construct models of how biological systems work at the whole cell or whole organism level. How to integrate multiple levels of information to achieve this task is not trivial, and we discuss some of the possible approaches below.

The simplest conceptual model of a biological system uses the analogy of the circuit diagram. In this case, different inputs and their modifiers lead to specific system outputs. Such models have been applied to compartmentalized cellular systems, specifically metabolic networks or specific transcriptional activators. Theoretically, we can model an entire cell 3-dimensionally as an overlay of 2-dimensional circuit diagrams. Such models require not just information on positive and negative regulators, but also reaction kinetics for each modifier of the circuit. It is at this point that such models break down in the context of looking at the entire system in question, as the amount of input data required to draw the circuit model is exponentially prohibitive. At a simple level, the existing integrated yeast protein interaction network is a graph of 5,321 nodes containing 78,390 connecting edges, which would require detailed kinetic information on a minimum of millions different reactions. In humans, we can expect a network of protein interactions with 120,000 nodes, and millions of connecting edges (see Ffig-networks). This degree of reductionism is unlikely to be obtainable with existing technology.

The application of theories of complex networks has brought many new insights into biological systems. The simplest of these is the complete PPI network viewed as a graph with individual proteins as nodes and their connections as edges. The structure of such a network reveals information about points of vulnerability, signaling pathways, and protein complexes. However, it does not reveal the direction of information flow, the kinetics of individual reactions, the consequences of positive or negative modulation on cellular outputs. Many interactions are transient, and thus they carry signal only as a response to specific stimuli. We still lack information about the context of these interactions. As a result, most proteins eventually will be connected to all other proteins in this static view of protein interactions.

Is there promise for biological understanding in other models of complex systems? The recent description of cellular automata by Stephen Wolfram is one example of novel theoretical frameworks for understanding any complex system [554]. The theory conjectures that any complex system and its behavior can be understood by a series of simple rules or programs that can dictate extreme complexity. One of the most startling conclusions from these ideas is that evolution does not advance by sampling all possible solutions to a problem or permutations of possible solutions, but instead randomly finds a solution that works and sticks with it. This would suggest however that

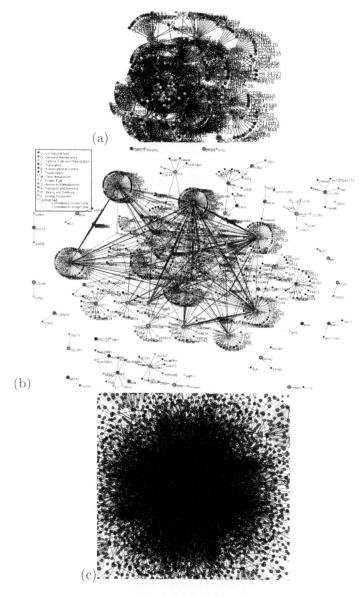

FIGURE 7.2: Although all PPI networks still comprise large number of false negatives, it is expected that with the increased complexity of the organism (moving from yeast to worm to human) we will get increased density of PPI networks. As a result, analysis and visualization of human PPI network will be a challenge, as we will need new heuristics in graph theory algorithms to scale to hundreds of thousands of nodes, and we will need new dynamic and high definition visualization to distinguish millions of edges. Although moving from visualizing the networks (a) to analyzing their structure-function properties (b) will be helpful, density of future networks (c) will require other approaches and especially contextual information about the interactions, such as presented in Figures 2.7 and 6.6. The context will tell us when, where and to what extent the proteins interact [48] (see Figure 2.7).

the working solution might be far from optimal. Similar problems have been found in studies of artificial intelligence, where solutions can be stuck in local minima unless additional algorithms are used to search the neighborhood to find globally optimal solutions.

Close examination of the organization and design of advanced technologies indirectly suggest universal principles that may be relevant to biology. If truly universal principles exist, they should manifest themselves in at least limited ways in scale-model systems. An interesting conceptual analogy for biological systems has been proposed by [98, 125, 439, 461] involving the ubiquitous Lego® system, with the signature feature of a patented snap connection for easy but precise and stable assembly of components. In this framework, the snap can be viewed as the basic Lego protocol, and Lego bricks are its basic modules.

This analogy suggests that protocols are far more important to biologic complexity than are modules [125]. In everyday usage, protocols are rules designed to manage relationships and processes smoothly and effectively. If modules are ingredients, parts, components, subsystems, and players, then protocols describe the corresponding recipes, architectures, rules, interfaces, etiquettes, and codes of conduct. Protocols here are rules that prescribe allowed interfaces between modules, permitting system functions that could not be achieved by isolated modules. Protocols also facilitate the addition of new protocols and organization into collections of mutually supportive protocol suites. A good protocol is one that supplies robustness and supports evolution [125].

Lego bricks and toys are reusable and robust to trauma. The snap is extremely versatile, permitting endless varieties of toys from an array of components. The toy, and what it creates, is consequently both robust and versatile. This makes both a given Lego collection and the entire toy system evolvable to changes in what one chooses to build, to the addition of new Lego-compatible parts, and to novel toy designs. The low cost of modules and the popularity of the system confer other forms of robustness and evolvability; lost parts are easily replaced, and enthusiasts continuously design new modules and toys. The Lego protocol also creates fragilities at every level. Superficially minuscule damage to the snap at a key interface may cause an entire toy to fail, yet noninterfacing parts of bricks may be heavily damaged with minimal impact beyond aesthetics. The success of Lego is supported by the fact that any new snap, even a superior one, would not be easily adopted. Selection pressures thus preserve a protocol in two ways, as protocols both facilitate evolution and are difficult to change [125].

As systems become more complex, protocols facilitate the layering of additional protocols, particularly involving feedback and signaling. As described in [125], suppose we want to make a Lego structure incrementally more useful and versatile by "evolving" it to be:

1. Mobile;

2. Motorized;

3. Able to avoid collisions in a maze of obstacles.

The first increment is easy to achieve, with Lego protocol-compatible axles and wheels. Motorizing toys involves a second increment in complexity, requiring protocols for motor and battery interconnection (or other power source) as well as a separate protocol for gears. All can be integrated into a motorized protocol suite to make modular subassemblies of batteries, motors, gears, axles, and wheels. These are available, inexpensive additions. The third increment increases cost and complexity by orders of magnitude, requiring layers of protocols and modules for sensing, actuation, and feedback controls plus subsidiary, but essential ones, for communications and computing. All are available, but it is here that we begin to see the true complexity of advanced technologies. Unfortunately, we also start to lose the easily described, intuitive story of the basic protocols. Minimal descriptions of advanced Lego features enabling sensing and feedback control literally fill books, but the protocols also facilitate the building of elaborate, robust toys, precisely because this complexity is largely hidden from users. This is consistent with the claim that biological complexity too is dominated not by minimal function, but by the protocols and regulatory feedback loops that provide robustness and evolvability (see Figure 7.3).

This added complexity also creates new and often extreme fragilities [125]. Removing a toy's control system might cause reversion to mere mobility, but a small change in an otherwise intact control system could cause wild, catastrophic behavior. For example, a small software bug might easily lead to collision seeking, a fragility absent in simpler toys. Similarly, large multicellular organisms are unaffected by the death of a single cell, but failure of one cell's control system can lead to fatal autoimmune diseases or cancer.

In biology, the identification of protocols is easiest when shared by many different modules, as in Lego. Thus, abstractions such as gene regulation, covalent modification, membrane potentials, metabolic and signal transduction pathways, action potentials, and even transcription–translation, the cell cycle, and DNA replication could all be reasonably described as protocols, with their attendant modular implementations in various activators and repressors, kinases and phosphatases, ion channels, receptors, heterotrimeric guanine nucleotide binding proteins (G proteins), and so on. The cardiovascular system has protocols for gas and nutrient exchange and transport, implemented in heart, lung, vascular networks, and blood modules. The immune system involves elaborate protocols for complement and cell-mediated activation, implemented in modules such as T cells, natural killer cells, major histocompatibility complex molecules, and antibodies. Metazoan development has highly conserved protocols. Appropriate temporal and spatial expression during development is regulated by enormous numbers of feedback strategies. These and many other protocols facilitate robust development and function in ways similar to Lego protocols, and they produce similar fragilities [125].

FIGURE 7.3: (See color insert following page 72.) Generic block can be used to construct simple wall (a) or more functional and diverse toys (b), more intriguing structures, which create forms different from the basic components (circular shape from rectangular pieces) (c) or when using specialized parts even highly mobile and controllable devices (d). Biological systems too share many basic components, and higher organisms achieve complex behavior by not only new "parts," but also by more complex use of individual components.

The protocol concept is also a useful abstraction for understanding the evolution of complexity [125]. In this sense, good protocols allow new functions to be built from existing components and allow new components to be added or to evolve from existing ones. More complex protocols enable modularity and robustness but are in turn sources of fragility. Successful protocols become highly conserved because they both facilitate evolution and are difficult to change. Even these simple toy examples show the robust yet fragile features of complex regulatory networks. Their outward signatures are constant, extremely regulated variables, and they exhibit extraordinary robustness to component variations despite the fact that rare but catastrophic cascading failures can and do happen. These apparently paradoxical combinations can easily be a source of confusion to experimentalists, clinicians, and theoreticians alike, but are intrinsic features of highly optimized feedback regulation. Since net robustness and fragility are constrained quantities, they must be manipulated and controlled within complex networks, even more so than energy and materials. The key to good control design, then, is to ensure that fragility is tolerable and occurs where uncertainties are relatively small [125].

7.7 Software Development for System Modeling

The computational modeling of biological systems is a critical technology for organizing and integrating vast amounts of biological information. Although biological simulation is now being done for a wide range of pathways, cells, and systems, the role of in silico biology in medical and pharmaceutical research is likely to become increasingly prominent as we seek to exploit the data generated through rapid gene sequencing and proteomic mapping. However, progress will be significantly enhanced by enabling larger numbers of researchers to use and verify models in the course of their everyday experimental work, as simulation and experiment must go together to maximize understanding [313]. Up to now it has been extremely difficult to transfer models between research centers, or to extend existing models so that more complex models can be constructed in an object-oriented or modular fashion. This process will be enhanced by the development of uniform standards for representing and communicating the content of models, and by the wide distribution of software tools that permit even nonmodelers to access, execute, and improve existing models. Increasingly, publication of models is accompanied by their availability on Web sites. In addition, the process of establishing standards of communication and languages is developing.

Once both the network structure and its functional properties are understood for a large number of regulatory circuits, studies on classifications and comparison of circuits will provide further insights into the richness of design patterns used and how design patterns of regulatory circuits have been modified or conserved through evolution. The hope is that intensive investigation will reveal a possible evolutionary family of circuits as well as a "periodic table" for functional regulatory circuits [313]. Although attempts have been made to build simulation software and to make use of the many analysis and computing packages originally designed for general engineering purposes, there is no common infrastructure or standard to enable integration of these resources. The Systems Biology Mark-up Language (SBML)[1], along with CellML, represent attempts to define a standard for an XML-based computer-readable model definition that enables models to be exchanged between software tools. Systems Biology Workbench (SBW) is built on SBML and provides a framework of modular open-source software for systems biology research. Both SBML and SBW are collective efforts of a number of research institutions sharing the same vision. Progress in any of the above areas toward effective modeling requires substantial breakthroughs in computational sciences, genomics, proteomics, measurement technology, and the integration with existing knowledge.

In a few years' time, we shall all wonder how we ever managed to do without integrated models in biological research. For drug development, there will certainly be a major change as these tools come on line and rapidly increase in their power. This will grow in a nonlinear way with the degree of biological detail that is incorporated, as the number of interactions modeled increases much faster than the number of components. Biology is set to become highly quantitative and a computer-intensive discipline in the 21*st* century [313].

7.8   Cancer as a System Failure

From many viewpoints, the specific problem of cancer biology can be viewed as a problem of system failure. The "hallmarks" of cancer include many diverse and seemingly nonoverlapping biological processes, including cell division, angiogenesis, migration and adhesion, DNA repair, and intracellular signaling [229]. Although some cancer subtypes are defined by a single genetic alteration leading to a primary defect in one of the above listed processes, most solid tumors responsible for the largest burden of human illness are heterogeneous lesions characterized by many if not all defects observable simultaneously. This includes lung, breast, prostate, colon, and central nervous system tumors among others. For many of these cancers where the specific inciting lesion is unknown, one hypothesis would be to view the defects as system failures either at a cellular or multicellular level. Defining such failures,

---

[1]http://sbml.org/index.psp

however, would require precise definition of network structure for individual tumor types, which is currently not feasible.

One line of evidence of cancer as a system failure is the evidence for the angiogenic phase of tumorigenesis being a critical component of tumor progression once lesions reach a certain size. In this framework, the acquisition of mutations due to genetic instability leads to the transformation of normal cells in our body into cancer cells. This phase is not inherently lethal, and generally results in a microscopic tumor. The second phase involves a switch to the angiogenic phenotype, involving the constant recruitment of new blood vessels to the tumor. This converts the nonlethal *in situ* lesions into an expanding mass that is potentially lethal. A number of critical factors have been identified with the capacity to launch angiogenesis and enter a lethal phase of cancer, including basic FGF, VEGF, PDGF, thrombospondin, tumstatin, canstatin, endostatin, and angiostatin. Interestingly, there is a very low incidence of solid tumors in patients with Down Syndrome, who circulate elevated levels of endostatin, an endogenous angiogenesis inhibitor, due to an extra copy of chromosome 21. Tumor progression clearly depends on the balance between the *in situ* tumor's total angiogenic output and an individual's total angiogenic defense.

In a similar vein, the balances holding an *in situ* lesion in check before it advances to a focal tumor involve a balance between a high rate of tumor cell division and cell death. A number of immune surveillance mechanisms are postulated to be involved in such a balance. For example, autopsies of individuals dying of trauma often reveal microscopic colonies of cancer cells, also known as *in situ* tumors. Virtually all autopsied individuals aged 50 to 70 have *in situ* carcinomas in their thyroid gland, whereas only 0.1% of individuals in this age group are diagnosed with thyroid cancer during this period of their life. It has long puzzled clinicians and scientists why cancer develops and progresses to be lethal only in a very small percentage of people. The realization that a lot of us carry *in situ* tumors, but do not develop the disease, suggests that these microscopic tumors are mostly dormant and need additional signals to grow [175].

One can theorize that it may be critical "protocol" rather than "modules" that are to blame for most cancers, and that the elucidation of such mechanisms is our only hope for truly successful drug targets. It may indeed be the case that such a systems view will provide the key insight to understand the biology of tumorigenesis, and to implement this knowledge in the search for successful therapeutics.

References

[1] J. Abello, A. Buchsbaum, and J. Westbrook. A functional approach to external graph algorithms. In *Lecture Notes in Comp Sci*, volume 1461, pages 332–343. Springer, Berlin, 1998.

[2] B. M. Acton, A. Jurisicova, I. Jurisica, and R. F. Casper. Alterations in mitochondrial membrane potential during preimplantation stages of mouse and human embryo development. *Mol Hum Reprod*, 10(1):23–32, 2004.

[3] T. B. Acton, K. Gunsalus, R. Xiao, L. Ma, J. Aramini, M. C. Baron, Y. Chiang, T. Clement, B. Cooper, N. Denissova, S. Douglas, J. K. Everett, D. Palacios, R. H. Paranji, R. Shastry, M. Wu, C.-H. Ho, L. Shih, G. V. T. Swapna, M. Wilson, M. Gerstein, M. Inouye, J. F. Hunt, and G.T. Montelione. Robotic cloning and protein production platform of the Northeast Structural Genomics Consortium. *Meth Enzymol*, 394:210–243, 2005.

[4] L. A. Adamic. The small world web. *Lecture Notes in Comp Sci*, 1696:443–454, 1999.

[5] M. D. Adams, S. E. Celniker, R. A. Holt, C. A. Evans, J. D. Gocayne, P. G. Amanatides, S. E. Scherer, P. W. Li, R. A. Hoskins, R. F. Galle, R. A. George, S. E. Lewis, S. Richards, M. Ashburner, S. N. Henderson, G. G. Sutton, J. R. Wortman, M. D. Yandell, Q. Zhang, L. X. Chen, R. C. Brandon, Y. H. Rogers, R. G. Blazej, M. Champe, B. D. Pfeiffer, K. H. Wan, C. Doyle, E. G. Baxter, G. Helt, C. R. Nelson, G. L. Gabor, J. F. Abril, A. Agbayani, H. J. An, C. Andrews-Pfannkoch, D. Baldwin, R. M. Ballew, A. Basu, J. Baxendale, L. Bayraktaroglu, E. M. Beasley, K. Y. Beeson, P. V. Benos, B. P. Berman, D. Bhandari, S. Bolshakov, D. Borkova, M. R. Botchan, J. Bouck, P. Brokstein, P. Brottier, K. C. Burtis, D. A. Busam, H. Butler, E. Cadieu, A. Center, I. Chandra, J. M. Cherry, S. Cawley, C. Dahlke, L. B. Davenport, P. Davies, B. de Pablos, A. Delcher, Z. Deng, A. D. Mays, I. Dew, S. M. Dietz, K. Dodson, L. E. Doup, M. Downes, S. Dugan-Rocha, B. C. Dunkov, P. Dunn, K. J. Durbin, C. C. Evangelista, C. Ferraz, S. Ferriera, W. Fleischmann, C. Fosler, A. E. Gabrielian, N. S. Garg, W. M. Gelbart, K. Glasser, A. Glodek, F. Gong, J. H. Gorrell, Z. Gu, P. Guan, M. Harris, N. L. Harris, D. Harvey, T. J. Heiman, J. R. Hernandez, J. Houck, D. Hostin, K. A. Houston, T. J. Howland, M. H. Wei, C. Ibegwam, et al. The

genome sequence of *Drosophila melanogaster*. *Science*, 287(5461):2185–95, 2000.

[6] R. Aebersold and M. Mann. Mass spectrometry-based proteomics. *Nature*, 422(6928):198–207, 2003.

[7] R. Agrawal, T. Imielinski, and A. N. Swami. Mining association rules between sets of items in large databases. *Proc ACM SIGMOD Int Conf on Management of Data*, 22(2):207–216, 1993.

[8] D. W. Aha. *The AAAI-99 KM/CBR Workshop: Summary of Contributions*. AAAI Press, Menlo Park, CA, 1999.

[9] W. Aiello, F. Chung, and L. Lu. A random graph model for power law graphs. *Exp Math*, 10:53–66, 2001.

[10] R. Albert and A. L. Barabasi. Topology of evolving networks: local events and universality. *Phys Rev Lett*, 85(24):5234–5237, 2000.

[11] R. Albert and A. L. Barabasi. Statistical mechanics of complex networks. *Rev Mod Phys*, 74:47–97, 2002.

[12] R. Albert, H. Jeong, and A. L. Barabasi. Diameter of the world-wide web. *Nature*, 401:387–392, 1999.

[13] R. Albert, H. Jeong, and A. L. Barabasi. Error and attack tolerance of complex networks. *Nature*, 406:378–382, 2000.

[14] C. J. Alpert and A. B. Kahng. Recent directions in netlist partitioning: a survey. *Integration*, 19:1–81, 1995.

[15] S. F. Altschul, W. Gish, W. Miller, and D. J. Lipman. Basic local alignment search tool. *J Mol Biol*, 215:403–410, 1990.

[16] S. F. Altschul, T. L. Madden, A. A. Schffer, J. Zhang, Z. Zhang, W. Miller, and D. J. Lipman. Gapped BLAST and PSI-BLAST: a new generation of protein database search programs. *Nucleic Acids Res*, 25:3389–3402, 1997.

[17] L. A. N. Amaral, A. Scala, M. Barthelemy, and H. E. Stanley. Classes of behavior of small-world networks. *Proc Natl Acad Sci USA*, 97:11149–11152, 2000.

[18] M. R. Anderberg. *Cluster Analysis for Applications*. Academic Press, New York, 1973.

[19] E. D. Angelini, Y. Wang, and A. Laine. Classification of micro array genomic images with Laplacian pyramidal filters and neural networks. In *Workshop on Genomic Signal Processing and Statistics (GENSIPS)*. Baltimore, MD, 2004.

[20] M. Anghel, Z. Toroczkai, K. E. Bassler, and G. Korniss. Competition-driven network dynamics: emergence of a scale-free leadership structure and collective efficiency. *Phys Rev Lett*, 92(5):058701, 2004.

[21] Anonymous. Help! The data are coming. *Nature*, 399(6736):505, 1999.

[22] R. Apweiler, T. K. Attwood, A. Bairoch, A. Bateman, E. Birney, M. Biswas, P. Bucher, L. Cerutti, F. Corpet, M. D. Croning, R. Durbin, L. Falquet, W. Fleischmann, J. Gouzy, H. Hermjakob, N. Hulo, I. Jonassen, D. Kahn, A. Kanapin, Y. Karavidopoulou, R. Lopez, B. Marx, N. J. Mulder, T. M. Oinn, M. Pagni, F. Servant, C. J. Sigrist, and E. M. Zdobnov. Interpro–an integrated documentation resource for protein families, domains and functional sites. *Bioinformatics*, 16(12):1145–1150, 2000.

[23] R. Apweiler, A. Bairoch, C. H. Wu, W. C. Barker, B. Boeckmann, S. Ferro, E. Gasteiger, H. Huang, R. Lopez, M. Magrane, M. J. Martin, D. A. Natale, C. O'Donovan, N. Redaschi, and L. S. Yeh. Uniprot: the universal protein knowledgebase. *Nucleic Acids Res*, 32(Database issue):D115–119, 2004.

[24] R. J. Arnold, P. Hrncirova, K. Annaiah, and M. V. Novotny. Fast proteolytic digestion coupled with organelle enrichment for proteomic analysis of rat liver. *J Proteome Res*, 3(3):653–657, 2004.

[25] M. Ashburner. FlyBase. *Genome News*, 13:19–20, 1993.

[26] M. Ashburner, C. A. Ball, J. A. Blake, D. Botstein, H. Butler, J. M. Cherry, A. P. Davis, K. Dolinski, S. S. Dwight, J. T. Eppig, M. A. Harris, D. P. Hill, L. Issel-Tarver, A. Kasarskis, S. Lewis, J. C. Matese, J. E. Richardson, M. Ringwald, G. M. Rubin, and G. Sherlock. Gene ontology: tool for the unification of biology. the gene ontology consortium. *Nat Genet*, 25(1):25–29, 2000.

[27] S. Asthana, O. D. King, F. D. Gibbons, and F. P. Roth. Predicting protein complex membership using probabilistic network reliability. *Genome Res*, 14(6):1170–1175, 2004.

[28] G. D. Bader, D. Betel, and C. W. V. Hogue. BIND: the biomolecular interaction network database. *Nucleic Acids Res*, 31(1):248–250, 2003.

[29] G. D. Bader and C. W. Hogue. Bind–a data specification for storing and describing biomolecular interactions, molecular complexes and pathways. *Bioinformatics*, 16(5):465–477, 2000.

[30] G. D. Bader and C. W. V. Hogue. Analyzing yeast protein–protein interaction data obtained from different sources. *Nat Biotechnol*, 20:991–997, 2002.

[31] G. D. Bader and C. W. V. Hogue. An automated method for finding molecular complexes in large protein interaction networks. *BMC Bioinformatics*, 4:2, 2003.

[32] J. S. Bader, A. Chaudhuri, J. M. Rothberg, and J. Chant. Gaining confidence in high-throughput protein interaction networks. *Nat Biotechnol*, 22(1):78–85, 2004.

[33] J. S. Bader, A. Chaudhuri, J. M. Rothberg, and J. Chant. Gaining confidence in high-throughput protein interaction networks. *Nat Biotechnol*, 22(1):78–85, 2004.

[34] A. Badia. Cooperative query answering with generalized quantifiers. *J Intell Inf Syst*, 12(1):75–97, 1999.

[35] A. Bairoch and R. Apweiler. The SWISS-PROT protein sequence database and its supplement TrEMBL in 2000. *Nucleic Acids Res*, 28:45–48, 2000.

[36] P. G. Baker, A. Brass, S. Bechhofer, C. Goble, N. Paton, and R. Stevens. TAMBIS: Transparent access to multiple bioinformatics information sources; an overview. In *Proc of the 6th Int Conf on Intelligent Systems for Molecular Biology (ISMB98)*, pages 25–34. AAAI Press, Menlo Park, CA, 1998.

[37] P. G. Baker, C. A. Goble, S. Bechhofer, N. W. Paton, R. Stevens, and A. Brass. An ontology for bioinformatics applications. *Bioinformatics*, (6):510–520, 1999.

[38] F. Ball, J. Mollison, and G. Scalio-Tomba. Epidemics with two levels of mixing. *Annals Appl Probab*, 7:46–89, 1997.

[39] Z. Bar-Joseph, G. K. Gerber, T. I. Lee, N. J. Rinaldi, J. Y. Yoo, F. Robert, D. B. Gordon, E. Fraenkel, T. S. Jaakkola, R. A. Young, and D. K. Gifford. Computational discovery of gene modules and regulatory networks. *Nat Biotechnol*, 21(11):1337–1342, 2003.

[40] A. L. Barabasi and R. Albert. Emergence of scaling in random networks. *Science*, 286(5439):509–512, 1999.

[41] A. L. Barabasi, R. Albert, and H. Jeong. Mean-field theory for scale-free random networks. *Physica A*, 272:173–197, 1999.

[42] A. L. Barabasi, Z. Dezso, E. Ravasz, Z.-H. Yook, and Z. N. Oltvai. Scale-free and hierarchical structures in complex networks. In *AIP Conf Proc*, volume 661, pages 1–16. Melville, New York, 2003.

[43] A. L. Barabasi and Z. N. Oltvai. Network biology: understanding the cell's functional organization. *Nat Rev Genet*, 5(2):101–113, 2004.

[44] A. L. Barabasi, E. Ravasz, and T. Vicsek. Deterministic scale-free networks. *Physica A*, 299:559–564, 2001.

[45] A. D. Barbour and G. Reinert. Small worlds. *Random Struct Algor*, 19:54–74, 2001.

[46] J. B. Bard and S. Y. Rhee. Ontologies in biology: design, applications and future challenges. *Nat Rev Genet*, 5(3):213–222, 2004.

[47] A. Barrat and M. Weigt. On the properties of small-world network models. *Eur Phys J B*, 13:547–560, 2000.

[48] M. Barrios-Rodiles, K. R. Brown, B. Ozdamar, R. Bose, Z. Liu, R. S. Donovan, F. Shinjo, Y. Liu, J. Dembowy, I. W. Taylor, V. Luga, N. Przulj, M. Robinson, H. Suzuki, Y. Hayashizaki, I. Jurisica, and J. L. Wrana. High-throughput mapping of a dynamic signaling network in mammalian cells. *Science*, 307(5715):1621–1625, 2005.

[49] M. Barthelemy and L. A. N. Amaral. Small-world networks: evidence for crossover picture. *Phys Rev Lett*, 82:3180–3183, 1999.

[50] V. Batagelj and A. Mrvar. Pajek — program for large network analysis. *Connections*, 2:47–57, 1998.

[51] A. D. Baxevanis. The molecular biology database collection: 2003 update. *Nucleic Acids Res*, 31(1):1–12, 2003.

[52] S. A. Beausoleil, M. Jedrychowski, D. Schwartz, J. E. Elias, J. Villen, J. Li, M. A. Cohn, L. C. Cantley, and S. P. Gygi. Large-scale characterization of hela cell nuclear phosphoproteins. *Proc Natl Acad Sci USA*, 101(33):12130–12135, 2004.

[53] N. J. Belkin and W. B. Croft. Retrieval techniques. In M. Williams, editor, *Annual Review of Information Science and Technology*, volume 22, pages 109–145. Elsevier, New York, 1987.

[54] A. Ben-Dor, R. Shamir, and Z. Yakhini. Clustering gene expression patterns. *J Comput Biol*, 6(3/4):281–297, 1999.

[55] E. A. Bender and E. R. Canfield. The asymptotic number of labeled graphs with given degree sequences. *J Comb Theory A*, 24:296–307, 1978.

[56] D. A. Benson, I. Karsch-Mizrachi, D. J. Lipman, J. Ostell, B. A. Rapp, and D. L. Wheeler. GenBank. *Nucleic Acids Res*, 30:17–20, 2002.

[57] R. Bergmann, S. Breen, M. Goker, M. Manago, and S. Wess. *Developing Industrial Case-Based Reasoning Applications: The INRECA Methodology*, volume 1612. Springer, Berlin, 1999.

[58] M. Bern, D. Goldberg, R. C. Stevens, and P. Kuhn. Automatic classification of protein crystallization images using a curve-tracking algorithm. *J Applied Crystallogr*, 37:279–287, 2004.

[59] D. Berrar, B. Sturgeon, I. Bradbury, and W. Dubitzky. Microarray data integration and machine learning techniques for lung cancer survival prediction. *CAMDA*, 2003.

[60] L. Bertalanffy. *Modern Theories of Development: An Introduction to Theoretical Biology*. Oxford University Press, New York, 1933.

[61] D. C. Berwick and J. M. Tavare. Identifying protein kinase substrates: hunting for the organ-grinder's monkeys. *Trends Biochem Sci*, 29(5):227–232, 2004.

[62] D. Betel, R. Isserlin, and C. W. Hogue. Analysis of domain correlations in yeast protein complexes. *Bioinformatics*, 20 Suppl 1:I55–62, 2004.

[63] C. Bettstetter. On the minimum node degree and connectivity of a wireless multihop network. In *Proc of the 3rd ACM Int Symposium on Mobile ad hoc Networking and Computing*, pages 80–91. 2002.

[64] M. Biswas, J. F. O'Rourke, E. Camon, G. Fraser, A. Kanapin, Y. Karavidopoulou, P. Kersey, E. Kriventseva, V. Mittard, N. Mulder, I. Phan, F. Servant, and R. Apweiler. Applications of interpro in protein annotation and genome analysis. *Brief Bioinform*, 3(3):285–295, 2002.

[65] B. Blagoev, S. E. Ong, I. Kratchmarova, and M. Mann. Temporal analysis of phosphotyrosine-dependent signaling networks by quantitative proteomics. *Nat Biotechnol*, pages 1139–1145, 2004.

[66] M. Blatt, S. Wiseman, and E. Domany. Superparamagnetic clustering of data. *Phys Rev Lett*, 76(18):3251–3254, 1996.

[67] J. R. Bock and D. A. Gough. Predicting protein–protein interactions from primary structure. *Bioinformatics*, 17(5):455–460, 2001.

[68] E. R. Bodenstaff, F. J. Hoedemaeker, M. E. Kuil, H. P. de Vrind, and J. P. Abrahams. The prospects of protein nanocrystallography. *Acta Crystallogr D Biol Crystallogr*, 58(Pt 11):1901–1906, 2002.

[69] B. Bollobas. *Random Graphs*. Academic, London, 1985.

[70] B. Bollobas and O. Riordan. The diameter of a scale-free random graph. *Combinatorica*, 24(1):5–24, 2004.

[71] H. Bono, T. Kasukawa, Y. Hayashizaki, and Y. Okazaki. Read: Riken expression array database. *Nucleic Acids Res*, 30(1):211–213, 2002.

[72] M. Boots and A. Sasaki. 'Small worlds' and the evolution of virulence: infection occurs locally and at a distance. *Proc R Soc London B, Biological Sciences*, 266:1933–1938, 1999.

[73] A. Borgida. Modeling class hierarchies with contradictions. In *Int Conf on Management of Data Archive*, pages 434–443. Chicago, Illinois, 1988.

[74] A. Borgida, S. Greenspan, and J. Mylopoulos. Knowledge representation as a basis for requirements specification. *IEEE Computer*, 18(4):82–101, 1985.

[75] P. Bork, L. J. Jensen, C. von Mering, A. K. Ramani, I. Lee, and E. M. Marcotte. Protein interaction networks from yeast to human. *Curr Opin Struct Biol*, 14(3):292–299, 2004.

[76] S. Bornholdt and H. Ebel. World-wide web scaling exponent from simon's 1955 model. *Phys Rev E*, 64:046401, 2001.

[77] M. Boulos, A. Roudsari, and E. Carson. Towards a semantic medical web: HealthCyberMap's tool for building an RDF metadata base of health information resources based on the qualified Dublin core metadata set. *Med Sci Monit*, 8(7):MT124–136, 2002.

[78] S. J. Boulton, A. Gartner, J. Reboul, P. Vaglio, N. Dyson, D. E. Hill, and M. Vidal. Combined functional genomic maps of the *C. elegans* DNA damage response. *Science*, 295(5552):127–131, 2002.

[79] R. Brachman. What Is-a is and isn't: An analysis of taxonomic links in semantic networks. *IEEE Computer*, (October), 1983.

[80] A. Brazma, P. Hingamp, J. Quackenbush, G. Sherlock, P. Spellman, C. Stoeckert, J. Aach, W. Ansorge, C. A. Ball, H. C. Causton, T. Gaasterland, P. Glenisson, F. C. Holstege, I. F. Kim, V. Markowitz, J. C. Matese, H. Parkinson, A. Robinson, U. Sarkans, S. Schulze-Kremer, J. Stewart, R. Taylor, J. Vilo, and M. Vingron. Minimum information about a microarray experiment (miame)-toward standards for microarray data. *Nat Genet*, 29(4):365–371, 2001.

[81] B.-J. Breitkreutz, C. Stark, and M. Tyers. The GRID: The general repository for interaction datasets. *Genome Biol*, 4:R23:R23.1–R23.3, 2003.

[82] B.-J. Breitkreutz, C. Stark, and M. Tyers. Osprey: a network visualization system. *Genome Biol*, 4:R22:R22.1–R22.4, 2003.

[83] A. Broder, R. Kumar, F. Maghoul, P. Raghavan, S. Rajagopalan, R. Stata, A. Tomkins, and J. Wiener. Graph structure of the web. *Comput Netw*, 33:309–320, 2000.

[84] K. Brown and I. Jurisica. Online predicted human interaction database. *Bioinformatics*, 21(9):2076–2082, 2005.

[85] D. Bu, Y. Zhao, L. Cai, H. Xue, X. Zhu, H. Lu, J. Zhang, S. Sun, L. Ling, N. Zhang, G. Li, and R. Chen. Topological structure analysis of the protein–protein interaction network in budding yeast. *Nucleic Acids Res*, 31(9):2443–2450, 2003.

[86] M. Buckland. The landscape of information science: The American Society for Information Science at 62. *J Am Soc Info Sci*, 50(11):970974, 1999.

[87] M. Bunge. *Treatise on Basic Philosophy: Ontology I – The Furniture of the Wworld.* Reidel, Dordrecht, 1977.

[88] C. Bystroff and Y. Shao. Fully automated ab initio protein structure prediction using I-SITES, HMMSTR and ROSETTA. *Bioinformatics*, 18 Suppl 1:S54–61, 2002.

[89] Norrie M. C. and M. Wunderli. Coordination system modelling. In *Proc 13th Int Conf on The Entity Relationship Approach.* Manchester, UK, 1994.

[90] G. Cagney and A. Emili. *De novo* peptide sequencing and quantitative profiling of complex protein mixtures using mass-coded abundance tagging. *Nat Biotechnol*, 20(2):163–170, 2002.

[91] D. S. Callaway, M. E. J. Newman, S. H. Strogatz, and D. J. Watts. Network robustness and fragility: percolation on random graphs. *Phys Rev Lett*, 85:5468–5471, 2000.

[92] E. Camon, M. Magrane, D. Barrell, V. Lee, E. Dimmer, J. Maslen, D. Binns, N. Harte, R. Lopez, and R. Apweiler. The gene ontology annotation (GOA) database: sharing knowledge in uniprot with gene ontology. *Nucleic Acids Res*, 32 Database issue:D262–266, 2004.

[93] W. R. Cannon and K. D. Jarman. Improved peptide sequencing using isotope information inherent in tandem mass spectra. *Rapid Commun Mass Sp*, 17(15):1793–1801, 2003.

[94] J. F. Canny. A computational approach to edge detection. *IEEE T Pattern Anal*, pages 679–698, 1986.

[95] R. Carnap. *The Logical Structure of the World & Pseudoproblems in Philosophy.* University of California Press, Berkeley, CA, 1967. Translated by Rolf A. George.

[96] P. Carter, J. Liu, and B. Rost. Pep: Predictions for entire proteomes. *Nucleic Acids Res*, 31(1):410–413, 2003.

[97] R. F. Chancho and R. V. Sole. The small world of human language. *Proc R Soc London B, Biological Sciences*, 268(1482):2261–2265, 2001.

[98] M. A. Changizi, M. A. McDannald, and D. Widders. Scaling of differentiation in networks: nervous systems, organisms, ant colonies, ecosystems, businesses, universities, cities, electronic circuits, and legos. *J Theor Biol*, 218(2):215–237, 2002.

[99] P. Chaurand, S. A. Schwartz, and R. M. Caprioli. Assessing protein patterns in disease using imaging mass spectrometry. *J Proteome Res*, 3(2):245–252, 2004.

[100] N. E. Chayen. The role of oil in macromolecular crystallization. *Structure*, 5(10):1269–1274, 1997.

[101] P. Cheeseman, J. Kelly, M. Self, J. Stutz, W. Taylor, and D. Freeman. Autoclass: A Bayesian classification system. In *Proc of the Int Conf on Machine Learning*, pages 54–64. Ann Arbor, MI, 1988.

[102] C. Chekuri, A. Goldberg, D. Karger, M. Levine, and C. Stein. Experimental study of minimum cut algorithms. In *ACM-SIAM Symposium on Discrete Algorithms (SODA 97)*, pages 324–333. 1997.

[103] D. Chelius and P. V. Bondarenko. Quantitative profiling of proteins in complex mixtures using liquid chromatography and mass spectrometry. *J Proteome Res*, 1:317–323, 2002.

[104] C. Chen. Searching for intellectual turning points: progressive knowledge domain visualization. *Proc Natl Acad Sci USA*, 101 Suppl 1:5303–5310, 2004.

[105] P. P. Chen. The entity-relationship model: Towards a unified view of data. *ACM T Database Syst*, 1(1):9–36, 1976.

[106] Y. Chen and D. Xu. Computational analyses of high-throughput protein–protein interaction data. *Curr Protein Pept Sci*, 4(3):159–181, 2003.

[107] S. A. Chervitz. Comparison of the complete protein sets of worm and yeast: orthology and divergence. *Science*, 282:2022–2028, 1998.

[108] M. Chicurel. Mathematical biology. life is a game of numbers. *Nature*, 408(6815):900–901, 2000.

[109] M. H. Choi, J. N. Kim, and B. C. Chung. Rapid hplc-electrospray tandem mass spectrometric assay for urinary testosterone and dihydrotestosterone glucuronides from patients with benign prostate hyperplasia. *Clin Chem*, 49(2):322–5, 2003.

[110] W. W. Chu, H. Yang, K. Chiang, M. Minock, G. Chow, and C. Larson. CoBase: A scalable and extensible cooperative information system. *J Intell Inf Syst*, 6(2/3):223–259, 1996.

[111] F. Chung and L. Lu. The diameter of sparse random graphs. *Adv Appl Math*, 26:257–279, 2001.

[112] J. J. Cimino. From data to knowledge through concept-oriented terminologies: Experience with the medical entities dictionary. *J Am Med Inf Assoc*, 7(3):288–297, 2000.

[113] B. N. Clark, J. C. Colbourn, and D. S. Johnson. Unit disk graphs. *Discrete Math*, 86:165–177, 1991.

[114] E. F. Codd. A relational model of data for large shared data banks. *Commun ACM*, 13(6):377–387, 1970.

[115] R. Cohen, K. Erez, D. ben Avraham, and S. Havlin. Resilience of the internet to random breakdowns. *Phys Rev Lett*, 85:4626–4628, 2000.

[116] E. M. Conlon, X. S. Liu, J. D. Lieb, and J. S. Liu. Integrating regulatory motif discovery and genome-wide expression analysis. *Proc Natl Acad Sci USA*, 100(6):3339–3344, 2003.

[117] D. Connolly, F. van Harmelen, I. Horrocks, D. L. McGuinness, P. F. Patel-Schneider, and L. A. Stein. DAML+OIL. reference description. Technical report, 2001.

[118] The C. elegans Sequencing Consortium. Genome sequence of the nematode *C. elegans*: a platform for investigating biology. The *C. elegans* sequencing consortium. *Science*, 282(5396):2012–2018, 1998.

[119] The Gene Ontology Consortium. Gene ontology: tool for the unification of biology. *Nat Gen*, 25:25–29, 2000.

[120] M. M. Cordero, T. J. Cornish, R. J. Cotter, and I. A. Lys. Sequencing peptides without scanning the reflectron: post-source decay with a curved-field reflectron time-of-flight mass spectrometer. *Rapid Commun Mass Sp*, 9(14):1356–1361, 1995.

[121] M. C. Costanzo, M. E. Crawford, J. E. Hirschman, J. E. Kranz, P. Olsen, L. S. Robertson, M. S. Skrzypek, B. R. Braun, K. L. Hopkins, P. Kondu, C. Lengieza, J. E. Lew-Smith, M. Tillberg, and J. I. Garrels. YPD, PombelPD and WorkPD: model organism volumes of the BioKnowledge library, and integrated resource for protein information. *Nucleic Acids Res*, 29:75–79, 2001.

[122] S. Coulomb, M. Bauer, D. Bernard, and M.-C. Marsolier-Kergoat. Mutational robustness is only weakly related to the topology of protein interaction networks. In *CNET 2004: Science of Complex Networks: from Biology to the Internet and WWW*. University of Aveiro, Aveiro, Portugal, 2004.

[123] C. Creighton and S. Hanash. Mining gene expression databases for association rules. *Bioinformatics*, 19(1):79–86, 2003.

[124] F. Crimins, R. Dimitri, T. Klein, N. Palmer, and L. Cowen. Higher dimensional approach for classification of lung cancer microarray data. *CAMDA*, 2003.

[125] M. E. Csete and J. C. Doyle. Reverse engineering of biological complexity. *Science*, 295(5560):1664–1669, 2002.

[126] C. A. Cumbaa and I. Jurisica. Automatic classification and pattern discovery in high-throughput protein crystallization trials. *J Struct and Funct Genomics*. In press.

[127] C. A. Cumbaa, A. Lauricella, N. Fehrman, C. Veatch, R. Collins, J. Luft, G. T. DeTitta, and I. Jurisica. Automatic classification of sub-microlitre protein-crystallization trials in 1536-well plates. *Acta Crystallogr D Biol Crystallogr*, 59(Pt 9):1619–1627, 2003.

[128] T. Dandekar, B. Snel, M. Huynen, and P. Bork. Conservation of gene order: a fingerprint of proteins that physically interact. *Trends Biochem Sci*, 23:324–328, 1998.

[129] A. D'Arcy, A. M. Sweeney, and A. Haber. Practical aspects of using the microbatch method in screening conditions for protein crystallization. *Methods*, 34(3):323–328, 2004.

[130] B. de Bruijn and J. Martin. Getting to the core of knowledge: mining biomedical literature. *Int J Med Inf*, 67(1–3):7–18, 2002.

[131] T. L. Dean and M. P. Wellman. *Planning and Control*. Morgan Kaufmann Publishers, San Mateo, CA, 1991.

[132] C. M. Deane, L. Salwinski, I. Xenarios, and D. Eisenberg. Protein interactions: Two methods for assessment of the reliability of high throughput observations. *Mol Cell Proteomics*, 1(5):349–356, 2002.

[133] M. Deng, S. Mehta, F. Sun, and T. Chen. Inferring domain-domain interactions from protein–protein interactions. *Genome Res*, 12(10):1540–1548, 2002.

[134] M. Deng, F. Sun, and T. Chen. Assessment of the reliability of protein–protein interactions and protein function prediction. *Pac Symp Biocomput*, pages 140–151, 2003.

[135] M. Deng, Z. Tu, F. Sun, and T. Chen. Mapping gene ontology to proteins based on protein–protein interaction data. *Bioinformatics*, 20(6):895–902, 2004.

[136] J. DeRisi, L. Penland, P. O. Brown, M. L. Bittner, P. S. Meltzer, M. Ray, Y. Chen, Y. A. Su, and J. M. Trent. Use of a cDNA microarray to analyse gene expression patterns in human cancer. *Nat Genet*, 14(4):457–460, 1996.

[137] J. Ding, D. Berleant, D. Nettleton, and E. Wurtele. Mining MEDLINE: abstracts, sentences, or phrases? *Pac Symp Biocomput*, pages 326–37, 2002.

[138] S. Doddi, A. Marathe, S. S. Ravi, and D. C. Torney. Discovery of association rules in medical data. *Med Inform Internet Med*, 26(1):25–33, 2001.

[139] S. N. Dorogovtsev, A. V. Goltsev, and J. F. F. Mendes. Pseudofractal scale-free web. *Phys Rev E*, 65:066122, 2002.

[140] S. N. Dorogovtsev and J. F. F. Mendes. Evolution of networks with aging of sites. *Phys Rev E*, 62:1842–1845, 2000.

[141] J. Drews. Research & development. basic science and pharmaceutical innovation. *Nat Biotechnol*, 17(5):406, 1999.

[142] J. Drews. Drug discovery: a historical perspective. *Science*, 287(5460):1960–1964, 2000.

[143] A. Ducruix and R. Giege. *Crystallization of Nucleic Acids and Proteins. A Practical Approach*. Oxford University Press, New York, 1992.

[144] R. O. Duda and P. E. Hart. *Pattern Classification and Scene Analysis*. Wiley, New York, 1973.

[145] E. Durr, J. Yu, K. M. Krasinska, L. A. Carver, J. R. Yates, III, J. E. Testa, P. Oh, and J. E. Schnitzer. Direct proteomic mapping of the lung microvascular endothelial cell surface *in vivo* and in cell culture. *Nat Biotechnol*, 22(8):985–992, 2004.

[146] S. S. Dwight, M. A. Harris, K. Dolinski, C. A. Ball, G. Binkley, K. R. Christie, D. G. Fisk, L. Issel-Tarver, M. Schroeder, G. Sherlok, A. Sethuraman, S. Weng, D. Botstein, and J. M. Cherry. Saccharomyces genome database (SGD) provides secondary gene annotation using the gene ontology (GO). *Nucleic Acids Res*, 30:69–72, 2002.

[147] A. Echalier, R. L. Glazer, V. Fulop, and M. A. Geday. Assessing crystallization droplets using birefringence. *Acta Crystallogr D Biol Crystallogr*, 60(Pt 4):696–702, 2004.

[148] J. P. Eckmann and E. Moses. Curvature of co-links uncovers hidden thematic layers in the world wide web. *Proc Natl Acad Sci USA*, 99(9):5825–5829, 2002.

[149] P. Edman. Phenylthiohydantoins in protein analysis. *Ann NY Acad Sci*, 88:602–610, 1960.

[150] P. Edman. Sequence determination. *Mol Biol Biochem Biophys*, 8:211–255, 1970.

[151] P. Edman and G. Begg. A protein sequenator. *Eur J Biochem*, 1(1):80–91, 1967.

[152] A. M. Edwards, B. Kus, R. Jansen, D. Greenbaum, J. Greenblatt, and M. Gerstein. Bridging structural biology and genomics: assessing protein interaction data with known complexes. *Trends Genet*, 18:529–536, 2002.

[153] M. B. Eisen, P. T. Spellman, P. O. Brown, and D. Botstein. Cluster analysis and display of genome-wide expression patterns. *Proc Natl Acad Sci USA*, 95(25):14863–14868, 1998.

[154] D. B. Emmert, P. J. Stoehr, G. Stoesser, and G. N. Cameron. The european bioinformatics institute (ebi) databases. *Nucleic Acids Res*, 22(17):3445–3449, 1994.

[155] J. K. Eng, A. L. McCormack, and J. R. Yates, III. An approach to correlate tandem mass-spectral data of peptides with amino-acid-sequences in a protein database. *J Am Soc Mass Spectrom*, 11:976–989, 1994.

[156] A. J. Enright. *Computational Analysis of Protein Function within Complete Genomes*. PhD thesis, University of Cambridge, United Kingdom, 2002.

[157] A. J. Enright, I. Iliopoulos, N. C. Kyrpides, and C. A. Ouzounis. Protein interaction maps for complete genomes based on gene fusion events. *Nature*, 402:86–90, 1999.

[158] A. J. Enright, S. Van Dongen, and C. A. Ouzounis. An efficient algorithm for large-scale detection of protein families. *Nucleic Acids Res*, 30(7):1575–1584, 2002.

[159] P. Erdos and A. Renyi. On random graphs. *Publ Math*, 6:290–297, 1959.

[160] P. Erdos and A. Renyi. On the evolution of random graphs. *Publ Math Inst Hung Acad Sci*, 5:17–61, 1960.

[161] P. Erdos and A. Renyi. On the strength of connectedness of a random graph. *Acta Math Sci*, 12:261–267, 1961.

[162] S. Even. *Graph Algorithms*. Computer Science Press, Rockville, MD, 1979.

[163] M. Faloutsos, P. Faloutsos, and C. Faloutsos. On power-law relationships of the internet topology. *Comput Commun Rev*, 29:251–262, 1999.

[164] R. G. Farr, A. L. Perryman, and C. T. Samuzdi. Re-clustering the database for crystallization of macromolecules. *J Cryst Growth*, 183(4):653–668, 1998.

[165] U. M. Fayyad, G. Piatetsky-Shapiro, and P. Smyth. From data mining to knowledge discovery: An overview. In U.M. Fayyad, G. Piatetsky-Shapiro, P. Smyth, and R. Uthurusamy, editors, *Advances in Knowledge Discovery and Data Mining*, pages 1–34. AAAI Press/The MIT Press, Menlo Park, CA, 1996.

[166] U. M. Fayyad, G. Piatetsky-Shapiro, P. Smyth, and R. Uthurusamy, editors. *Advances in Knowledge Discovery and Data Mining*. AAAI/MIT Press, Menlo Park, CA, 1996.

[167] D. Fell and A. Wagner. The small world of metabolism. *Nat Biotechnol*, 19:1121–1122, 2000.

[168] J. B. Fenn, M. Mann, C. K. Meng, S. F. Wong, and C. M. Whitehouse. Electrospray ionization for mass spectrometry of large biomolecules. *Science*, 246(4926):64–71, 1989.

[169] J. B. Fenn, M. Mann, C. K. Meng, S. F. Wong, and C. M. Whitehouse. Electrospray ionization for mass spectrometry of large biomolecules. *Science*, 246(4926):64–71, 1989.

[170] D. Fenyo and R. C. Beavis. A method for assessing the statistical significance of mass spectrometry-based protein identifications using general scoring schemes. *Anal Chem*, 75(4):768–774, 2003.

[171] S. B. Ficarro, M. L. McCleland, P. T. Stukenberg, D. J. Burke, M. M. Ross, J. Shabanowitz, D. F. Hunt, and F. M. White. Phosphoproteome analysis by mass spectrometry and its application to *Saccharomyces cerevisiae*. *Nat Biotechnol*, 20(3):301–305, 2002.

[172] S. Fields and Johnston M. Genomics: A crisis in postgenomic nomenclature. *Science*, 296(5568):671–672, 2002.

[173] S. Fields and O. Song. A novel genetic system to detect protein–protein interactions. *Nature*, 340(6230):245–246, 1989.

[174] L. Florens, M. P. Washburn, J. D. Raine, R. M. Anthony, M. Grainger, J. D. Haynes, J. K. Moch, N. Muster, J. B. Sacci, D. L. Tabb, A. A. Witney, D. Wolters, Y. Wu, M. J. Gardner, A. A. Holder, R. E. Sinden, J. R. Yates, III, and D. J. Carucci. A proteomic view of the plasmodium falciparum life cycle. *Nature*, 419(6906):520–526, 2002.

[175] J. Folkman and R. Kalluri. Cancer without disease. *Nature*, 427(6977):787, 2004.

[176] A. J. Forbes, S. M. Patrie, G. K Taylor, Y. B. Kim, L. Jiang, and N. L. Kelleher. Targeted analysis and discovery of posttranslational modifications in proteins from methanogenic archaea by top-down MS. *Proc Natl Acad Sci USA*, 101(9):2678–2683, 2004.

[177] K. D. Forbus, D. Gentner, and K. Law. MAC/FAC: A model of similarity-based retrieval. *Cognitive Science*, 19(2):141–205, 1995.

[178] A. G. Fraser, R. S. Kamath, P. Zipperlen, M. Martinez-Campos, M. Sohrmann, and J. Ahringer. Functional genomic analysis of *C. elegans* chromosome I by systematic RNA interference. *Nature*, 408(6810):325–330, 2000.

[179] H. B. Fraser and A. E. Hirsh. Evolutionary rate depends on number of protein–protein interactions independently of gene expression level. *BMC Evol Biol*, 4(1):13, 2004.

[180] J. Frawley and G. Piatetsky-Shapiro. *Knowledge Discovery in Databases*. AAAI Press, Menlo Park, CA, 1991.

[181] C. Friedman, P. Kra, H. Yu, M. Krauthammer, and A. Rzhetsky. Genies: a natural-language processing system for the extraction of molecular pathways from journal articles. *Bioinformatics*, 17 Suppl. 1:74–82, 2001.

[182] N. Friedman, M. Linial, I. Nachman, and D. Pe'er. Using Bayesian networks to analyze expression data. *J Comput Biol*, 7(3–4):601–620, 2000.

[183] A. M. Frisch and P. Haddawy. Anytime deduction for probabilistic logic. *Artif Intell*, 69:93–122, 1994.

[184] E. E. Furlong, E. C. Andersen, B. Null, K. P. White, and M. P. Scott. Patterns of gene expression during drosophila mesoderm development. *Science*, 293(5535):1629–1633, 2001.

[185] T. Gaasterland. Restricting query relaxation through user constraints. In *Proc on Int Conf on Intelligent and Cooperative Information Systems*, pages 359–366. Rotterdam, 1993.

[186] M. E. Garber, O. G. Troyanskaya, K. Schluens, S. Petersen, Z. Thaesler, M. Pacyna-Gengelbach, M. van de Rijn, G. D. Rosen, C. M. Perou, R. I. Whyte, R. B. Altman, P. O. Brown, D. Botstein, and I. Petersen. Diversity of gene expression in adenocarcinoma of the lung. *Proc Natl Acad Sci USA*, 98(24):13784–13789, 2001.

[187] K. Garwood, T. McLaughlin, C. Garwood, S. Joens, N. Morrison, C. F. Taylor, K. Carroll, C. Evans, A. D. Whetton, S. Hart, D. Stead, Z. Yin, A. J. Brown, A. Hesketh, K. Chater, L. Hansson, M. Mewissen, P. Ghazal, J. Howard, K. S. Lilley, S. J. Gaskell, A. Brass, S. J. Hubbard, S. G. Oliver, and N. W. Paton. Pedro: a database for storing, searching and disseminating experimental proteomics data. *BMC Genomics*, 5(1):68, 2004.

[188] T. K. Garyantes. 1536-well assay plates: When do they make sense? *Drug Discov Today*, 7(9):489–490, 2002.

[189] S. P. Gaucher, S. W. Taylor, E. Fahy, B. Zhang, D. E. Warnock, S. S. Ghosh, and B. W. Gibson. Expanded coverage of the human heart mitochondrial proteome using multidimensional liquid chromatography coupled with tandem mass spectrometry. *J Proteome Res*, 3(3):495–505, 2004.

[190] A. C. Gavin, M. Bosche, R. Krause, P. Grandi, M. Marzioch, A. Bauer, J. Schultz, J. M. Rick, A. M. Michon, C. M. Cruciat, M. Remor, C. Hofert, M. Schelder, M. Brajenovic, H. Ruffner, A. Merino, K. Klein, M. Hudak, D. Dickson, T. Rudi, V. Gnau, A. Bauch, S. Bastuck,

B. Huhse, C. Leutwein, M. A. Heurtier, R. R. Copley, A. Edelmann, E. Querfurth, V. Rybin, G. Drewes, M. Raida, T. Bouwmeester, P. Bork, B. Seraphin, B. Kuster, G. Neubauer, and G. Superti-Furga. Functional organization of the yeast proteome by systematic analysis of protein complexes. *Nature*, 415(6868):141–147, 2002.

[191] H. Ge, Z. Liu, G. M. Church, and M. Vidal. Correlation between transcriptome and interactome mapping data from *Saccharomyces cerevisiae*. *Nat Genet*, 29(4):482–486, 2001.

[192] Y. Ge, B. G. Lawhorn, M. ElNaggar, E. Strauss, J. H. Park, T. P. Begley, and F. W. McLafferty. Top down characterization of larger proteins (45 kda) by electron capture dissociation mass spectrometry. *J Am Chem Soc*, 124(4):672–678, 2002.

[193] M. R. Genesereth. Knowledge interchange format. In James Allen, Richard Fikes, and Erik Sandewall, editors, *Proc of the 2nd Int Conf on Principles of Knowledge Representation and Reasoning*, pages 599–600. Morgan Kaufmann Publishers, San Mateo, CA, 1991.

[194] S. A. Gerber, J. Rush, O. Stemman, M. W. Kirschner, and S. P. Gygi. Absolute quantification of proteins and phosphoproteins from cell lysates by tandem ms. *Proc Natl Acad Sci USA*, 100(12):6940 6945, 2003.

[195] M. Gerstein, N. Lan, and R. Jansen. Proteomics. Integrating interactomes. *Science*, 295(5553):284–287, 2002.

[196] S. Ghaemmaghami, W. K. Huh, K. Bower, R. W. Howson, A. Belle, N. Dephoure, E. K. O'Shea, and J. S. Weissman. Global analysis of protein expression in yeast. *Nature*, 425(6959):737–741, 2003.

[197] G. Giaever, A. M. Chu, L. Ni, C. Connelly, L. Riles, S. Veronneau, S. Dow, A. Lucau-Danila, K. Anderson, B. Andre, A. P. Arkin, A. Astromoff, M. El-Bakkoury, R. Bangham, R. Benito, S. Brachat, S. Campanaro, M. Curtiss, K. Davis, A. Deutschbauer, K. D. Entian, P. Flaherty, F. Foury, D. J. Garfinkel, M. Gerstein, D. Gotte, U. Guldener, J. H. Hegemann, S. Hempel, Z. Herman, D. F. Jaramillo, D. E. Kelly, S. L. Kelly, P. Kotter, D. LaBonte, D. C. Lamb, N. Lan, H. Liang, H. Liao, L. Liu, C. Luo, M. Lussier, R. Mao, P. Menard, S. L. Ooi, J. L. Revuelta, C. J. Roberts, M. Rose, P. Ross-Macdonald, B. Scherens, G. Schimmack, B. Shafer, D. D. Shoemaker, S. Sookhai-Mahadeo, R. K. Storms, J. N. Strathern, G. Valle, M. Voet, G. Volckaert, C. Y. Wang, T. R. Ward, J. Wilhelmy, E. A. Winzeler, Y. Yang, G. Yen, E. Youngman, K. Yu, H. Bussey, J. D. Boeke, M. Snyder, P. Philippsen, R. W. Davis, and M. Johnston. Functional profiling of the *Saccharomyces cerevisiae* genome. *Nature*, 418(6896):387–391, 2002.

[198] R. A. Gibbs, G. M. Weinstock, M. L. Metzker, D. M. Muzny, E. J. Sodergren, S. Scherer, G. Scott, D. Steffen, K. C. Worley, P. E. Burch, G. Okwuonu, S. Hines, L. Lewis, C. DeRamo, O. Delgado, S. Dugan-Rocha, G. Miner, M. Morgan, A. Hawes, R. Gill, Celera, R. A. Holt, M. D. Adams, P. G. Amanatides, H. Baden-Tillson, M. Barnstead, S. Chin, C. A. Evans, S. Ferriera, C. Fosler, A. Glodek, Z. Gu, D. Jennings, C. L. Kraft, T. Nguyen, C. M. Pfannkoch, C. Sitter, G. G. Sutton, J. C. Venter, T. Woodage, D. Smith, H. M. Lee, E. Gustafson, P. Cahill, A. Kana, L. Doucette-Stamm, K. Weinstock, K. Fechtel, R. B. Weiss, D. M. Dunn, E. D. Green, R. W. Blakesley, G. G. Bouffard, P. J. De Jong, K. Osoegawa, B. Zhu, M. Marra, J. Schein, I. Bosdet, C. Fjell, S. Jones, M. Krzywinski, C. Mathewson, A. Siddiqui, N. Wye, J. McPherson, S. Zhao, C. M. Fraser, J. Shetty, S. Shatsman, K. Geer, Y. Chen, S. Abramzon, W. C. Nierman, P. H. Havlak, R. Chen, K. J. Durbin, A. Egan, Y. Ren, X. Z. Song, B. Li, Y. Liu, X. Qin, S. Cawley, A. J. Cooney, L. M. D'Souza, K. Martin, J. Q. Wu, M. L. Gonzalez-Garay, A. R. Jackson, K. J. Kalafus, M. P. McLeod, A. Milosavljevic, D. Virk, A. Volkov, D. A. Wheeler, Z. Zhang, J. A. Bailey, E. E. Eichler, E. Tuzun, et al. Genome sequence of the brown norway rat yields insights into mammalian evolution. *Nature*, 428(6982):493–521, 2004.

[199] D. Gibson, J. Kleinberg, and P. Raghavan. Inferring web communities from link toplogy. In *Proc of the 9th ACM Conf on Hypertext and Hypermedia*. ACM Press, New York, NY, 1998.

[200] E. N. Gilbert. Random plane networks. *SIAM J*, 9:533–543, 1961.

[201] G. L. Gilliland, M. Tung, and J. E. Ladner. The biological macromolecule crystallization database: crystallization procedures and strategies. *Acta Crystallogr D Biol Crystallogr*, 58(1):916–920, 2002.

[202] K. Ginalski and L. Rychlewski. Protein structure prediction of CASP5 comparative modeling and fold recognition targets using consensus alignment approach and 3D assessment. *Proteins*, 53 Suppl 6:410–417, 2003.

[203] P. Giorgini, J. Mylopoulos, E. Nicchiarelli, and R. Sebastiani. Formal reasoning techniques for goal models. *J Data Semantics*, (October):1–20, 2003.

[204] L. Giot, J. S. Bader, C. Brouwer, A. Chaudhuri, B. Kuang, Y. Li, Y. L. Hao, C. E. Ooi, B. Godwin, E. Vitols, G. Vijayadamodar, P. Pochart, H. Machineni, M. Welsh, Y. Kong, B. Zerhusen, R. Malcolm, Z. Varrone, A. Collis, M. Minto, S. Burgess, L. McDaniel, E. Stimpson, F. Spriggs, J. Williams, K. Neurath, N. Ioime, M. Agee, E. Voss, K. Furtak, R. Renzulli, N. Aanensen, S. Carrolla, E. Bickelhaupt, Y. Lazovatsky, A. DaSilva, J. Zhong, C. A. Stanyon, Jr., Finley, R. L., K. P. White, M. Braverman, T. Jarvie, S. Gold, M. Leach, J. Knight, R. A. Shimkets,

M. P. McKenna, J. Chant, and J. M. Rothberg. A protein interaction map of *Drosophila melanogaster*. *Science*, 302(5651):1727–1736, 2003.

[205] M. Girvan and M. E. Newman. Community structure in social and biological networks. *Proc Natl Acad Sci USA*, 99(12):7821–7826, 2002.

[206] H. C. Godfray. Challenges for taxonomy. *Nature*, 417(6884):17–19, 2002.

[207] B. Goethals. Survey on frequent pattern mining. Technical report, University of Helsinki, http://www.cs.helsinki.fi/u/goethals/publications/survey.pdf.

[208] D. S. Goldberg and F. P. Roth. Assessing experimentally derived interactions in a small world. *Proc Natl Acad Sci USA*, 100(8):4372–4376, 2003.

[209] P. Gonczy, C. Echeverri, K. Oegema, A. Coulson, S. J. Jones, R. R. Copley, J. Duperon, J. Oegema, M. Brehm, E. Cassin, E. Hannak, M. Kirkham, S. Pichler, K. Flohrs, A. Goessen, S. Leidel, A. M. Alleaume, C. Martin, N. Ozlu, P. Bork, and A. A. Hyman. Functional genomic analysis of cell division in *c. elegans* using RNAi of genes on chromosome III. *Nature*, 408(6810):331–336, 2000.

[210] V. Gopalakrishnan, G. Livingston, D. Hennessy, B. Buchanan, and J. M. Rosenberg. Machine-learning techniques for macromolecular crystallization data. *Acta Crystallogr D Biol Crystallogr*, D60:1705–1716, 2004.

[211] N. I. Govorukhina, A. Keizer-Gunnink, A. G. van der Zee, S. de Jong, H. W. de Bruijn, and R. Bischoff. Sample preparation of human serum for the analysis of tumor markers. comparison of different approaches for albumin and gamma-globulin depletion. *J Chromatogr A*, 1009(1-2):171–178, 2003.

[212] B. D. Grant and H. A. Wilkinson. Functional genomic maps in caenorhabditis elegans. *Curr Opin Cell Biol*, 15(2):206–212, 2003.

[213] C. Gravel, L. Stergiou, S. N. Gagnon, and S. Desnoyers. The *C. elegans* gene pme-5: molecular cloning and role in the DNA-damage response of a tankyrase orthologue. *DNA Repair (Amst)*, 3(2):171–182, 2004.

[214] T. J. Griffin, C. M. Lock, X. J. Li, A. Patel, I. Chervetsova, H. Lee, M. E. Wright, J. A. Ranish, S. S. Chen, and R. Aebersold. Abundance ratio-dependent proteomic analysis by mass spectrometry. *Anal Chem*, 75(4):867–874, 2003.

[215] W. J. Griffiths. Nanospray mass spectrometry in protein and peptide chemistry. *Exs*, 88:69–79, 2000.

[216] A. Grigoriev. A relationship between gene expression and protein interactions on the proteome scale: analysis of the bacteriophage T7 and the yeast *Saccharomyces cerevisiae*. *Nucleic Acids Res*, 29(17):3513–3519, 2001.

[217] D. Groth, H. Lehrach, and S. Hennig. Goblet: a platform for gene ontology annotation of anonymous sequence data. *Nucleic Acids Res*, 32(Web server issue):W313–317, 2004.

[218] T. R. Gruber. Ontolingua: A mechanism to support portable ontologies. In *Technical Report KSL-91-66*. Stanford University, Knowledge Systems Laboratory, 1992.

[219] T. R. Gruber. A translation approach to portable ontology specification. *Knowl Acquis*, 5(2):199–220, 1993.

[220] H. Gu, Y. Perl, J. Geller, M. Halper, and M. Singh. A methodology for partitioning a vocabulary hierarchy into trees. *Artif Intell Med*, 15:77–98, 1999.

[221] N. Guelzim, S. Bottani, P. Bourgine, and F. Kepes. Topological and causal structure of the yeast transcriptional regulatory network. *Nat Genet*, 31(1):60–63, 2002.

[222] P. Gupta and P. Kumar. Critical power for asymptotic connectivity in wireless networks. In *Stochastic Analysis, Control, Optimization and Applications: A Volume in Honor of W.H. Fleming*, pages 547–566. Birkhauser, Boston, 1998.

[223] S. P. Gygi, B. Rist, S. A. Gerber, F. Turecek, M. H. Gelb, and R. Aebersold. Quantitative analysis of complex protein mixtures using isotope-coded affinity tags. *Nat Biotechnol*, 17(10):994–999, 1999.

[224] R. Hafner. The asymptotic distribution of random clumps. *Computing*, 10:335–351, 1972.

[225] U. Hahn, M. Romacker, and S. Schulz. Creating knowledge repositories from biomedical reports: the MEDSYNDIKATE text mining system. *Pac Symp Biocomput*, pages 338–349, 2002.

[226] R. Halaban, J. S. Rubin, Y. Funasaka, M. Cobb, T. Boulton, D. Faletto, E. Rosen, A. Chan, K. Yoko, W. White, and et al. Met and hepatocyte growth factor/scatter factor signal transduction in normal melanocytes and melanoma cells. *Oncogene*, 7(11):2195–2206, 1992.

[227] J. D. Han, N. Bertin, T. Hao, D. S. Goldberg, G. F. Berriz, L. V. Zhang, D. Dupuy, A. J. Walhout, M. E. Cusick, F. P. Roth, and M. Vidal. Evidence for dynamically organized modularity in the yeast protein–protein interaction network. *Nature*, 430(6995):88–93, 2004.

[228] K. Han, B. Park, H. Kim, J. Hong, and J. Park. HPID: The human protein interaction database. *Bioinformatics*, 20(15):2466–2470, 2004.

[229] D. Hanahan and R. A. Weinberg. The hallmarks of cancer. *Cell*, 100(1):57–70, 2000.

[230] D. J. Hand, H. Mannila, and P. Smyth. *Principles of Data Mining*. MIT Press, Cambridge, MA, 2001.

[231] B. J. Hanson, B. Schulenberg, W. F. Patton, and R. A. Capaldi. A novel subfractionation approach for mitochondrial proteins: a three-dimensional mitochondrial proteome map. *Electrophoresis*, 22(5):950–959, 2001.

[232] J. Hao and J. Orlin. A faster algorithm for finding the minimum cut in a directed graph. *J Algorithm*, 17(3):424–446, 1994.

[233] R. M. Haralick, K. Shanmugam, and I. Dinstein. Textural features for image classification. *IEEE T Syst Man Cy*, 3(6):610–621, 1973.

[234] M. A. Harris, J. Clark, A. Ireland, J. Lomax, M. Ashburner, R. Foulger, K. Eilbeck, S. Lewis, B. Marshall, C. Mungall, J. Richter, G. M. Rubin, J. A. Blake, C. Bult, M. Dolan, H. Drabkin, J. T. Eppig, D. P. Hill, L. Ni, M. Ringwald, R. Balakrishnan, J. M. Cherry, K. R. Christie, M. C. Costanzo, S. S. Dwight, S. Engel, D. G. Fisk, J. E. Hirschman, E. L. Hong, R. S. Nash, A. Sethuraman, C. L. Theesfeld, D. Botstein, K. Dolinski, B. Feierbach, T. Berardini, S. Mundodi, S. Y. Rhee, R. Apweiler, D. Barrell, E. Camon, E. Dimmer, V. Lee, R. Chisholm, P. Gaudet, W. Kibbe, R. Kishore, E. M. Schwarz, P. Sternberg, M. Gwinn, L. Hannick, J. Wortman, M. Berriman, V. Wood, N. de la Cruz, P. Tonellato, P. Jaiswal, T. Seigfried, and R. White. The gene ontology (go) database and informatics resource. *Nucleic Acids Res*, 32 Database issue:D258–261, 2004.

[235] E. Hartuv and R. Shamir. An algorithm for clustering cDNA fingerprints. *Genomics*, 66(3):249–256, 2000.

[236] E. Hartuv and R. Shamir. A clustering algorithm based on graph connectivity. *Inform Process Lett*, 76(4–6):175–181, 2000.

[237] T. R. Hazbun, L. Malmstrom, S. Anderson, B. J. Graczyk, B. Fox, M. Riffle, B. A. Sundin, J. D. Aranda, W. H. McDonald, C. H. Chiu, B. E. Snydsman, P. Bradley, E. G. Muller, S. Fields, D. Baker, J. R. Yates, III, and T. N. Davis. Assigning function to yeast proteins by integration of technologies. *Mol Cell*, 12(6):1353–1365, 2003.

[238] P. D. Hebert, M. Y. Stoeckle, T. S. Zemlak, and C. M. Francis. Identification of birds through DNA barcodes. *PLoS Biol*, 2(10):e312, 2004.

[239] L. V. Hedges and I. Okin. *Statistical Methods for Meta-Analysis*. Academic Press, 1985.

[240] G. Van Heijst, S. Falasconi, A. Abu-Hanna, G. Schreiber, and M. Stefanelli. A cese study in ontology library construction. *Artif Intell Med*, 7:227–255, 1995.

[241] M. Heller, H. Mattou, C. Menzel, and X. Yao. Trypsin catalyzed 16o-to-18o exchange for comparative proteomics: tandem mass spectrometry comparison using maldi-tof, esi-qtof, and esi-ion trap mass spectrometers. *J Am Soc Mass Spectrom*, 14(7):704–718, 2003.

[242] D. Hennessy, B. Buchanan, D. Subramanian, P. A. Wilkosz, and J. M. Rosenberg. Statistical methods for the objective design of screening procedures for macromolecular crystallization. *Acta Crystallogr D Biol Crystallogr*, 56 (Pt 7):817–827, 2000.

[243] D. Hennessy, V. Gopalakrishnan, B. G. Buchanan, J. M. Rosenberg, and D. Subramanian. Induction of rules for biological macromolecule crystallization. *Proc 2nd Int Conf Intell Syst Mol Biol*, pages 179–187, 1994.

[244] H. Hermjakob, L. Montecchi-Palazzi, G. Bader, J. Wojcik, L. Salwinski, A. Ceol, S. Moore, S. Orchard, U. Sarkans, C. von Mering, B. Roechert, S. Poux, E. Jung, H. Mersch, P. Kersey, M. Lappe, Y. Li, R. Zeng, D. Rana, M. Nikolski, H. Husi, C. Brun, K. Shanker, S. G. Grant, C. Sander, P. Bork, W. Zhu, A. Pandey, A. Brazma, B. Jacq, M. Vidal, D. Sherman, P. Legrain, G. Cesareni, I. Xenarios, D. Eisenberg, B. Steipe, C. Hogue, and R. Apweiler. The HUPO PSI's molecular interaction format — a community standard for the representation of protein interaction data. *Nat Biotechnol*, 22(2):177–183, 2004.

[245] H. Hermjakob, L. Montecchi-Palazzi, C. Lewington, S. Mudali, S. Kerrien, S. Orchard, M. Vingron, B. Roechert, P. Roepstorff, A. Valencia, H. Margalit, J. Armstrong, A. Bairoch, G. Cesareni, D. Sherman, and R. Apweiler. Intact: an open source molecular interaction database. *Nucleic Acids Res*, 32 Database issue:D452–455, 2004.

[246] D. M. Hillis. The emergence of systematic biology. *Syst Biol*, 50(3):301–303, 2001.

[247] Y. Ho, A. Gruhler, A. Heilbut, G. D. Bader, L. Moore, S. L. Adams, A. Millar, P. Taylor, K. Bennett, K. Boutilier, L. Yang, C. Wolting, I. Donaldson, S. Schandorff, J. Shewnarane, M. Vo, J. Taggart, M. Goudreault, B. Muskat, C. Alfarano, D. Dewar, Z. Lin, K. Michalickova, A. R. Willems, H. Sassi, P. A. Nielsen, K. J. Rasmussen, J. R. Andersen, L. E. Johansen, L. H. Hansen, H. Jespersen, A. Podtelejnikov, E. Nielsen, J. Crawford, V. Poulsen, B. D. Sorensen, J. Matthiesen, R. C. Hendrickson, F. Gleeson, T. Pawson, M. F. Moran, D. Durocher, M. Mann, C. W. Hogue, D. Figeys, and M. Tyers. Systematic identification of protein complexes in *Saccharomyces cerevisiae* by mass spectrometry. *Nature*, 415(6868):180–183, 2002.

[248] R. Hoffmann and A. Valencia. A gene network for navigating the literature. *Nat Genet*, 36(7):664, 2004.

[249] H. Holley and M. Karplus. Protein secondary structure prediction with a neural network. *Proc Nat Acad Sci USA*, 86:152–156, 1989.

[250] D. M. Horn, R. A. Zubarev, and F. W. McLafferty. Automated *de novo* sequencing of proteins by tandem high-resolution mass spectrometry. *Proc Natl Acad Sci USA*, 97(19):10313–10317, 2000.

[251] I. Horrocks, P. F. Patel-Schneider, and F. van Harmelen. Reviewing the design of DAML+OIL: An ontology language for the semantic web. In *Presented at 18th National Conf on Artif Intell (AAAI'02)*. Edmonton, Alberta, 2002.

[252] T. W. Huang, A. C. Tien, W. S. Huang, Y. C. Lee, C. L. Peng, H. H. Tseng, C. Y. Kao, and C. Y. Huang. Point: a database for the prediction of protein–protein interactions based on the orthologous interactome. *Bioinformatics*, 20(17):3273–3276, 2004.

[253] T. Hubbard, D. Barker, E. Birney, G. Cameron, Y. Chen, L. Clark, T. Cox, J. Cuff, V. Curwen, T. Down, R. Durbin, E. Eyras, J. Gilbert, M. Hammond, L. Huminiecki, A. Kasprzyk, H. Lehvaslaiho, P. Lijnzaad, C. Melsopp, E. Mongin, R. Pettett, M. Pocock, S. Potter, A. Rust, E. Schmidt, S. Searle, G. Slater, J. Smith, W. Spooner, A. Stabenau, J. Stalker, E. Stupka, A. Ureta-Vidal, I. Vastrik, and M. Clamp. The Ensembl genome database project. *Nucleic Acids Res*, 30:38–41, 2002.

[254] T. Hubbard, D. Barker, E. Birney, G. Cameron, Y. Chen, L. Clark, T. Cox, J. Cuff, V. Curwen, T. Down, R. Durbin, E. Eyras, J. Gilbert, M. Hammond, L. Huminiecki, A. Kasprzyk, H. Lehvaslaiho, P. Lijnzaad, C. Melsopp, E. Mongin, R. Pettett, M. Pocock, S. Potter, A. Rust, E. Schmidt, S. Searle, G. Slater, J. Smith, W. Spooner, A. Stabenau, J. Stalker, E. Stupka, A. Ureta-Vidal, I. Vastrik, and M. Clamp. The Ensembl genome database project. *Nucleic Acids Res*, 30(1):38–41, 2002.

[255] T. R. Hughes, M. J. Marton, A. R. Jones, C. J. Roberts, R. Stoughton, C. D. Armour, H. A. Bennett, E. Coffey, H. Dai, Y. D. He, M. J. Kidd, A. M. King, M. R. Meyer, D. Slade, P. Y. Lum, S. B. Stepaniants, D. D. Shoemaker, D. Gachotte, K. Chakraburtty, J. Simon, M. Bard, and S. H. Friend. Functional discovery via a compendium of expression profiles. *Cell*, 102(1):109–126, 2000.

[256] W. K. Huh, J. V. Falvo, L. C. Gerke, A. S. Carroll, R. W. Howson, J. S. Weissman, and E. K. O'Shea. Global analysis of protein localization in budding yeast. *Nature*, 425(6959):686–691, 2003.

[257] B. L. Humphreys and D. A. Lindberg. The unified medical language system: an informatics research collaboration. *J Am Med Informat Assoc*, 5(1):1–11, 1998.

[258] L. Hunter. Ontologies for programs, not people. *Genome Biol*, 3(6):IN-TERACTIONS:1002, 2002.

[259] M. Huynen and P. Bork. Measuring genome evolution. *Proc Natl Acad Sci USA*, 95:5849–5856, 1998.

[260] F. Iragne, A. Barre, N. Goffard, and A. De Daruvar. Aliasserver: a web server to handle multiple aliases used to refer to proteins. *Bioinformatics*, 20(14):2331–2332, 2004.

[261] T. Ito, T. Chiba, R. Ozawa, M. Yoshida, M. Hattori, and Y. Sakaki. A comprehensive two-hybrid analysis to explore the yeast protein interactome. *Proc Natl Acad Sci USA*, 98(8):4569–4574, 2001.

[262] T. Ito, K. Tashiro, S. Muta, R. Ozawa, T. Chiba, M. Nishizawa, K. Yamamoto, S. Kuhara, and Y. Sakaki. Toward a protein–protein interaction map of the budding yeast: A comprehensive system to examine two-hybrid interactions in all possible combinations between the yeast proteins. *Proc Natl Acad Sci USA*, 97(3):1143–1147, 2000.

[263] S. Itzkovitz and U. Alon. Subgraphs and network motifs in geometric networks. *Condensed Matter*, 2004.

[264] S. Itzkovitz, R. Milo, N. Kashtan, G. Ziv, and U. Alon. Subgraphs in random networks. *Phys Rev E*, 68, 2003.

[265] A. K. Jain, M. N. Murty, and P. J. Flynn. Data clustering: a review. *ACM Comput Surv*, 31(3):264–323, 1999.

[266] J. Jancarik and S. H. Kim. Spare matrix sampling: a screening method for crystallization of proteins. *J Applied Crystallogr*, 24(409), 1991.

[267] R. Jansen, D. Greenbaum, and M. Gerstein. Relating whole-genome expression data with protein–protein interactions. *Genome Res*, 12(1):37–46, 2002.

[268] R. Jansen, H. Yu, D. Greenbaum, Y. Kluger, N. J. Krogan, S. Chung, A. Emili, M. Snyder, J. F. Greenblatt, and M. Gerstein. A bayesian networks approach for predicting protein–protein interactions from genomic data. *Science*, 302(5644):449–453, 2003.

[269] F. V. Jensen. *An Introduction to Bayesian Networks*. Springer Verlag, New York, NY, 1996.

[270] F. V. Jensen. *Bayesian Networks and Decision Graphs*. Springer Verlag, New York, NY, 2001.

[271] H. Jeong, A. L. Barabasi B. Tombor, and Z. N. Oltvai. The global organization of cellular networks. In *Proc of the 2nd Workshop on Computation of Biochemical Pathways and Genetic Networks.* Heidelberg, Germany, 2001.

[272] H. Jeong, S. P. Mason, A. L. Barabasi, and Z. N. Oltvai. Lethality and centrality in protein networks. *Nature,* 411(6833):41–42, 2001.

[273] H. Jeong, Z. N. Oltvai, and A. L. Barabasi. Prediction of protein essentiality based on genomic data. *ComPlexUs,* 1:19–28, 2003.

[274] H. Jeong, B. Tombor, R. Albert, Z. N. Oltvai, and A. L. Barabasi. The large-scale organization of metabolic networks. *Nature,* 407(6804):651–654, 2000.

[275] R. S. Johnson, S. A. Martin, K. Biemann, J. T. Stults, and J. T. Watson. Novel fragmentation process of peptides by collision-induced decomposition in a tandem mass spectrometer: differentiation of leucine and isoleucine. *Anal Chem,* 59(21):2621–2625, 1987.

[276] S.C. Johnson. Hierarchical clustering schemes. *Psychometrika,* 2:241–254, 1967.

[277] S. Jones and J. M. Thornton. protein–protein interactions: a review of protein dimer structures. *Prog Biophys Mol Biol,* 63:31–65, 1995.

[278] S. Jones and J. M. Thornton. Principles of protein–protein interactions. *Proc Natl Acad Sci USA,* 93:13–20, 1996.

[279] I. Jurisica and J. Glasgow. Improving performance of case-based classification using context-based relevance. *Int J Artif Intell Tools, Special Issue of IEEE ITCAI-96 Best Papers,* 6(4):511–536, 1997.

[280] I. Jurisica and J. Glasgow. Extending case-based reasoning by discovering and using image features in IVF. In *ACM Symp on Applied Computing (SAC 2000).* Como, Italy, 2000.

[281] I. Jurisica and J. Glasgow. Application of case-based reasoning in molecular biology. *AI Mag, Special issue on Bioinformatics,* 25(1):85–95, 2004.

[282] I. Jurisica, J. Glasgow, and J. Mylopoulos. Incremental iterative retrieval and browsing for efficient conversational CBR systems. *Int J Appl Intell,* 12(3):251–268, 2000.

[283] I. Jurisica, J. Mylopoulos, J. Glasgow, H. Shapiro, and R. F. Casper. Case-based reasoning in IVF: prediction and knowledge mining. *Artif Intell Med,* 12(1):1–24, 1998.

[284] I. Jurisica, J. Mylopoulos, and E. Yu. Ontologies for knowledge management: An information systems perspective. *An Int J Knowl Inform Syst, Special issue on Ontologies,* 6(4):380–401, 2004.

[285] I. Jurisica, P. Rogers, J. Glasgow, R. Collins, J. Wolfley, J. Luft, and G. T. DeTitta. Improving objectivity and scalability in protein crystallization: Integrating image analysis with knowledge discovery. *IEEE Intell Syst J, Special Issue on Intelligent Systems in Biology*, 16(6):26–34, 2001.

[286] I. Jurisica, P. Rogers, J. Glasgow, S. Fortier, J. Luft, M. Bianca, and G. T. DeTitta. Image-feature extraction for protein crystallization: Integrating image analysis and case-based reasoning. In *The 13th Innovative Applications of Artificial Intelligence Conf (IAAI-01)*, pages 73–80. AAAI Press, Menlo Park, CA, 2001.

[287] I. Jurisica, P. Rogers, J. Glasgow, S. Fortier, J. Luft, J. Wolfley, M. Bianca, D. Weeks, and G. T. DeTitta. Intelligent decision support for protein crystal growth. *IBM Syst J, Special Issue on Deep Computing for Life Sciences*, 40(2):394–409, 2001.

[288] F. C. Kafatos and T. Eisner. Unification in the century of biology. *Science*, 303(5662):1257, 2004.

[289] C. Kahn. An internet-based ontology editor for medical appropriateness criteria. *Comput Meth Prog Bio*, 56:3136, 1998.

[290] R. S. Kamath, A. G. Fraser, Y. Dong, G. Poulin, R. Durbin, M. Gotta, A. Kanapin, N. Le Bot, S. Moreno, M. Sohrmann, D. P. Welchman, P. Zipperlen, and J. Ahringer. Systematic functional analysis of the *Caenorhabditis elegans* genome using RNAi. *Nature*, 421(6920):231–237, 2003.

[291] K. A. Kantardjieff, M. Jamshidian, and B. Rupp. Reply to the letter: Distributions of pI provide prior information for the design of crystallization screening experiments: Response to a comment on "protein isoelectric point as a predictor for increased crystallization screening efficiency". *Bioinformatics*, 20(4):2171–2174, 2004.

[292] U. Karaoz, T. M. Murali, S. Letovsky, Y. Zheng, C. Ding, C. R. Cantor, and S. Kasif. Whole-genome annotation by using evidence integration in functional-linkage networks. *Proc Natl Acad Sci USA*, 101(9):2888–2893, 2004.

[293] M. Karas and F. Hillenkamp. Laser desorption ionization of proteins with molecular masses exceeding 10,000 daltons. *Anal Chem*, 60(20):2299–2301, 1988.

[294] N. Kashtan, S. Itzkovitz, R. Milo, and U. Alon. Efficient sampling algorithm for estimating subgraph concentrations and detecting network motifs. *Bioinformatics*, 20:1746–1758, 2004.

[295] N. Kashtan, S. Itzkovitz, R. Milo, and U. Alon. Topological generalizations of network motifs. *Phys Rev E*, 70:031909, 2004.

[296] S. Kauffman. *At Home in the Universe*. Oxford, New York, 1995.

[297] S. A. Kauffman. Metabolic stability and epigenesis in randomly constructed genetic nets. *J Theor Biol*, 22:437–467, 1969.

[298] J. J. Keeling. The effects of local spatial structure on epidemiological invasions. *Proc R Soc London B, Biological Sciences*, 266:859–867, 1999.

[299] N. L. Kelleher, R. A. Zubarev, K. Bush, B. Furie, B. C. Furie, F. W. McLafferty, and C. T. Walsh. Localization of labile posttranslational modifications by electron capture dissociation: The case of gamma-carboxyglutamic acid. *Anal Chem*, 71(19):4250–4253, 1999.

[300] A. Keller, A.I. Nesvizhskii, E. Kolker, and R. Aebersold. Empirical statistical model to estimate the accuracy of peptide identifications made by ms/ms and database search. *Anal Chem*, 74(20):5383–5392, 2002.

[301] P. Kemmeren, N. L. van Berkum, J. Vilo, T. Bijma, R. Donders, A. Brazma, and F. C. Holstege. Protein interaction verification and functional annotation by integrated analysis of genome-scale data. *Mol Cell*, 9(5):1133–1143, 2002.

[302] A. Kemper and C. Wiesner. Building scalable electronic market places using hyperquery-based distributed query processing. *World Wide Web*, 8(1):27–60, 2005.

[303] J. O. Kephart and S. R. White. Directed-graph epidemiological models of computer viruses. *Proc 1991 IEEE Comput Soc Symp Res Security Privacy*, pages 343–359, 1991.

[304] P. J. Kersey, J. Duarte, A. Williams, Y. Karavidopoulou, E. Birney, and R. Apweiler. The int protein index: an integrated database for proteomics experiments. *Proteomics*, 4(7):1985–1988, 2004.

[305] B. T. Kile, K. E. Hentges, A. T. Clark, H. Nakamura, A. P. Salinger, B. Liu, N. Box, D. W. Stockton, R. L. Johnson, R. R. Behringer, A. Bradley, and M. J. Justice. Functional genetic analysis of mouse chromosome 11. *Nature*, 425(6953):81–86, 2003.

[306] S. Kim and T. Szyperski. GFT NMR, a new approach to rapidly obtain precise high-dimensional NMR spectral information. *J Am Chem Soc*, 125(5):1385–1393, 2003.

[307] S. K. Kim, J. Lund, M. Kiraly, K. Duke, M. Jiang, J. M. Stuart, A. Eizinger, B. N. Wylie, and G. S. Davidson. A gene expression map for *Caenorhabditis elegans*. *Science*, 293(5537):2087–2092, 2001.

[308] M. S. Kimber, F. Vallee, S. Houston, A. Necakov, T. Skarina, E. Evdokimova, S. Beasley, D. Christendat, A. Savchenko, C. H. Arrowsmith, M. Vedadi, M. Gerstein, and A. M. Edwards. Data mining crystallization databases: knowledge-based approaches to optimize protein crystal screens. *Proteins*, 51(4):562–568, 2003.

[309] A. D. King, N. Przulj, and I. Jurisica. Protein complex prediction via cost-based clustering. *Bioinformatics*, 20(17):3013–3020, 2004.

[310] O. D. King, R. E. Foulger, S. S. Dwight, J. V. White, and F. P. Roth. Predicting gene function from patterns of annotation. *Genome Res*, 13(5):896–904, 2003.

[311] M. Kirschner, J. Gerhart, and T. Mitchison. Molecular "vitalism". *Cell*, 100(1):79–88, 2000.

[312] T. Kislinger, K. Rahman, D. Radulovic, B. Cox, J. Rossant, and A. Emili. Prism, a generic large scale proteomic investigation strategy for mammals. *Mol Cell Proteomics*, 2(2):96–106, 2003.

[313] H. Kitano. Systems biology: a brief overview. *Science*, 295(5560):1662–1664, 2002.

[314] J. Kleinberg. Authoritative sources in a hyper-linked environment. In *Proc of the 9th ACM Conf on Hypertext and Hypermedia*. ACM Press, New York, NY, 1998.

[315] J. M. Kleinberg. Navigation in a small world. *Nature*, 406:845, 2000.

[316] T. Kohonen. *Self-Organizing Maps*. Springer, Berlin, 2001.

[317] T. Kokeny. Constraint satisfaction problems with order-sorted domains. *Int J Artif Intell Tools*, 4(1 & 2):55–72, 1995.

[318] A. Koller, M. P. Washburn, B. M. Lange, N. L. Andon, C. Deciu, P. A. Haynes, L. Hays, D. Schieltz, R. Ulaszek, J. Wei, D. Wolters, and J. R. Yates, III. Proteomic survey of metabolic pathways in rice. *Proc Natl Acad Sci USA*, 99(18):11969–11974, 2002.

[319] J. L. Kolodner. *Case-Based Reasoning*. Morgan Kaufmann Publishers, San Mateo, CA, 1993.

[320] P. L Krapivsky and S. Redner. Organization of growing random networks. *Phys Rev E*, 63:066123, 2001.

[321] P. L Krapivsky, S. Redner, and F. Leyvraz. Connectivity of growing random networks. *Phys Rev Lett*, 85:4629–4632, 2000.

[322] M. Krauthammer, P. Kra, I. Iossifov, S. M. Gomez, G. Hripcsak, V. Hatzivassiloglou, C. Friedman, and A. Rzhetsky. Of truth and pathways: chasing bits of information through myriads of articles. *Bioinformatics*, 18 Suppl. 1:249–257, 2002.

[323] J. Krijgsveld, R. F. Ketting, T. Mahmoudi, J. Johansen, M. Artal-Sanz, C. P. Verrijzer, R. H. Plasterk, and A. J. Heck. Metabolic labeling of *C. elegans* and *D. melanogaster* for quantitative proteomics. *Nat Biotechnol*, 21:927–931, 2003.

[324] M. E. Kuil, E. R. Bodenstaff, F. J. Hoedemaeker, and J. P. Abrahams. Protein nano-crystallogenesis. *Enzyme Microb Technol*, 30(3):262–265, 2002.

[325] A. Kumar, S. Agarwal, J. A. Heyman, S. Matson, M. Heidtman, S. Piccirillo, L. Umansky, A. Drawid, R. Jansen, Y. Liu, K. H. Cheung, P. Miller, M. Gerstein, G. S. Roeder, and M. Snyder. Subcellular localization of the yeast proteome. *Genes Dev*, 16(6):707–719, 2002.

[326] N. Lan, G. T. Montelione, and M. Gerstein. Ontologies for proteomics: towards a systematic definition of structure and function that scales to the genome level. *Curr Opin Chem Biol*, 7(1):44–54, 2003.

[327] E. S. Lander, L. M. Linton, B. Birren, C. Nusbaum, M. C. Zody, J. Baldwin, K. Devon, K. Dewar, M. Doyle, W. FitzHugh, R. Funke, D. Gage, K. Harris, A. Heaford, J. Howland, L. Kann, J. Lehoczky, R. LeVine, P. McEwan, K. McKernan, J. Meldrim, J. P. Mesirov, C. Miranda, W. Morris, J. Naylor, C. Raymond, M. Rosetti, R. Santos, A. Sheridan, C. Sougnez, N. Stange-Thomann, N. Stojanovic, A. Subramanian, D. Wyman, J. Rogers, J. Sulston, R. Ainscough, S. Beck, D. Bentley, J. Burton, C. Clee, N. Carter, A. Coulson, R. Deadman, P. Deloukas, A. Dunham, I. Dunham, R. Durbin, L. French, D. Grafham, S. Gregory, T. Hubbard, S. Humphray, A. Hunt, M. Jones, C. Lloyd, A. McMurray, L. Matthews, S. Mercer, S. Milne, J. C. Mullikin, A. Mungall, R. Plumb, M. Ross, R. Shownkeen, S. Sims, R. H. Waterston, R. K. Wilson, L. W. Hillier, J. D. McPherson, M. A. Marra, E. R. Mardis, L. A. Fulton, A. T. Chinwalla, K. H. Pepin, W. R. Gish, S. L. Chissoe, M. C. Wendl, K. D. Delehaunty, T. L. Miner, A. Delehaunty, J. B. Kramer, L. L. Cook, R. S. Fulton, D. L. Johnson, P. J. Minx, S. W. Clifton, T. Hawkins, E. Branscomb, P. Predki, P. Richardson, S. Wenning, T. Slezak, N. Doggett, J. F. Cheng, A. Olsen, S. Lucas, C. Elkin, E. Uberbacher, M. Frazier, et al. Initial sequencing and analysis of the human genome. *Nature*, 409(6822):860–921, 2001.

[328] C. P. Langlotz and S. A. Caldwell. The completeness of existing lexicons for representing radiology report information. *J Digit Imaging*, 15 Suppl 1:201–205, 2002.

[329] M. Lappe and L. Holm. Unraveling protein interaction networks with near-optimal efficiency. *Nat Biotechnol*, 22(1):98–103, 2004.

[330] L. F. Largo-Fernandez, R. Huerta, F. Corbancho, and J. Siguenza. Fast response and temporal coherent oscillations in small-world networks. *Phys Rev Lett*, 84:2758–2761, 2000.

[331] D. Leake, editor. *Case-Based Reasoning: Experience, Lessons, and Future Directions*. AAAI Press, Menlo Park, CA, 1996.

[332] Y. G. Leclerc. Continuous terrain modeling from image sequences with applications to change detection. In *Workshop on Image Understanding.* New Orleans, LA, 1997.

[333] B. Lehner and A. G. Fraser. A first-draft human protein interaction map. *Genome Biol,* 5(R63):R63.1–R63.9, 2004.

[334] D. B. Lenat and R. V. Guha. *Building Large Knowledge-Based Systems: Representation and Inference in the CYC Project.* Addison-Wesley, Reading, Massachusetts, 1990.

[335] S. Letovsky and S. Kasif. Predicting protein function from protein/protein interaction data: a probabilistic approach. *Bioinformatics,* 19 Suppl 1:i197–204, 2003.

[336] J. Li, H. Steen, and S. P. Gygi. Protein profiling with cleavable isotope coded affinity tag (cICAT) reagents: The yeast salinity stress response. *Mol Cell Proteomics,* 2(11):1198–1204, 2003.

[337] S. Li, C. M. Armstrong, N. Bertin, H. Ge, S. Milstein, M. Boxem, P. O. Vidalain, J. D. Han, A. Chesneau, T. Hao, D. S. Goldberg, N. Li, M. Martinez, J. F. Rual, P. Lamesch, L. Xu, M. Tewari, S. L. Wong, L. V. Zhang, G. F. Berriz, L. Jacotot, P. Vaglio, J. Reboul, T. Hirozane-Kishikawa, Q. Li, H. W. Gabel, A. Elewa, B. Baumgartner, D. J. Rose, H. Yu, S. Bosak, R. Sequerra, A. Fraser, S. E. Mango, W. M. Saxton, S. Strome, S. Van Den Heuvel, F. Piano, J. Vandenhaute, C. Sardet, M. Gerstein, L. Doucette-Stamm, K. C. Gunsalus, J. W. Harper, M. E. Cusick, F. P. Roth, D. E. Hill, and M. Vidal. A map of the interactome network of the metazoan *C. elegans. Science,* 303(5657):540–543, 2004.

[338] S. Li and C. Dass. Iron(III)-immobilized metal ion affinity chromatography and mass spectrometry for the purification and characterization of synthetic phosphopeptides. *Anal Biochem,* 270(1):9–14, 1999.

[339] S. Z. Li. Invariant representation, matching and pose estimation of 3D space curves under similarity transformation. *Pattern Recogn,* 30(3):447–458, 1997.

[340] X. J. Li, P. G. Pedrioli, J. Eng, D. Martin, E. C. Yi, H. Lee, and R. Aebersold. A tool to visualize and evaluate data obtained by liquid chromatography-electrospray ionization-mass spectrometry. *Anal Chem,* 76(13):3856–3860, 2004.

[341] N. Lin, B. Wu, R. Jansen, M. Gerstein, and H. Zhao. Information assessment on predicting protein–protein interactions. *BMC Bioinformatics,* 5(1):154, 2004.

[342] H. Liu and C. Friedman. Mining terminological knowledge in large biomedical corpora. *Pac Symp Biocomput,* pages 415–26, 2003.

[343] H. Liu, R. G. Sadygov, and J. R. Yates, III. A model for random sampling and estimation of relative protein abundance in shotgun proteomics. *Anal Chem*, 76(14):4193–4201, 2004.

[344] L. Liu, M. Halper, J. Geller, and Y. Perl. Controlled vocabularies in OODBs: Modeling issues and implementation. *Parallel Distrib Databases*, 7(1):37–65, 1999.

[345] L. Liu, M. Halper, J. Geller, and Y. Perl. Using OOBD modeling to partition a vocabulary into structurally and semantically uniform concept groups. *IEEE T Knowl Data En*, 14(4):850866, 2002.

[346] G. Livingson, X. Li, G. Li, L. Hao, and J. Zhou. Using rule induction methods to analyze gene expression data. *Proc IEEE Comp Soc Conf on Bioinformatics*, page 439, 2003.

[347] L. S. Lock, I. Royal, M. A. Naujokas, and M. Park. Identification of an atypical grb2 carboxyl-terminal sh3 domain binding site in gab docking proteins reveals grb2-dependent and -independent recruitment of gab1 to receptor tyrosine kinases. *J Biol Chem*, 275(40):31536–31545, 2000.

[348] P. W. Lord, R. D. Stevens, A. Brass, and C. A. Goble. Semantic similarity measures as tools for exploring the gene ontology. *Pac Symp Biocomput*, pages 601–612, 2003.

[349] L. Lu, A. K. Arakaki, H. Lu, and J. Skolnick. Multimeric threading-based prediction of protein–protein interactions on a genomic scale: application to the saccharomyces cerevisiae proteome. *Genome Res*, 13(6A):1146–1154, 2003.

[350] T. Luczak. Component behavior near the critical point of the random graph process. *Random Struct Algor*, 1:287, 1990.

[351] J. Luft, J. Wolfley, I. Jurisica, J. Glasgow, S. Fortier, and G. T. DeTitta. Macromolecular crystallization in a high throughput laboratory — the search phase. *J Cryst Growth*, 232:591–595, 2001.

[352] J. R. Luft, R. J. Collins, N. A. Fehrman, A. M. Lauricella, C. K. Veatch, and G. T. DeTitta. A deliberate approach to screening for initial crystallization conditions of biological macromolecules. *J Struct Biol*, 142(1):170–179, 2003.

[353] H. Ma and A.-P. Zheng. Structure and evolution analysis of metabolic networks based on genomic data. In *4th Biopathways Consortium Meeting*. Edmonton, AB, 2002.

[354] M. J. MacCoss, C. C. Wu, and J. R. Yates, III. Probability-based validation of protein identifications using a modified sequest algorithm. *Anal Chem*, 74(21):5593–5599, 2002.

[355] R. M. MacGregor. Representing reified relations in loom. *J Exp Theor Artif Intell*, 5:179–183, 1993.

[356] D. Maglott, J. Ostell, K. D. Pruitt, and T. Tatusova. Entrez Gene: gene-centered information at NCBI. *Nucleic Acids Res*, 33 Database Issue:D54–58, 2005.

[357] W. H. Majoros, G. M. Subramanian, and M. D. Yandell. Identification of key concepts in biomedical literature using a modified markov heuristic. *Bioinformatics*, 19(3):402–407, 2003.

[358] S. Mangan and U. Alon. Structure and function of the feed-forward loop network motif. *Proc Natl Acad Sci USA*, 100:11980–11985, 2003.

[359] S. Mangan, A. Zaslaver, and U. Alon. The coherent feedforward loop serves as a sign-sensitive delay element in transcription networks. *J Mol Biol*, 334/2:197–204, 2003.

[360] M. Mann, R. C. Hendrickson, and A. Pandey. Analysis of proteins and proteomes by mass spectrometry. *Annu Rev Biochem*, 70:437–473, 2001.

[361] M. Mann and M. Wilm. Error-tolerant identification of peptides in sequence databases by peptide sequence tags. *Anal Chem*, 66(24):4390–4399, 1994.

[362] E. M. Marcotte, M. Pellegrini, M. J. Thompson, T. O. Yeates, and D. Eisenberg. A combined algorithm for genomewide prediction of protein function. *Nature*, 402:83–86, 1999.

[363] E. Marshall. Getting the noise out of gene arrays. *Science*, 306(5696):630–631, 2004.

[364] E. Marshall. Taxonomy. Will DNA bar codes breathe life into classification? *Science*, 307(5712):1037, 2005.

[365] S. G. Martinez-Garza, A. Nunez-Salazar, A. L. Calderon-Garciduenas, F. J. Bosques-Padilla, A. Niderhauser-Garcia, and H. A. Barrera-Saldana. Frequency and clinicopathology associations of k-ras mutations in colorectal cancer in a northeast mexican population. *Dig Dis*, 17(4):225–229, 1999.

[366] S. Maslov and K. Sneppen. Specificity and stability in topology of protein networks. *Science*, 296(5569):910–913, 2002.

[367] D. W. Matula. The cohesive strength of graphs. In G. Chartrand and S. F. Kapoor, editors, *The Many Facets of Graph Theory*, pages 215–221. Springer, Berlin, 1969.

[368] D. W. Matula. Cluster analysis via graph theoretic techniques. In R. C. Mullin, K. B. Reid, and D. P. Roselle, editors, *Proc Louisiana Conf on Combinatorics, Graph Theory and Computing*, pages 199–212. University of Manitoba, Winnipeg, 1970.

[369] D. W. Matula. k-components, clusters and slicing in graphs. *SIAM J Applied Math*, 22(3):459–480, 1972.

[370] D. W. Matula. Graph theoretic techniques for cluster analysis algorithms. In J. van Ryzin, editor, *Classification and Clustering*, pages 95–129. Academic Press, New York, 1977.

[371] D. W. Matula. Determining edge connectivity in O(nm) time. In *28th IEEE Symposium on Foundations of Computer Science*, pages 249–251. 1987.

[372] R. M. May. *Stability and Complexity in Model Ecosystems*. Princeton University Press, Princeton, 1973.

[373] P. McBrien and A. Poulovassilis. Schema evolution in heterogeneous database architectures, a schema transformation approach. In A. B. Pidduck, J. Mylopoulos, C. C. Woo, and M. T. Özsu, editors, *Lecture Notes in Comp Sci, (CAiSE-02)*, volume 2348, pages 484–499. 2002.

[374] P. B. McGarvey, H. Huang, W. C. Barker, B. C. Orcutt, J. S. Garavelli, G. Y. Srinivasarao, L. S. Yeh, C. Xiao, and C. H. Wu. PIR: a new resource for bioinformatics. *Bioinformatics*, 16:290–291, 2000.

[375] T. McInerney and D. Terzopoulos. T-snakes: topology adaptive snakes. *Med Image Anal*, 4(2):73–91, 2000.

[376] D. McShan, S. Rao, and I. Shah. Microbial metabolic pathway inference by heuristic search. In *4th Biopathways Consortium Meeting*. Edmonton, AB, 2002.

[377] K. F. Medzihradszky, X. Zhang, R. J. Chalkley, S. Guan, M. A. McFarland, M. J. Chalmers, A. G. Marshall, R. L. Diaz, C. D. Allis, and A. L. Burlingame. Characterization of tetrahymena histone h2b variants and posttranslational populations by electron capture dissociation (ecd) fourier transform ion cyclotron mass spectrometry (ft-icr ms). *Mol Cell Proteomics*, pages 872–886, 2004.

[378] K. Mehlhorn and S. Naher. *LEDA: A Platform for Combinatorial and Geometric Computing*. Cambridge University Press, Cambridge, 1999.

[379] J. C. Mellor, I. Yanai, K. H. Clodfelter, J. Mintseris, and C. DeLisi. Predictome: a database of putative functional links between proteins. *Nucleic Acids Res*, 30(1):306–309, 2002.

[380] H. W. Mewes, K. Albermann, M. Bahr, D. Frishman, A. Gleissner, J. Hani, K. Heumann, K. Kleine, A. Maierl, S. G. Oliver, F. Pfeiffer, and A. Zollner. Overview of the yeast genome. *Nature*, 387(6632 Suppl):7–65, 1997.

[381] H. W. Mewes, D. Frishman, U. Guldener, G. Mannhaupt, K. Mayer, M. Mokrejs, B. Morgenstern, M. Munsterkotter, S. Rudd, and B. Weil.

MIPS: a database for genomes and protein sequences. *Nucleic Acids Res*, 30(1):31–34, 2002.

[382] K. Michalickova, G. D. Bader, M. Dumontier, H. Lieu, D. Betel, R. Isserlin, and C. W. Hogue. SeqHound: biological sequence and structure database as a platform for bioinformatics research. *BMC Bioinformatics*, 3(1):32, 2002.

[383] G. L. Miklos and R. Maleszka. Microarray reality checks in the context of a complex disease. *Nat Biotechnol*, 22(5):615–621, 2004.

[384] S. Milgram. The small world problem. *Psychol Today*, 2:60–67, 1967.

[385] R. C. Miller and B. A. Myers. Outlier finding: focusing user attention on possible errors. In *UIST*, pages 81–90. 2001.

[386] R. Milo, S. Itzkovitz, N. Kashtan, R. Levitt, S. Shen-Orr, I. Ayzenshtat, M. Sheffer, and U. Alon. Superfamilies of evolved and designed networks. *Science*, (303):1538–1542, 2004.

[387] R. Milo, S. S. Shen-Orr, S. Itzkovitz, N. Kashtan, D. Chklovskii, and U. Alon. Network motifs: simple building blocks of complex networks. *Science*, 298:824–827, 2002.

[388] G. W. Mineau and Godin R. Automatic structuring of knowledge bases by conceptual clustering. *IEEE T Knowl Data En*, 7(5):824–829, 1992.

[389] M. Molloy and B. Reed. A critical point of random graphs with a given degree sequence. *Random Struct Algor*, 6:161–180, 1995.

[390] M. Molloy and B. Reed. The size of the largest component of a random graph on a fixed degree sequence. *Comb Probab Comput*, 7:295–306, 1998.

[391] J. M. Montoya and R. V. Sole. Small world patterns of food webs. *Working paper 00-10-059, Santa Fe Institute*, 2001.

[392] V. K. Mootha, J. Bunkenborg, J. V. Olsen, M. Hjerrild, J. R. Wisniewski, E. Stahl, M. S. Bolouri, H. N. Ray, S. Sihag, M. Kamal, N. Patterson, E. S. Lander, and M. Mann. Integrated analysis of protein composition, tissue diversity, and gene regulation in mouse mitochondria. *Cell*, 115(5):629–640, 2003.

[393] Y. Moreau, S. Aerts, B. De Moor, B. De Strooper, and M. Dabrowski. Comparison and meta-analysis of microarray data: from the bench to the computer desk. *Trends Genet*, 19(10):570–577, 2003.

[394] D. W. Morris, C. Y. Kim, and A. McPherson. Automation of protein crystallization trials: use of a robot to deliver reagents to a novel multichamber vapor diffusion plate. *Biotechniques*, 7(5):522–527, 1989.

[395] A. A. Mosi and G. K. Eigendorf. Current mass spectrometric methods in organic chemistry. *Curr Org Chem*, 2(2):145–172, 1998.

[396] K. Murphy, Y. Weiss, and M. Jordan. Loopy-belief propagation for approximate inference: An empirical study. In *Uncertainty in AI*, pages 467–475. AAAI Press, Menlo Park, CA, 1999.

[397] M. A. Musen. Medical informatics: Searching for underlying components. *Methods Inf Med*, 41(1):12–19, 2002.

[398] J. Mylopoulos. Information modeling in the time of the revolution. *Inform Syst*, 23(3/4):127–155, 1998.

[399] J. Mylopoulos, A. Borgida, M. Jarke, and M. Koubarakis. Telos: Representing knowledge about information systems. *ACM T Inform Syst*, 8(4):325–362, 1990.

[400] J. Mylopoulos, I. Jurisica, and E. Yu. Computational mechanisms for knowledge organization. In *5th Int Conf of the Int Society of Knowledge Organization (ISKO 5)*, pages 125–132. Lille, France, 1998.

[401] H. Nagamochi and T. Ibaraki. Computing edge connectivity in multigraphs and capacitated graphs. *SIAM J Discrete Math*, 5:54–66, 1992.

[402] R. Neches, R. Fikes, T. Finin, T. Gruber, R. Patil, T. Senator, and W. R. Swarton. Enabling technology for knowledge sharing. *AI Mag*, pages 37–51, March 1991.

[403] G. Nenadic, I. Spasic, and S. Ananiadou. Terminology-driven mining of biomedical literature. *Bioinformatics*, 19(8):938–943, 2003.

[404] M. E. Newman. Scientific collaboration networks: I. network construction and fundamental results. *Phys Rev E*, 64:016131, 2001.

[405] M. E. Newman. Ego-centered networks and the ripple effect. *Social Networks*, 25:83–95, 2003.

[406] M. E. Newman and D. J. Watts. Renormalization group analysis in the small-world network model. *Phys Lett A*, 263:341–346, 1999.

[407] M. E. Newman and D. J. Watts. Scaling and percolation in the small-world network model. *Phys Rev E*, 60:7332–7342, 1999.

[408] M. E. J. Newman. Models of the small world: a review. *J Stat Phys*, 101:819–841, 2000.

[409] M. E. J. Newman. The structure and function of networks. *Comput Phys Commun*, 147:44–45, 2001.

[410] M. E. J. Newman. Random graphs as models of networks. In S. Bornholdt and H. G. Schuster, editors, *Handbook of Graphs and Networks*. Wiley-VHC, Berlin, 2002.

[411] M. E. J. Newman. The structure and function of complex networks. *SIAM Rev*, 45(2):167–256, 2003.

[412] M. E. J. Newman, C. Mooire, and D. J. Watts. Mean-field solution of the small-world network model. *Phys Rev Lett*, 84:3201–3204, 2000.

[413] M. E. J. Newman, S. H. Strogatz, and D. J. Watts. Random graphs with arbitrary degree distributions and their applications. *Phys Rev E*, 64:026118, 2001.

[414] A. Y. Ng, M. I. J., and Y. Weiss. On spectral clustering: Analysis and an algorithm. In G. Dieterich, S. Becker, and Z Ghahramani, editors, *Advances in Neural Information Processing Systems 14*. MIT Press, Cambridge, MA, 2002.

[415] S. Ng and M. Wong. Toward routine automatic pathway discovery from on-line scientific text abstracts. *Genome Informatics*, 10:104–112, 1999.

[416] D. Noble. Modeling the heart–from genes to cells to the whole organ. *Science*, 295(5560):1678–1682, 2002.

[417] P. Novak, M. M. Young, J. S. Schoeniger, and G. H. Kruppa. A top-down approach to protein structure studies using chemical cross-linking and fourier transform mass spectrometry. *Eur J Mass Spectrom*, 9(6):623–631, 2003.

[418] Y. Oda, T. Nagasu, and B. T. Chait. Enrichment analysis of phosphorylated proteins as a tool for probing the phosphoproteome. *Nat Biotechnol*, 19(4):379–382, 2001.

[419] T. Ohkawa, D. Namihira, N. Komoda, A. Kidera, and H. Nakamura. Protein structure classification by structural transformations. In *Proc of the IEEE Int Joint Symposia on Intelligence and Systems*, pages 23–29. Rockville, MD, 1996.

[420] D. E. Oliver. Synchronization of diverging versions of a controlled medical terminology. In *Technical Report SMI 98-0741*. Stanford University School of Medicine, Stanford, CA, 1998.

[421] D. E. Oliver, D. L. Rubin, J. M. Stuart, M. Hewett, T. E. Klein, and R. B. Altman. Ontology development for a pharmacogenetics knowledge base. *Pac Symp Biocomput*, pages 65–76, 2002.

[422] M. C. O'Neill and L. Song. Neural network analysis of lymphoma microarray data: prognosis and diagnosis near-perfect. *BMC Bioinformatics*, 4(1):13, 2003.

[423] S. E. Ong, B. Blagoev, I. Kratchmarova, D. B. Kristensen, H. Steen, A. Pandey, and M. Mann. Stable isotope labeling by amino acids in cell culture, silac, as a simple and accurate approach to expression proteomics. *Mol Cell Proteomics*, 1(5):376–386, 2002.

[424] S. Orchard, H. Hermjakob, R. K. Julian, Jr., K. Runte, D. Sherman, J. Wojcik, W. Zhu, and R. Apweiler. Common interchange standards for proteomics data: Public availability of tools and schema. *Proteomics*, 4(2):490–491, 2004.

[425] H. F. Orthne, J. R. Scherrer, and R. Dahlen. Sharing and communicating health care information: summary and recommendations. *Int J Biomed Comput*, 34(1-4):303–318, 1994.

[426] R. Overbeek, N. Larsen, G. D. Pusch, M. D'Souza, E. Selkov, Jr., N. Kyrpides, M Fonstein, N. Maltsev, and E. Selkov. WIT: integrated system for high-throughput genome sequence analysis and metabolic reconstruction. *Nucleic Acids Res*, 28(1):123–125, 2000.

[427] T. Oyama, K. Kitano, K. Satou, and T. Ito. Extraction of knowledge on protein–protein interaction by association rule discovery. *Bioinformatics*, 18(5):705–714, 2002.

[428] B. Ozdamar, R. Bose, M. Barrios-Rodiles, H. R. Wang, Y. Zhang, and J. L. Wrana. Regulation of the polarity protein Par6 by TGFbeta receptors controls epithelial cell plasticity. *Science*, 307(5715):1603–1609, 2005.

[429] R. Page and R. C. Stevens. Crystallization data mining in structural genomics: using positive and negative results to optimize protein crystallization screens. *Methods*, 34(3):373–389, 2004.

[430] Y. Pan, T. Kislinger, A. O. Gramolini, E. Zvaritch, E. G. Kranias, D. H. MacLennan, and A. Emili. Identification of biochemical adaptations in hyper- or hypocontractile hearts from phospholamban mutant mice by expression proteomics. *Proc Natl Acad Sci USA*, 101(8):2241–2246, 2004.

[431] A. Pandey and M. Mann. Proteomics to study genes and genomes. *Nature*, 405:837–846, 2000.

[432] A. Pandey, A. V. Podtelejnikov, B. Blagoev, X. R. Bustelo, M. Mann, and H. F. Lodish. Analysis of receptor signaling pathways by mass spectrometry: identification of vav-2 as a substrate of the epidermal and platelet-derived growth factor receptors. *Proc Natl Acad Sci USA*, 97(1):179–184, 2000.

[433] R. Pastor-Satorras and A. Vespignani. Epidemic spreading in scale-free networks. *Phys Rev Lett*, 86:3200–3203, 2001.

[434] P. Pavlidis, J. Weston, J. Cai, and W. N. Grundy. Gene functional classification from heterogenous data. *RECOMB*, 2001.

[435] V. Pavlovic, A. Garg, and S. Kasif. A bayesian framework for combining gene predictions. *Bioinformatics*, 18(1):19 27, 2002.

[436] T. Pawson and P. Nash. Assembly of cell regulatory systems through protein interaction domains. *Science*, 300(5618):445–452, 2003.

[437] J. Pearl. *Probabilistic Reasoning in Intelligent Systems*. Morgan Kaufmann Publishers, San Mateo, CA, 1988.

[438] M. Penrose. *Geometric Random Graphs*. Oxford University Press, Oxford, 2003.

[439] R. N. Perham. Structural aspects of biomolecular recognition and self-assembly. *Biosens Bioelectron*, 9(9–10):753–60, 1994.

[440] S. Peri, J. D. Navarro, T. Z. Kristiansen, R. Amanchy, V. Surendranath, B. Muthusamy, T. K. Gandhi, K. N. Chandrika, N. Deshpande, S. Suresh, B. P. Rashmi, K. Shanker, N. Padma, V. Niranjan, H. C. Harsha, N. Talreja, B. M. Vrushabendra, M. A. Ramya, A. J. Yatish, M. Joy, H. N. Shivashankar, M. P. Kavitha, M. Menezes, D. R. Choudhury, N. Ghosh, R. Saravana, S. Chandran, S. Mohan, C. K. Jonnalagadda, C. K. Prasad, C. Kumar-Sinha, K. S. Deshpande, and A. Pandey. Human protein reference database as a discovery resource for proteomics. *Nucleic Acids Res*, 32 Database issue:D497–501, 2004.

[441] D. N. Perkins, D. J. Pappin, D. M. Creasy, and J. S. Cottrell. Probability-based protein identification by searching sequence databases using mass spectrometry data. *Electrophoresis*, 20(18):3551–3567, 1999.

[442] E. G. M. Petrakis and C. Faloutsos. Similarity searching in medical image databases. *IEEE T Knowl Data En*, 9(3):435–447, 1997.

[443] E. F. Petricoin, A. M. Ardekani, B. A. Hitt, P. J. Levine, V. A. Fusaro, S. M. Steinberg, G. B. Mills, C. Simone, D. A. Fishman, E. C. Kohn, and L. A. Liotta. Use of proteomic patterns in serum to identify ovarian cancer. *Lancet*, 359(9306):572–577, 2002.

[444] A. Pirotte, E. Zimányi, D. Massart, and T. Yakusheva. Materialization: a powerful and ubiquitous abstraction pattern. In J. B. Bocca, M. Jarke, and C. Zaniolo, editors, *20th Int Conf on Very Large Data Bases*, pages 630–641. Morgan Kaufmann Publishers, Los Altos, CA, 1994.

[445] M. C. Posewitz and P. Tempst. Immobilized gallium(III) affinity chromatography of phosphopeptides. *Anal Chem*, 71(14):2883–2892, 1999.

[446] J. H. Prestegard, H. Valafar, J. Glushka, and F. Tian. Nuclear magnetic resonance in the era of structural genomics. *Biochem*, 40:8677–8685, 2001.

[447] N. Przulj, D. G. Corneil, and I. Jurisica. Modeling interactome: Scale-free or geometric? *Bioinformatics*, 20(18):3508–3515, 2004.

[448] N. Przulj and I. Jurisica. A call graph analysis. In *CASCON*. IBM Toronto Lab, Markham, ON, 2003.

[449] N. Przulj, G. Lee, and I. Jurisica. Functional analysis of large software networks. In *IBM Academy: Proactive Problem Prediction, Avoidance and Diagnosis*. IBM T. J. Watson Research Center, 2003.

[450] N. Przulj, D. A. Wigle, and I. Jurisica. Functional topology in a network of protein interactions. *Bioinformatics*, 20(3):340–348, 2004.

[451] M. L. Pusey, Z. J. Liu, W. Tempel, J. Praissman, D. Lin, B. C. Wang, J. A. Gavira, and J. D. Ng. Life in the fast lane for protein crystallization and x-ray crystallography. *Prog Biophys Mol Biol*, 88(3):359–86, 2005.

[452] J. R. Quinlan. *C4.5: Programs for Machine Learning*. Morgan Kaufmann Publishers, San Mateo, CA, 1993.

[453] D. Radulovic, S. Jelveh, S. Ryu, T. G. Hamilton, E. Foss, Y. Mao, and A. Emili. Informatics platform for global proteomic profiling and biomarker discovery using liquid-chromatography-tandem mass spectrometry. *Mol Cell Proteomics*, pages 984–997, 2004.

[454] J. C. Rain, L. Selig, H. De Reuse, V. Battaglia, C. Reverdy, S. Simon, G. Lenzen, F. Petel, J. Wojcik, V. Schachter, Y. Chemama, A. Labigne, and P. Legrain. The protein–protein interaction map of helicobacter pylori. *Nature*, 409(6817):211–215, 2001.

[455] R. Ramakrishnan and J. Gehrke. *Database Management Systems*. McGraw-Hill, 2003.

[456] S. Ramaswamy and T. R. Golub. DNA microarrays in clinical oncology. *J Clin Oncol*, 20(7):1932–1941, 2002.

[457] E. Ravasz, A. L. Somera, D. A. Mongru, Z. N. Oltvai, and A. L. Barabasi. Hierarchical organization of modularity in metabolic networks. *Science*, 297:1551–1555, 2002.

[458] D. Rebholz-Schuhmann, H. Kirsch, and F. Couto. Facts from text—is text mining ready to deliver? *PLoS Biol*, 3(2):e65, 2005.

[459] D. R. Rhodes, T. R. Barrette, M. A. Rubin, D. Ghosh, and A. M. Chinnaiyan. Meta-analysis of microarrays: interstudy validation of gene expression profiles reveals pathway dysregulation in prostate cancer. *Cancer Res*, 62(15):4427–4433, 2002.

[460] G. Rigaut, A. Shevchenko, B. Rutz, M. Wilm, M. Mann, and B. Seraphin. A generic protein purification method for protein complex characterization and proteome exploration. *Nat Biotechnol*, 17:1030–1032, 1999.

[461] G. Rijnders and D. H. Blank. Materials science: build your own superlattice. *Nature*, 433(7024):369–370, 2005.

[462] E. Rillof. An empirical study of automated dictionary construction for information extraction in three domains. *Artif Intell Med*, 85, 1996.

[463] C. J. Roberts, B. Nelson, M. J. Marton, R. Stoughton, M. R. Meyer, H. A. Bennett, Y. D. He, H. Dai, W. L. Walker, T. R. Hughes, M. Tyers, C. Boone, and S. H. Friend. Signaling and circuitry of multiple mapk pathways revealed by a matrix of global gene expression profiles. *Science*, 287(5454):873–880, 2000.

[464] J. F. Roddick. Schema evolution in database systems — An annotated bibliography. *SIGMOD Record*, 21(4):35–40, 1992.

[465] P. Rodriguez-Viciana, P. H. Warne, R. Dhand, B. Vanhaesebroeck, I. Gout, M. J. Fry, M. D. Waterfield, and J. Downward. Phosphatidylinositol-3-oh kinase as a direct target of ras. *Nature*, 370(6490):527–532, 1994.

[466] G. M. Rubin, M. D. Yandell, J. R. Wortman, G. L. G. Miklos, C. R. Nelson, I. K. Hariharan, M. E. Fortini, P. W. Li, R. Apweiler, W. Fleischmann, J. M. Cherry, S. Henikoff, M. P. Skupski, S. Misra, M. Ashburner, E. Birney, M. S. Boguski, T. Brody, P. Brokstein, S. E. Celniker, S. A. Chervitz, D. Coates, A. Cravchik, A. Gabrielian, R. F. Galle, W. M. Gelbart, R. A. George, L. S. B. Goldstein, F. Gong, P. Guan, N. L. Harris, B. A. Hay, R. A. Hoskins, J. Li, Z. Li, R. O. Hynes, S. J. M. Jones, P. M. Kuehl, B. Lemaitre, J. T. Littleton, D. K. Morrison, C. Mungall, P. H. O'Farrell, O. K. Pickeral, C. Shue, L. B. Vosshall, J. Zhang, Q. Zhao, X. H. Zheng, F. Zhong, W. Zhong, R. Gibbs, J. C. Venter, M. D. Adams, and S. Lewis. Comparative genomics of the eukaryotes. *Science*, 287:2204–2215, 2000.

[467] B. Rupp. Maximum-likelihood crystallization. *J Struct Biol*, 142(1):162–169, 2003.

[468] S. A. Russell, W. Old, K. A. Resing, and L. Hunter. Proteomic informatics. *Int Rev Neurobiol*, 61:127–157, 2004.

[469] S. J. Russell and P. Norvig. *Artif Intell: a Modern Approach; 2nd edition*. Prentice Hall, 2003.

[470] M. Safran, I. Solomon, O. Shmueli, M. Lapidot, S. Shen-Orr, A. Adato, U. Ben-Dor, N. Esterman, N. Rosen, I. Peter, T. Olender, V. Chalifa-Caspi, and D. Lancet. Genecards 2002: towards a complete, object-oriented, human gene compendium. *Bioinformatics*, 18(11):1542–1543, 2002.

[471] G. Salton. Relevance feedback and the optimization of retrieval effectiveness. In G. Salton, editor, *The SMART Retrieval System: Experiments in Automatic Document Processing*, pages 324–336. Prentice-Hall, Englewood Cliffs, NJ, 1971.

[472] K. Satou, G. Shibayama, T. Ono, Y. Yamamura, E. Furuichi, S. Kuhara, and T. Takagi. Finding association rules on heterogeneous genome data. *Pac Symp Biocomput*, pages 397–408, 1997.

[473] A. Savchenko, A. Yee, A. Khachatryan, T. Skarina, E. Evdokimova, M. Pavlova, A. Semesi, J. Northey, S. Beasley, N. Lan, R. Das, M. Gerstein, C. H. Arrowmith, and A. M. Edwards. Strategies for structural proteomics of prokaryotes: quantifying the advantages of studying orthologous proteins and of using both NMR and x-ray crystallography approaches. *Proteins*, 50(3):392–399, 2003.

[474] C. H. Schilling, D. Letscher, and B. O. Palsson. Theory for the systematic definition of metabolic pathways and their use in interpreting metabolic function from a pathway-oriented perspective. *J Theor Biol*, 203:229–248, 2000.

[475] E. C. Schirmer, L. Florens, T. Guan, J. R. Yates, III, and L. Gerace. Nuclear membrane proteins with potential disease links found by subtractive proteomics. *Science*, 301(5638):1380–1382, 2003.

[476] A. Schmidt, J. Kellermann, and F. Lottspeich. A novel strategy for quantitative proteomics using isotope-coded protein labels. *Proteomics*, 5(1):4–15, 2005.

[477] S. Schulze-Kremer. Ontologies for molecular biology. *Pac Symp Biocomput*, pages 695–706, 1998.

[478] S. A. Schwartz, R. J. Weil, M. D. Johnson, S. A. Toms, and R. M. Caprioli. Protein profiling in brain tumors using mass spectrometry: feasibility of a new technique for the analysis of protein expression. *Clin Cancer Res*, 10(3):981–987, 2004.

[479] B. Schwikowski, P. Uetz, and A. Fields. A network of protein–protein interactions in yeast. *Nat Biotechnol*, 18:1257–1261, 2000.

[480] E. Segal, M. Shapira, A. Regev, D. Pe'er, D. Botstein, D. Koller, and N. Friedman. Module networks: identifying regulatory modules and their condition-specific regulators from gene expression data. *Nat Genet*, 34(2):166–176, 2003.

[481] E. Segal, B. Taskar, A. Gasch, N. Friedman, and D. Koller. Rich probabilistic models for gene expression. *Bioinformatics*, 17 Suppl 1:243–252, 2001.

[482] J. Selbig, T. Mevissen, and T. Lengauer. Decision tree-based formation of consensus protein secondary structure prediction. *Bioinformatics*, 15(12):1039–1046, 1999.

[483] R. Service. Structural biology. structural genomics, round 2. *Science*, 307(5715):1554–1558, 2005.

[484] S. P. Shah, Y. Huang, T. Xu, M. M. Yuen, J. Ling, and B. F. Ouellette. Atlas — a data warehouse for integrative bioinformatics. *BMC Bioinformatics*, 6(1):34, 2005.

[485] B. E. Shakhnovich, N. V. Dokholyan, C. DeLisi, and E. I. Shakhnovich. Functional fingerprints of folds: evidence for correlated structure-function evolution. *J Mol Biol*, 326(1):1–9, 2003.

[486] R. Shamir and R. Sharan. Algorithmic approaches to clustering gene expression data. In T. Jiang, T. Smith, Y. Xu, and M. Zhang, editors, *Current Topics in Computational Biology*. MIT Press, 2001.

[487] R. Sharan, T. Ideker, B. P. Kelley, R. Shamir, and R. M. Karp. Identification of protein complexes by comparative analysis of yeast and bacterial protein interaction data. In *Proc of the 8th Annual Int Conf on Computational Molecular Biology (RECOMB'04)*. 2004.

[488] R. Sharan and R. Shamir. CLICK: A clustering algorithm with applications to gene expression analysis. In *Int Conf on Intelligent Systems for Molecular Biology (ISMB)*, pages 307–316. 2000.

[489] S. S. Shen-Orr, R. Milo, S. Mangan, and U. Alon. Network motifs in the transcriptional regulation network of escherichia coli. *Nat Gen*, 31:64–68, 2002.

[490] A. Shevchenko, I. Chernushevic, M. Wilm, and M. Mann. *De novo* sequencing of peptides recovered from in-gel digested proteins by nano-electrospray tandem mass spectrometry. *Mol Biotechnol*, 20(1):107–118, 2002.

[491] Y. Shiio, S. Donohoe, E. C. Yi, D. R. Goodlett, R. Aebersold, and R. N. Eisenman. Quantitative proteomic analysis of myc oncoprotein function. *Embo J*, 21(19):5088–5096, 2002.

[492] H. A. Simon. On a class of skew distribution functions. *Biometrika*, 42:425–440, 1955.

[493] H. A. Simon. Why should machines learn? In R. S. Michalski, J. G. Carbonell, and T. M. Mitchell, editors, *Machine Learning. An Artif Intell Approach*, volume 1, pages 25–38. Morgan Kaufmann Publishers, San Mateo, CA, 1983.

[494] J. Skolnick, D. Kihara, and Y. Zhang. Development and large scale benchmark testing of the PROSPECTOR3 threading algorithm. *Proteins*, 56(3):502–518, 2004.

[495] J. R. Smith and S.-Fu Chang. Tools and techniques for color image retrieval. In *Storage and Retrieval for Image and Video Databases (SPIE)*, pages 426–437. 1996.

[496] V. Spirin, D. Zhao, and L. A. Mirny. Discovery of protein complexes in the network of protein interactions. In *3rd Int Conf on Systems Biology (ICSB)*. Karolinska Institutet, Stockholm, Sweden, 2002.

[497] O. Sporns, G. Tononi, and G. M. Edelman. Theoretical neuroanatomy: relating anatomical and functional connectivity in graphs and cortical connection matrices. *Cereb Cortex*, 10:127–141, 2000.

[498] G. Spraggon, S. A. Lesley, A. Kreusch, and J. P. Priestle. Computational analysis of crystallization trials. *Acta Crystallogr D Biol Crystallogr*, D58:1915–1923, 2002.

[499] E. Sprinzak, S. Sattath, and H. Margalit. How reliable are experimental protein–protein interaction data? *J Mol Biol*, 327(5):919–923, 2003.

[500] H. Steen and A. Pandey. Proteomics goes quantitative: measuring protein abundance. *Trends Biotechnology*, 20(9):361–364, 2002.

[501] M. Steffen, A. Petti, J. Aach, P. D'haeseleer, and G. Church. Automated modeling of signal transduction networks. *BMC Bioinformatics*, 2002.

[502] J. Stelling, S. Klamt, K. Bettenbrock, S. Schuster, and E. D. Gilles. Metabolic network structure determines key aspects of functionality and regulation. *Nature*, 420:190–193, 2002.

[503] K. E Stephan. Computational analysis of functional connectivity between areas of primate visual cortex. *Phil T R Soc London B*, 355:111–126, 2000.

[504] R. E. Stepp and R. S. Michalski. Conceptual clustering: Inventing goal oriented classifications of structured objects. In J. G. Carbonell R. S. Michalski and T. M. Mitchell, editors, *Machine Learning: An Artif Intell Approach, Volume II*, pages 471–498. Morgan Kaufmann Publishers, Los Altos, CA, 1986.

[505] G. Steve, A. Gangemi, and D. M. Pisanelli. Integrating medical terminologies with ONIONS methodology. In *http://www.saussure.irmkant.rm.cnr.it/onto/publ/onions97/onions97.pdf*. 1997.

[506] R. Stevens, C. Goble, I. Horrocks, and S. Bechhofer. Building a bioinformatics ontology using OIL. *IEEE T Inf Technol Biomed*, 6(2):135–141, 2002.

[507] I. I. Stewart, T. Thomson, and D. Figeys. 18O labeling: a tool for proteomics. *Rapid Commun Mass Sp*, 15(24):2456–2465, 2001.

[508] M. Stoer and F. Wagner. A simple min-cut algorithm. *J ACM*, 44(4):585–591, 1997.

[509] G. Stoesser, W. Baker, A. van den Broek, E. Camon, M. Garia-Pastor, C. Kanz, T. Kulikova, R. Leinonen, Q. Lin, V. Lombard, R. Leopez,

N. Redaschi, P. Stoehr, M. A. Tuli, K. Tzouvara, and R. Vaughan. The EMBL nucleotide sequence database. *Nucleic Acids Res*, 30:21–26, 2002.

[510] N. Strang, M. Cucherat, and J. P. Boissel. Which coding system for therapeutic information in evidence-based medicine. *Comput Methods Programs Biomed*, 68(1):73–85, 2002.

[511] S. H. Strogatz. Exploring complex networks. *Nature*, 410:268–276, 2001.

[512] J. M. Stuart, E. Segal, D. Koller, and S. K. Kim. A gene-coexpression network for global discovery of conserved genetic modules. *Science*, 302(5643):249–255, 2003.

[513] A. I. Su, T. Wiltshire, S. Batalov, H. Lapp, K. A. Ching, D. Block, J. Zhang, R. Soden, M. Hayakawa, G. Kreiman, M. P. Cooke, J. R. Walker, and J. B. Hogenesch. A gene atlas of the mouse and human protein-encoding transcriptomes. *Proc Natl Acad Sci USA*, 101(16):6062–6067, 2004.

[514] M. Sultan, D. Wigle, C. A. Cumbaa, M Maziarz, J. Glasgow, M. S. Tsao, and I. Jurisica. Binary tree-structured vector quantization approach to clustering and visualizing microarray data. *Bioinformatics. Special Issue of ISMB'02*, 18(Supplement 1):S111–119, 2002.

[515] Y. Tateno and T. Gojobori. DNA Data Bank of Japan in the age of information biology. *Nucleic Acids Res*, 25(1):14–17, 1997.

[516] Y. Tateno, T. Imanishi, S. Miyazaki, K. Fukami-Kobayashi, N. Saitou, H. Sugawara, and T. Gojobori. DAN data bank of japan (DDBJ). *Nucleic Acids Res*, 30:27–30, 2002.

[517] M. Tewari, P. J. Hu, J. S. Ahn, N. Ayivi-Guedehoussou, P. O. Vidalain, S. Li, S. Milstein, C. M. Armstrong, M. Boxem, M. D. Butler, S. Busiguina, J. F. Rual, N. Ibarrola, S. T. Chaklos, N. Bertin, P. Vaglio, M. L. Edgley, K. V. King, P. S. Albert, J. Vandenhaute, A. Pandey, D. L. Riddle, G. Ruvkun, and M. Vidal. Systematic interactome mapping and genetic perturbation analysis of a *C. elegans* TGF-beta signaling network. *Mol Cell*, 13(4):469–482, 2004.

[518] P. R. Thagard, K. J. Holyoak, G. Nelson, and D. Gotchfeld. Analog retrieval by constraint satisfaction. *Artif Intell*, 46:259–310, 1990.

[519] A. H. Tong, B. Drees, G. Nardelli, G. D. Bader, B. Brannetti, L. Castagnoli, M. Evangelista, S. Ferracuti, B. Nelson, S. Paoluzi, M. Quondam, A. Zucconi, C. W. Hogue, S. Fields, C. Boone, and G. Cesareni. A combined experimental and computational strategy to define protein interaction networks for peptide recognition modules. *Science*, 295(5553):321–324, 2002.

[520] A. H. Tong, M. Evangelista, A. B. Parsons, H. Xu, G. D. Bader, N. Page, M. Robinson, S. Raghibizadeh, C. W. Hogue, H. Bussey, B. Andrews, M. Tyers, and C. Boone. Systematic genetic analysis with ordered arrays of yeast deletion mutants. *Science*, 294(5550):2364–2368, 2001.

[521] S. Tortola, E. Marcuello, I. Gonzalez, G. Reyes, R. Arribas, G. Aiza, F. J. Sancho, M. A. Peinado, and G. Capella. p53 and k-ras gene mutations correlate with tumor aggressiveness but are not of routine prognostic value in colorectal cancer. *J Clin Oncol*, 17(5):1375–1381, 1999.

[522] O. G. Troyanskaya, K. Dolinski, A. B. Owen, R. B. Altman, and D. Botstein. A Bayesian framework for combining heterogeneous data sources for gene function prediction in *Saccharomyces cerevisiae*. *Proc Natl Acad Sci USA*, 100(14):8348–8353, 2003.

[523] S. W. Tu, H. Eriksson, J. H. Gennari, Y. Shahar, and M. A. Musen. Ontology-based configuration of problemsolving methods and generation of knowledge-acquisition tools: application of PROTEGE-II to protocol-based decision support. *Artif Intell Med*, 7:257–289, 1995.

[524] Y. Tu. How robust is the internet? *Nature*, 406:353–354, 2000.

[525] P. D. Turney. Cost-sensitive classification: Empirical evaluation of a hybrid genetic decision tree induction algorithm. *J Artif Intell Res*, 2:369–409, 1995.

[526] P. Uetz, L. Giot, G. Cagney, T. A. Mansfield, R. S. Judson, J. R. Knight, D. Lockshon, V. Narayan, M. Srinivasan, P. Pochart, A. Qureshi-Emili, Y. Li, B. Godwin, D. Conover, T. Kalbfleish, G. Vijayadamodar, M. Yang, M. Johnston, S. Fields, and J. M. Rothberg. A comprehensive analysis of protein–protein interactions in *Saccharomyces cerevisiae*. *Nature*, 403:623–627, 2000.

[527] R. V. Vadali, Y. Shi, S. Kumar, L. V. Kale, M. E. Tuckerman, and G. J. Martyna. Scalable fine-grained parallelization of plane-wave-based ab initio molecular dynamics for large supercomputers. *J Comput Chem*, 25(16):2006–2022, 2004.

[528] S. M. van Dongen. *Graph Clustering by Flow Simulation*. PhD thesis, University of Utrecht, The Netherlands, 2000.

[529] J. C. Venter, M. D. Adams, E. W. Myers, P. W. Li, R. J. Mural, G. G. Sutton, H. O. Smith, M. Yandell, C. A. Evans, R. A. Holt, J. D. Gocayne, P. Amanatides, R. M. Ballew, D. H. Huson, J. R. Wortman, Q. Zhang, C. D. Kodira, X. H. Zheng, L. Chen, M. Skupski, G. Subramanian, P. D. Thomas, J. Zhang, G. L. Gabor Miklos, C. Nelson, S. Broder, A. G. Clark, J. Nadeau, V. A. McKusick, N. Zinder, A. J. Levine, R. J. Roberts, M. Simon, C. Slayman,

M. Hunkapiller, R. Bolanos, A. Delcher, I. Dew, D. Fasulo, M. Flanigan, L. Florea, A. Halpern, S. Hannenhalli, S. Kravitz, S. Levy, C. Mobarry, K. Reinert, K. Remington, J. Abu-Threideh, E. Beasley, K. Biddick, V. Bonazzi, R. Brandon, M. Cargill, I. Chandramouliswaran, R. Charlab, K. Chaturvedi, Z. Deng, V. Di Francesco, P. Dunn, K. Eilbeck, C. Evangelista, A. E. Gabrielian, W. Gan, W. Ge, F. Gong, Z. Gu, P. Guan, T. J. Heiman, M. E. Higgins, R. R. Ji, Z. Ke, K. A. Ketchum, Z. Lai, Y. Lei, Z. Li, J. Li, Y. Liang, X. Lin, F. Lu, G. V. Merkulov, N. Milshina, H. M. Moore, A. K. Naik, V. A. Narayan, B. Neelam, D. Nusskern, D. B. Rusch, S. Salzberg, W. Shao, B. Shue, J. Sun, Z. Wang, A. Wang, X. Wang, J. Wang, M. Wei, R. Wides, C. Xiao, C. Yan, et al. The sequence of the human genome. *Science*, 291(5507):1304–1351, 2001.

[530] C. von Mering, M. Huynen, D. Jaeggi, S. Schmidt, P. Bork, and B. Snel. STRING: a database of predicted functional associations between proteins. *Nucleic Acids Res*, 31(1):258–261, 2003.

[531] C. von Mering, R. Krause, B. Snel, M. Cornell, S. G. Oliver, S. Fields, and P. Bork. Comparative assessment of large-scale data sets of protein–protein interactions. *Nature*, 417(6887):399–403, 2002.

[532] A. Wagner and D. Fell. The small world inside large metabolic networks. *Proc R Soc London B*, 268:1803–1810, 2001.

[533] J. Wallinga, K. J. Edmunds, and M. Kretzschmar. Perspective: human contact patterns and the spread of airborne infectious diseases. *Trends Microbiol*, 7:372–377, 1999.

[534] T. Walsh. Search in small world. *Proc 16th Int Joint Conf Artif Intell*, pages 1172–1177, 1999.

[535] J. T.-Li Wang, M. J. Zaki, H. Toivonen, and D. Shasha, editors. *Data Mining in Bioinformatics*. Springer, 2005.

[536] M. P. Washburn, R. Ulaszek, C. Deciu, D. M. Schieltz, and J. R. Yates, III. Analysis of quantitative proteomic data generated via multidimensional protein identification technology. *Anal Chem*, 74(7):1650–1657, 2002.

[537] M. P. Washburn, D. Wolters, and J. R. Yates, III. Large-scale analysis of the yeast proteome by multidimensional protein identification technology. *Nat Biotechnol*, 19(3):242–247, 2001.

[538] R. H. Waterston, K. Lindblad-Toh, E. Birney, J. Rogers, J. F. Abril, P. Agarwal, R. Agarwala, R. Ainscough, M. Alexandersson, P. An, S. E. Antonarakis, J. Attwood, R. Baertsch, J. Bailey, K. Barlow, S. Beck, E. Berry, B. Birren, T. Bloom, P. Bork, M. Botcherby, N. Bray, M. R. Brent, D. G. Brown, S. D. Brown, C. Bult, J. Burton, J. Butler,

R. D. Campbell, P. Carninci, S. Cawley, F. Chiaromonte, A. T. Chinwalla, D. M. Church, M. Clamp, C. Clee, F. S. Collins, L. L. Cook, R. R. Copley, A. Coulson, O. Couronne, J. Cuff, V. Curwen, T. Cutts, M. Daly, R. David, J. Davies, K. D. Delehaunty, J. Deri, E. T. Dermitzakis, C. Dewey, N. J. Dickens, M. Diekhans, S. Dodge, I. Dubchak, D. M. Dunn, S. R. Eddy, L. Elnitski, R. D. Emes, P. Eswara, E. Eyras, A. Felsenfeld, G. A. Fewell, P. Flicek, K. Foley, W. N. Frankel, L. A. Fulton, R. S. Fulton, T. S. Furey, D. Gage, R. A. Gibbs, G. Glusman, S. Gnerre, N. Goldman, L. Goodstadt, D. Grafham, T. A. Graves, E. D. Green, S. Gregory, R. Guigo, M. Guyer, R. C. Hardison, D. Haussler, Y. Hayashizaki, L. W. Hillier, A. Hinrichs, W. Hlavina, T. Holzer, F. Hsu, A. Hua, T. Hubbard, A. Hunt, I. Jackson, D. B. Jaffe, L. S. Johnson, M. Jones, T. A. Jones, A. Joy, M. Kamal, E. K. Karlsson, et al. Initial sequencing and comparative analysis of the mouse genome. *Nature*, 420(6915):520–562, 2002.

[539] I. D. Watson. *Applying Case-Based Reasoning: Techniques for Enterprise Systems*. Morgan Kaufmann Publishers, San Francisco, CA, 1997.

[540] D. J. Watts. *Small Worlds*. Princeton University Press, Princeton, 1999.

[541] D. J. Watts and S. H. Strogatz. Collective dynamics of 'small-world' networks. *Nature*, 393:440–442, 1998.

[542] E. W. Weisstein. Radon transform. In *Eric Weisstein's World of Mathematics*. 2003.

[543] D. B. West. *Introduction to Graph Theory*. Prentice Hall, Upper Saddle River, NJ, 1996.

[544] J. D. Westbrook and P. E. Bourne. STAR/mmCIF: an ontology for macromolecular structure. *Bioinformatics*, 16(2):159–168, 2000.

[545] D. Wettschereck and G. T. Dietterich. An experimental comparison of the nearest neighbor and nearest hyperrectangle algorithmns. *Mach Learn*, 19(1):5–27, 1995.

[546] D. L. Wheeler, C. Chappey, A. E. Lash, D. D. Leipe, T. L. Madden, G. D. Schuler, T. A. Tatusova, and B. A. Rapp. Database resources of the national center for biotechnology information. *Nucleic Acids Res*, 28(1):10–14, 2000.

[547] D. A. Wigle, M. Tsao, and I. Jurisica. Making sense of lung-cancer gene-expression profiles. *Genome Biol*, 5(2):309, 2004.

[548] H. S. Wilf. *Generating Functionology*. Academic, Boston, 1990.

[549] R. J. Williams, E. L. Berlow, J. A. Dunne, A. L. Barabasi, and N. D. Martinez. Two degrees of separation in complex food webs. *Proc Natl Acad Sci USA*, 99:12913–12916, 2002.

[550] M. Wilm and M. Mann. Analytical properties of the nanoelectrospray ion source. *Anal Chem*, 68(1):1–8, 1996.

[551] J. Wilson. Towards the automated evaluation of crystallization trials. *Acta Crystallogr D Biol Crystallogr*, D58:1907–1904, 2002.

[552] S. K. Wismath, H. P. Soong, and S. G. Akl. Feature selection by interactive clustering. *Pattern Recogn*, 14:75–80, 1981.

[553] I. H. Witten and E. Frank. *Data Mining. Practical Machine Learning Tools and Techniques with Java Implementations.* Morgan Kaufmann Publishers, 1999.

[554] S. Wolfram. *A New Kind of Science.* Wolfram Media, Champaign, IL, 2002.

[555] S. L. Wong, L. V. Zhang, A. H. Y. Tong, Z. Li, D. S. Goldberg, O. D. King, G. Lesage, M. Vidal, B. Andrews, H. Bussey, C. Boone, and F. P. Roth. Combining biological networks to predict genetic interactions. *Proc Natl Acad Sci USA*, 101(44):15682–15687, 2004.

[556] R. M. Woodsmall and D. A. Benson. Information resources at the National Center for Biotechnology Information. *Bull Med Libr Assoc*, 81(3):282–284, 1993.

[557] C. C. Wu, M. J. MacCoss, K. E. Howell, and J. R. Yates, III. A method for the comprehensive proteomic analysis of membrane proteins. *Nat Biotechnol*, 21(5):532–538, 2003.

[558] C. C. Wu, M. J. MacCoss, K. E. Howell, D. E. Matthews, and J. R. Yates, III. Metabolic labeling of mammalian organisms with stable isotopes for quantitative proteomic analysis. *Anal Chem*, 76(17):4951–4959, 2004.

[559] C. C. Wu, M. J. MacCoss, G. Mardones, C. Finnigan, S. Mogelsvang, J. R. Yates, III, and K. E. Howell. Organellar proteomics reveals golgi arginine dimethylation. *Mol Biol Cell*, 15(6):2907–2919, 2004.

[560] S. Wuchty. Evolution and topology in the yeast protein interaction network. *Genome Res*, 14(7):1310–1314, 2004.

[561] I. Xenarios, L. Salwinski, X. J. Duan, P. Higney, S. M. Kim, and Eisenberg D. Dip: the database of interacting proteins. *Nucleic Acids Res*, 28(1):289–291, 2000.

[562] I. Xenarios, L. Salwinski, X. J. Duan, P. Higney, S. M. Kim, and D. Eisenberg. Dip, the database of interacting proteins: a research tool for studying cellular networks of protein interactions. *Nucleic Acids Res*, 30(1):303–305, 2002.

[563] H. Xie, A. Wasserman, Z. Levine, A. Novik, V. Grebinskiy, A. Shoshan, and L. Mintz. Large-scale protein annotation through gene ontology. *Genome Res*, 12(5):785–794, 2002.

[564] K. Yanagisawa, Y. Shyr, B. J. Xu, P. P. Massion, P. H. Larsen, B. C. White, J. R. Roberts, M. Edgerton, A. Gonzalez, S. Nadaf, J. H. Moore, R. M. Caprioli, and D. P. Carbone. Proteomic patterns of tumour subsets in non-small-cell lung cancer. *Lancet*, 362(9382):433–439, 2003.

[565] E. Yeger-Lotem and H. Margalit. Detection of regulatory circuits by integration of protein–protein and protein–DNA interaction data. In *4th Biopathways Consortium Meeting*. Edmonton, AB, 2002.

[566] E. Yeger-Lotem, S. Sattath, N. Kashtan, S. Itzkovitz, R. Milo, R. Y. Pinter, U. Alon, and H. Margalit. Network motifs in integrated cellular networks of transcription-regulation and protein–protein interaction. *Proc Natl Acad Sci USA*, 101(16):5934–5939, 2004.

[567] S.-H. Yook, H. Jeong, and A. L. Barabasi. Modeling the internet's large-scale topology. *Proc Natl Acad Sci USA*, 99:13382–6, 2002.

[568] A. Zanzoni, L. Montecchi-Palazzi, M. Quondam, G. Ausiello, Helmer-Citterich M., and G. Cesareni. MINT: a molecular interaction database. *FEBS Lett*, 513(1):135–140, 2002.

[569] B. R. Zeeberg, W. Feng, G. Wang, M. D. Wang, A. T. Fojo, M. Sunshine, S. Narasimhan, D. W. Kane, W. C. Reinhold, S. Lababidi, K. J. Bussey, J. Riss, J. C. Barrett, and J. N. Weinstein. Gominer: a resource for biological interpretation of genomic and proteomic data. *Genome Biol*, 4(4):R28, 2003.

[570] H. Zhang, X. J. Li, D. B. Martin, and R. Aebersold. Identification and quantification of n-linked glycoproteins using hydrazide chemistry, stable isotope labeling and mass spectrometry. *Nat Biotechnol*, 21(6):660–666, 2003.

[571] L. V. Zhang, S. L. Wong, O. D. King, and F. P. Roth. Predicting co-complexed protein pairs using genomic and proteomic data integration. *BMC Bioinformatics*, 5, 2004.

[572] Y. Zhen, N. Xu, B. Richardson, R. Becklin, J. R. Savage, K. Blake, and J. M. Peltier. Development of an lc-maldi method for the analysis of protein complexes. *J Am Soc Mass Spectrom*, 15(6):803–822, 2004.

[573] G. Zheng, E. O. George, and G. Narasimhan. Neural network classifiers and gene selection methods for microarray data on human lung adenocarcinoma. *CAMDA*, 2003.

[574] H. Zhou, J. A. Ranish, J. D. Watts, and R. Aebersold. Quantitative proteome analysis by solid-phase isotope tagging and mass spectrometry. *Nat Biotechnol*, 20(5):512–515, 2002.

[575] Z. Zhou, L. J. Licklider, S. P. Gygi, and R. Reed. Comprehensive proteomic analysis of the human spliceosome. *Nature*, 419(6903):182–185, 2002.

[576] X. Zhu, S. Sun, S. E. Cheng, and M. Bern. Classification of protein crystallization imagery. In *26th Annual Int Conf of IEEE Engineering in Medicine and Biology Society (EMBS-04)*. San Francisco, CA, 2004.

[577] W. M. Zuk and K. B. Ward. Methods of analysis of protein crystal images. *J Cryst Growth*, 110:148–155, 1991.

Index

A

Ab initio prediction and protein structure determination, 129
Absolute quantitation of proteins (AQUA), 58, 65
Abstraction and data base design, 17
Abstraction mechanisms in conceptual information models, 24, 25–29
Accuracy and data integration, 219–220
Accuracy vs. efficiency and KM approaches, 22–23
Adaptation algorithm and CBR, 182
Adaptation and robustness of a system, 247
Advanced Research Projects Agency (ARPA), 30
Aeschylus, 38
Affymetrix, 241
Aggregation as an abstraction mechanism, 26
Algorithms, two main groups of knowledge-discovery, 3; see also individual algorithms and subject headings
AliasServer, 214
Antibodies, 16
Application program interface (API), web-based, 203–204
ARPA, see Advanced Research Projects Agency
Artificial intelligence (AI), 1, 2, 11
Association mining and data integration, 230–233
Association rule and image analysis in HTP crystallization, 197
Atlas, 216
Attributes and relational data model, 17–18
Automation and data acquisition/ontology creation, 37

B

Barabasi-Albert network, 88–89
Basic Local Alignment Search Tool (BLAST), 8, 206, 212, 216–217
Bayesian framework and data integration, 23, 201, 221–228
Bayes net segmentation model, 155
Bias, PPI analysis and research, 96

Biochemical methods used to identify proteins and PPIs, 91
Bioinformatics, 37, 200
Biological designs of HTP platforms, challenges arising from the, 200–201
Biological Macromolecular Crystallization Database (BMCD), 132–133
Biology, 20, 24, 76–77; see also Systems biology
Biomolecular Interaction Network Database (BIND), 73, 95, 96, 205, 207, 213
Biopathways Consortium, 101
BLAST tool, 8, 206, 212, 216–217
Blur and image analysis in HTP crystallization, 147
Bottom-up clustering, 5
Bottom up sequencing strategy, 48

C

Cancer and systems biology, 241, 256–257
Canny edge detection and image analysis in HTP crystallization, 148, 149
Capillary electrophoresis (CE) and mass spectrometry, 41
Case-based reasoning (CBR), 23, 139, 144–146; see also Crystallization approaches, HTP protein
CBR, see Case-based reasoning
Certainty and discovered knowledge, 2, 5
Chromatography, multidimensional, 53–57
Chromosome 11 and functional genomics, 242
Classification as an abstraction mechanism, 25; see also Extracted features, image classification using; Images, classifying crystallization
Clique graph, 75–76
Cluster affinity search technique (CAST) algorithm, 121–122
Cluster identification via connectivity kernels (CLICK), 118, 119–121
Clusters/cluster analysis
    conceptual clustering methods, 4–5
    global data analysis, 69–70